Applying Artificial Intelligence to Computational Biology and Medical Informatics

Applying Artificial Intelligence to Computational Biology and Medical Informatics

Edited by

Mohammad "Sufian" Badar

Department of Bioengineering, University of California, Riverside, CA, United States

Universal Scientific Education and Research Network (USERN), Tehran, Iran

SPI Darbhanga, Darbhanga, Bihar, India

Department of Computer Science and Engineering (Bioinformatics), School of Engineering Sciences and Technology (SEST), Jamia Hamdard University, New Delhi, India

ELSEVIER

ACADEMIC PRESS

An imprint of Elsevier

Academic Press is an imprint of Elsevier
125 London Wall, London EC2Y 5AS, United Kingdom
50 Hampshire Street, 5th Floor, Cambridge, MA 02139, United States

For accessibility purposes, images in electronic versions of this book are accompanied by alt text descriptions independently provided by Elsevier, generated with the aid of AI. For more information, see https://www.elsevier.com/about/accessibility or Radarweg 29, 1043 NX Amsterdam, Netherlands.

Books and Journals published by Elsevier comply with applicable product safety requirements. For any product safety concerns or queries, please contact our authorised representative, Elsevier B.V., at productsafety@elsevier.com.

Publisher's note: Elsevier takes a neutral position with respect to territorial disputes or jurisdictional claims in its published content, including in maps and institutional affiliations.

Notices

Knowledge and best practice in this field are constantly changing. As new research and experience broaden our understanding, changes in research methods, professional practices, or medical treatment may become necessary.

Practitioners and researchers must always rely on their own experience and knowledge in evaluating and using any information, methods, compounds, or experiments described herein. In using such information or methods they should be mindful of their own safety and the safety of others, including parties for whom they have a professional responsibility.

To the fullest extent of the law, neither the Publisher nor the authors, contributors, or editors, assume any liability for any injury and/or damage to persons or property as a matter of products liability, negligence or otherwise, or from any use or operation of any methods, products, instructions, or ideas contained in the material herein.

ISBN: 978-0-443-44915-4

For information on all Academic Press publications visit our website at https://www.elsevier.com/books-and-journals

Publisher: Megan Ball
Acquisitions Editor: Linda Buschman
Editorial Project Manager: Priyanshi Peelwan
Production Project Manager: Aishwarya R
Cover Designer: Arumugam Kothandan

Typeset by TNQ Tech

Dedication

The Prophetic Dua for Healing: "O Allah, Lord of mankind, remove the disease and cure him. You are the one who cures, there is no cure but Yours, a cure that leaves no disease behind."

Sahih al-Bukhari

I am grateful to my parents, (Amman, Marhooma Nurunnisa, Abba Marhoom Mohd. Badre Alam), wife, Rana Kamal, daughters; Sarah Sufian Badar and Aisha Sufian Badar, and my brothers, sister Shahnaz Badar, nieces, nephews, and newly borne grandchildren, Fatimah Aasim, Musa Aasim, Amara Modabbir, and Fehma Yaqoot and . their support, encouragement, guidance, and constant love have sustained and encouraged me.

Contents

Contributors

Mohd Afaque Department of Respiratory Medicine, Hamdard Institute of Medical Sciences and HAHC Hospital, New Delhi, India

Afzal Ahmad Department of Electronics, School of Engineering Sciences and Technology (SEST), Jamia Hamdard, New Delhi, India

Khairol Amali bin Ahmad National Defence University of Malaysia (NDUM), Kuala Lumpur, Malaysia

Syed Ansar Ahmad Department of Maxillofacial Surgery, Faculty of Dentistry, Jamia Millia Islamia, New Delhi, India

Jatin Ahuja Infectious Diseases & Travel Health Specialist in Indraprastha Apollo Hospital, New Delhi, India; AIIMS, New Delhi, India; Travel Medicine, ISTM, Alpharetta, GA, United States

Bashir Alam Department of Computer Engineering, Faculty of Engineering and Technology, Jamia Millia Islamia, New Delhi, India

Faiq Alam Bachelor's of Technology, Zakir Husain College of Engineering and Technology, Aligarh Muslim University (AMU), Uttar Pradesh, India

Iqubal Alam Emergency Medicine Department, Dr Sulaiman Al Habib Hospital, Riyadh, Saudi Arabia

Mansaf Alam Department of Computer Science, Jamia Millia Islamia (JMI), New Delhi, India

Mohd Shadab Alam Department of Computer Science and Engineering, School of Engineering Sciences and Technology, Jamia Hamdard, New Delhi, India

Shamsheer Alam ESIC Hospital & PGIMSR, New Delhi, India

Sundus Mohammad Fakhre Alam Centre for Physiotherapy and Rehabilitation Sciences, Jamia Millia Islamia (JMI), New Delhi, India

Omar A. Alamoudi Department of Electrical Engineering, Faculty of Engineering, King Abdulaziz University, Jeddah, Saudi Arabia

Akhtar Ali Centre for Federal Studies, Public Policy and Governance, Jamia Hamdard University, New Delhi, India

Mohammad Subhan Ali WHO, Katihar, Bihar, India

Syed Ausaf Ali Hamdard Public School, New Delhi, India

Adil Amraz Department of Computer Science, Jamia Millia Islamia, New Delhi, India

Jishnu Anilkumar NaaS (NetOps & Automation), HPE (Xevyte Technologies), Bengaluru, Karnataka, India

Md Arshad Department of Computer Science, JMI, New Delhi, India

Belal Bin Asaf Department of Chest Surgery and Chest Oncosurgery and Lung Transplantation, Medanta Hospital Noida, Noida, Uttar Pradesh, India

Farheen Asaf Max Hospital, New Delhi, India

Aquil Atif Abedin Clinic, Phulwari Sharif, Patna, Bihar, India

Syed Fawaz Ayazi Bachelor's of Technology, Zakir Husain College of Engineering and Technology, Aligarh Muslim University (AMU), Uttar Pradesh, India

Abul Kalam Azad Max Health Care, Max Superspecialty Hospital, New Delhi, India

Labeebah Rizwan Badar Badar Medical Centre (BMC), New Delhi, India

Mohammad "Sufian" Badar Department of Bioengineering, University of California, Riverside, CA, United States; Universal Scientific Education and Research Network (USERN), Tehran, Iran; SPI Darbhanga, Darbhanga, Bihar, India; Department of Computer Science and Engineering (Bioinformatics), School of Engineering Sciences and Technology (SEST), Jamia Hamdard University, New Delhi, India

Mohammad Rehan Badar Apollo Hospital, Badar Medical Centre (BMC), Salmon Arm, BC, Canada

Sarah Sufian Badar Sharda University, Greater Noida, Uttar Pradesh, India

Ananta Dutta Choudhury Department of Computer Science, Jamia Millia Islamia, New Delhi, India

Chhavi Dudeja Jamia Hamdard University, School of Engineering Sciences & Technology, New Delhi, India

Ayesha Fatima Industrial Chemistry, Faculty of Science, AMU, Uttar Pradesh, India

Sadaf Fatima Department of Gastroenterology, Medanta Hospital, Sector 50, Noida, Uttar Pradesh, India

T.N. Giri Vocational Education, NIOS, Noida, Uttar Pradesh, India

Mohd Asrarul Haque Department of Forensic Medicine, Jawaharlal Nehru Medical College, Aligarh Muslim University (AMU), Aligarh, Uttar Pradesh, India

Fakroul Ridzuan Hashim Artificial Intelligent, Biomedical Engineering Universiti Pertahanan Nasional Malaysia, Kuala Lumpur, Malaysia

Syed Zafer Hashmi Department of Critical Care Medicine, Max Noida Sector 128, Noida, Uttar Pradesh, India

Aamna Hassan Department of Botany, Jamia Hamdard, New Delhi, India

Tabindah Hesam Graham Terrace Neath, Wales, United Kingdom

Asghar Hussain CapitalOne Bank, Mclean, VA, United States

Imran Hussain Department of Computer Science & Engineering (CSE), School of Engineering Sciences and Technology (SEST), Jamia Hamdard University, New Delhi, India

Md Zaid Hussain Siddique Department of Biotechnology, Jamia Hamdard, New Delhi, India

Syed Zeeshan Hussain Department of Computer Science, Jamia Millia Islamia (JMI), New Delhi, India

Mohammad Idrees Faculty of Engineering & Technology, AMU Aligarh, Aligarh, Uttar Pradesh, India

Aisha Idris Department of Biotechnology, Jamia Hamdard, New Delhi, India

Nida Iftekhar Department of Computer Science and Engineering, School of Engineering Sciences and Technology (SEST), Jamia Hamdard University, New Delhi, India

Nabeela Ilyas ENHANCE Ontario, Toronto, ON, Canada

Zubin Ingah Global Network Technologies, Sunnyvale, CA, United States

Hamid Iqbal Dr. B. R. Ambedkar National Institute of Technology (NITJ), Jalandhar, Punjab, India

Ehtesham Jameel Medicinal & Process Chemistry Division, CSIR-Central Drug Research Institute, Lucknow, Uttar Pradesh, India

Md Kaleem Department of Physics, Jahangirabad Institute of Technology, Barabanki, Uttar Pradesh, India

Azfar Kamal District Hospital, Sambhal, Uttar Pradesh, India

Sajjad Karim Department of Bioinformatics and Artificial Intelligence, Institute of Genomic Medicine Sciences, King Abdulaziz University, Jeddah, Saudi Arabia

Syed Sibtain Khalid Department of Computer Science and Engineering, School of Engineering Sciences and Technology (SEST), Jamia Hamdard University, New Delhi, India

Aabida Khan Department of Computer Science, Jamia Millia Islamia, New Delhi, India

Iqra Khan Department of Computer Science, Jamia Millia Islamia, New Delhi, India

Kamran Khan Hind Institute of Medical Sciences, Sitapur, Uttar Pradesh, India

Rehan Saeed Khan Department of Computer Science and Engineering, SEST, Jamia Hamdard, New Delhi, India

Uzma Khan Department of Computer Science, Jamia Millia Islamia, New Delhi, India

Yaqoot Taban Khan Rama Medical College Hospital & Research Centre, Kanpur, Uttar Pradesh, India

Jitendra Kumar Department of Forensic Medicine, Maulana Azad Medical College, New Delhi, India

Mohammed Abaan Kurikkal Department of Computer Science, Jamia Millia Islamia, New Delhi, India

Azhar Latif Meta, Menlo Park, CA, United States

Asma Maheen New Delhi, India

Arslan Mansoor Amazon, Bengaluru, Karnataka, India

Hammad Mansoor Telus Digital, Candor Tech Space, Sector 135, Noida, Uttar Pradesh, India

Sameen Masroor Jamia Hamdard University, School of Engineering Sciences & Technology, New Delhi, India

Anima Sushim Mathur Desert Medicine Research Centre (Indian Council of Medical Research), World Health Organization Project, Jodhpur, Rajasthan, India; Manav Rachna International Institute of Research and Studies, Faridabad, Haryana, India; Scientific Hub Research Writing Services, Jodhpur, Rajasthan, India

Sushim Harish Mathur New Delhi Municipal Council, Dr Ram Manohar Lohia Hospital, New Delhi, India

Hussam Bin Mehare School of Electrical and Mechanical Engineering, University of Adelaide, Adelaide, SA, Australia

Zeenat Mirza King Fahd Medical Research Centre, King Abdulaziz University, Jeddah, Saudi Arabia

Priyanka Mishra Centre of Excellence for Microbiome Research (CCMR) Lab, Department of Biochemistry, All India Institute of Medical Science, Bhubaneswar, Odisha, India

Syed Shad Mohsin Apollo Cradle & Children's Hospital, Ghaziabad, Uttar Pradesh, India

Hatwib Mugasa National Information Technology Authority, Kampala, Uganda

Md. Mustafa All India Institute of Medical Sciences (AIIMS), New Delhi, India

Sobia Naaz Department of Computer Science and Engineering (Bioinformatics), School of Engineering Sciences and Technology (SEST), Jamia Hamdard University, New Delhi, India

Sara Nadeem Department of Computer Science and Engineering, School of Engineering Sciences and Technology (SEST), Jamia Hamdard University, New Delhi, India

Md. Tabrez Nafis Department of Computer Science & Engineering, Jamia Hamdard (Deemed University), New Delhi, India

Aamir Nehal Touch n Heal Physiotherapy and Rehabilitation Centre, Motihari, Bihar, India

Noora Department of Pathology, Madhubani Medical College, Madhubani, Bihar, India

Hamid Omar University of Wisconsin Madison, Madison, WI, United States

Asma Orooj Department of Community Medicine, Agartala Government Medical College and GBP Hospital, Agartala, Tripura, India

Siddhant Panigrahi Robotics Lab, Department of Engineering Design, Indian Institute of Technology Madras, Chennai, Tamil Nadu, India

Shaista Parveen Dental Consultant, Patna, India

Md E.A. Raghib Khan Department of Computer Science, Jamia Millia Islamia, New Delhi, India

Naseem Rao Department of Computer Science and Engineering, School of Engineering Sciences and Technology (SEST), Jamia Hamdard University, New Delhi, India

Ashif Raza Department of Computer Science, Jamia Millia Islamia, New Delhi, India

Danish Reza Centre for Federal Studies, Public Policies and Governance, School of Humanities and Social Sciences, Jamia Hamdard, New Delhi, India

Fatemah Rizvi Jamia Millia Islamia University, New Delhi, India

Yukti Sabikhi Jamia Hamdard University, School of Engineering Sciences & Technology, New Delhi, India

Abdus Salam Sabir Charles Schwab, Lonetree, CO, United States

Marya Saeed Cardiothoracic and Vascular Surgery (CTVS) and Cardiac Critical Care Unit, Medanta Hospital, Noida, Uttar Pradesh, India

Muneerah Saeed Department of Pathology, R. G. Medical College, Hathras, Uttar Pradesh, India

Mohhamed Salahuddin Information Specialist, Core42, Abu Dhabi, United Arab Emirates

Aaliya Sameen Salfia Paramedical Institute (SPI), Darbhanga, India

Abdus Sattar Agrometeorology, Centre for Advanced Studies on Climate Change, Dr. Rajendra Prasad Central Agricultural University, Samastipur, Bihar, India

Syed Nazia Shah Department of Computer Science, Jamia Millia Islamia, New Delhi, India

Syed Khurram Shahabuddin Department of Biosciences, Jamia Millia Islamia (JMI), New Delhi, India

Mohd Shahzad Faculty of Law, Hamdard Institute of Legal Studies and Research (HILSR), Jamia Hamdard University, New Delhi, India

Shamim Ahmed Shamim Department of Nuclear Medicine, AIIMS, New Delhi, India

Taiba Shamim Department of Computer Science, Jamia Millia Islamia, New Delhi, India

Huda Shams Aligarh Muslim University (AMU), Aligarh, Uttar Pradesh, India

Shazmeen Shamsi Department of Computer Science, JMI, New Delhi, India

Farheen Siddiqui Department of Computer Science and Engineering, School of Engineering Sciences and Technology (SEST), Jamia Hamdard University, New Delhi, India

Shakeel Ahmad Siddiqui Department of Computer Science and Engineering, Jamia Hamdard, New Delhi, India

Shireen Siddiqui Department of Laboratory Medicine, Medanta Hospital Noida, Noida, Uttar Pradesh, India

Yousuf Siddiqui JNMC, AMU, Aligarh, Uttar Pradesh, India

Anshika Singh Jamia Hamdard University, School of Engineering Sciences & Technology, New Delhi, India

Prakriti Singh Department of Computer Science, Jamia Millia Islamia, New Delhi, India

Nahed Solouma Department of Biomedical Engineering, King Faisal University, Hofuf, Eastern Province, Kingdom of Saudi Arabia

Aafia Tasneem Badar Medical Center (BMC), New Delhi, India

Mohd Maksuf Ul Haque Department of Biotechnology, Jaypee Institute of Information Technology, Noida, Uttar Pradesh, India

Fakhr Ul-Islam FSGlobal, LLC, Marietta, GA, United States

Arif Wahab Indraprastha Apollo Hospital, Sarita Vihar, New Delhi, India

Shabbir Ahmad Warsi AURA Cosmetic Surgery Klinik, Paras HMRI Hospital, Patna, Bihar, India

Aasim Zafar Aligarh Muslim University (AMU), Aligarh, Uttar Pradesh, India

Sania Zahid Department of Computer Science, Jamia Millia Islamia, New Delhi, India

S.M. Zainul Abedin Abedin Clinic, Phulwari Sharif, Patna, Bihar, India

Emad Zeni Lamar University, Beaumont, TX, United States

Preface

Currently, AI and ML are being applied in all areas of health science. The causes, transmission, diagnosis, treatment, and management of all diseases can be carried out using AI and ML. Students and researchers of all branches of health science need a book that explains the application of AI and ML in computational biology and medical informatics. Very few books are available in the market for that purpose.

Purpose: Generally, books are written for those with a computer science, computational biology, or a medical informatics background. Thus, this book is written in a way that is coherent and easy to understand for those who do not have this background. Furthermore, the names of the chapters and the language of the contents are straightforward.

This is one of the few books for new academicians or researchers who can understand the cause and mechanism of any disease and are interested in learning how to integrate the emerging technology of AI and ML into their work.

This book discusses the integration of AI and ML in the treatment of many common diseases, so it is necessary to answer the questions that will arise. Some of these diseases may have been faced by the reader themselves or their family members. Keeping this in mind, we must address common problems and their solutions through case studies. We will include case studies where it is necessary.

Our AI and ML-designed model performs best if there is continuous inflow of data. We will provide online sources so that the readers can see how the model performs when features or parameters are changed. This will allow the authors to fulfil the needs of the readers in these innovative fields and areas.

Chapter 1: Introduction to AI/ML, Computational Biology, and Medical Informatics
Chapter 2: AI/ML Applications in Medical Imaging
Chapter 3: Natural Language Technologies in Biomedical Domain
Chapter 4: AI/ML in Chemoinformatics

These four chapters serve as the foundation in understanding how the fields of AI, ML, and computational biology can be applied in the various domains of medicine. It explains how these tools can help overcome problems faced by practicing physicians in the given fields.

Chapter 5: Deep Learning Methods for Network Biology
Chapter 6: Probabilistic Optimization of ML for Heart Disease Prediction

Chapter 7: The Need for Interpretable and Explainable Deep Learning Data in Health Care

The next three chapters explore the role of deep learning in healthcare data and medical networks. It also explores its application in predicting a rapidly growing health emergency, heart disease.

Chapter 8: Using hybrid models in healthcare data
Chapter 9: Improving Disease Prediction by Integrating Multiple ML and Optimization Techniques

Chapters 8 and 9 cover an important aspect of AI and ML application in healthcare. The integration of ML and optimization techniques with disease prediction is covered, along with how it improves disease prediction.

Chapter 10: Ethical, Societal, and Legal Issues in AI/ML for Healthcare
Chapter 11: Deep Learning in Gait Abnormality Detection: Principles and Illustrations
Chapter 12: Broad Applications of Network Embedding in Computational Biology, Genomics, Medicine, and Health

Chapters 10–12 cover an important aspect of AI and ML application in healthcare, the ethical and legal issues. This is one of the major drawbacks in the implementation of this technology in the healthcare sector. These chapters will tackle these concerns.

Chapter 13: AI Use in Clinical Prediction
Chapter 14: AI/ML for Medical Informatics and Public Health
Chapter 15: AI in Medical Imaging for Developing Countries: Challenges and Opportunities
Chapter 16: AI Applications in Disease Diagnosis and Treatment: New Applications

The last four chapters, 13, 14, 15, and 16 examine the role of AI and ML in medical informatics and its impact on public health. It discusses how AI can aid in medical imaging, disease diagnosis, and treatment, especially for developing countries, along with the challenges and opportunities it presents.

Three key features

1. This is one of the few books that are ideal for new academicians/researchers who do not have a background in computer science or health science but want to learn both.
2. The language is precise and easy to comprehend.
3. The chapters are designed in such a way that it maintains the flow and coherency, starting from level zero and going up to a higher level of understanding.

Acknowledgment

First and foremost, I would like to praise and thank Allah SWT, the almighty, who has granted me countless blessings, knowledge, and opportunities so that I can edit this book. Alhamdulillah.

I would like to mention Naana-Abdul Ghafoor, Naani-Quraisha, Mamun-Md Aslam, Mamani-Naushaba, Khala: Qamrun Nisa, Badrun Nisa, Khaloo: Abdul Hafeez, and Mohd. Shakir.

I would like to mention Chacha who was very close to my Abba, Engr Md Qamrul Hoda, Chachi Noor Jahan, and their children, Md. Zeyaul Hoda and his wife Fakhera Uzma, Mohammad Rafeul Hoda and his wife Fauzia Shaheen, Md. Ekramul Hoda and his wife Shaista Perween, Reyaz Fatima and her husband Irshad Ahmad, Nishat Fatima and her husband Sultan Zafar, Shahnaz Fatma and her husband Iqbal Ahmad, Iffat Fatma and her husband Faisal Manzar, and all members of Ahle Nawaab Jaan and Ahle Nazra.

Also, I would like to mention Phoophu Amra Jamal, Phoopha Dr. Firoz Ahmad, Sabih Mahmood, Dr. Noora, and their children Manal Sabih, Izma Sabih, Bushra Sabih, Rahmah Sabih, Fasih Mehmood, Nikhat Parween, and their Children Aisha Fasih, Aidah Fasih, Shahidah Fasih.

1

Introduction to AI/ML, computational biology, and medical informatics

Mohammad "Sufian" Badar[1, 2, 3, 4], Ananta Dutta Choudhury[5], Iqra Khan[5], Yaqoot Taban Khan[6], Nabeela Ilyas[7], Azfar Kamal[8] and Nahed Solouma[9]

[1]DEPARTMENT OF BIOENGINEERING, UNIVERSITY OF CALIFORNIA, RIVERSIDE, CA, UNITED STATES; [2]UNIVERSAL SCIENTIFIC EDUCATION AND RESEARCH NETWORK (USERN), TEHRAN, IRAN; [3]SPI DARBHANGA, DARBHANGA, BIHAR, INDIA; [4]DEPARTMENT OF COMPUTER SCIENCE AND ENGINEERING (BIOINFORMATICS), SCHOOL OF ENGINEERING SCIENCES AND TECHNOLOGY (SEST), JAMIA HAMDARD UNIVERSITY, NEW DELHI, INDIA; [5]DEPARTMENT OF COMPUTER SCIENCE, JAMIA MILLIA ISLAMIA, NEW DELHI, INDIA; [6]RAMA MEDICAL COLLEGE HOSPITAL & RESEARCH CENTRE, KANPUR, UTTAR PRADESH, INDIA; [7]ENHANCE ONTARIO, TORONTO, ON, CANADA; [8]DISTRICT HOSPITAL, SAMBHAL, UTTAR PRADESH, INDIA; [9]DEPARTMENT OF BIOMEDICAL ENGINEERING, KING FAISAL UNIVERSITY, HOFUF, EASTERN PROVINCE, KINGDOM OF SAUDI ARABIA

E-MAIL ADDRESSES: MSBADAR@SALFIAINSTITUTE.EDU.IN, MSBADAR@JAMIAHAMDARD.AC.IN, ANANTA.DC.04@GMAIL.COM, IIQRAKHHAN@GMAIL.COM, YAQOOTTABAN247@GMAIL.COM, NABEELA.ILYAS@GMAIL.COM, AZFARKAMAL1@GMAIL.COM, NSOLOUMA@KFU.EDU.SA.

What you will learn

In this chapter, you will learn:

- Foundational concepts of artificial intelligence, machine learning, computational biology, and medical informatics.
- The historical evolutions of these disciplines along with their convergence.

Applying Artificial Intelligence to Computational Biology and Medical Informatics. https://doi.org/10.1016/B978-0-443-44915-4.00001-0

- Leveraging algorithms and models in computational biology to solve biological challenges.
- Integrating AI/ML with computational biology and medical informatics to enhance healthcare research and practices.
- Real-world applications in healthcare and life sciences, including genomics, drug discovery, and precision medicine.
- Challenges in AI in healthcare, such as bias in data, privacy issues, interoperability, and precision medicine.

1.1 Overview

1.1.1 Scope of the chapter

The scope of this chapter is threefold:

1. *Foundational understanding*—To give readers an understanding of fundamental concepts of artificial knowledge (AI), machine learning (ML), computational biology, medical informatics, including their definitions, key methodologies, and interdisciplinary nature.
2. *Integration of disciplines*—To highlight how these domains interconnect to form an ecosystem that leverages computational modeling, predictive analytics, and biomedical data to drive innovation in life sciences.
3. *Real-world relevance*—To demonstrate the practical application of these technologies in areas such as disease modeling, drug target identification, precision medicine, and healthcare informatics, while addressing ethical and technical challenges.

This chapter is designed such that researchers, clinicians, data scientists, and life sciences professionals who aim to understand how computational intelligence transforms biomedical research and healthcare delivery.

1.1.2 Historical context and evolution

The evolution of AI and ML has unfolded over several decades, with each era marked by unique milestones. Parallel to this, the fields of computational biology and medical informatics have advanced significantly, often benefiting directly from progress in AI (see Fig. 1.1).

It started in the 1950s with symbolic reasoning, early logic-based systems, and medical decision-support experiments like MYCIN. By the 1980s, expert systems gained popularity, though their scalability was limited. The 1990s marked a shift toward statistical and data-driven methods such as decision trees and support vector machines (ML models), while the Human Genome Project accelerated computational biology and bioinformatics. Around the same time, medical informatics advanced

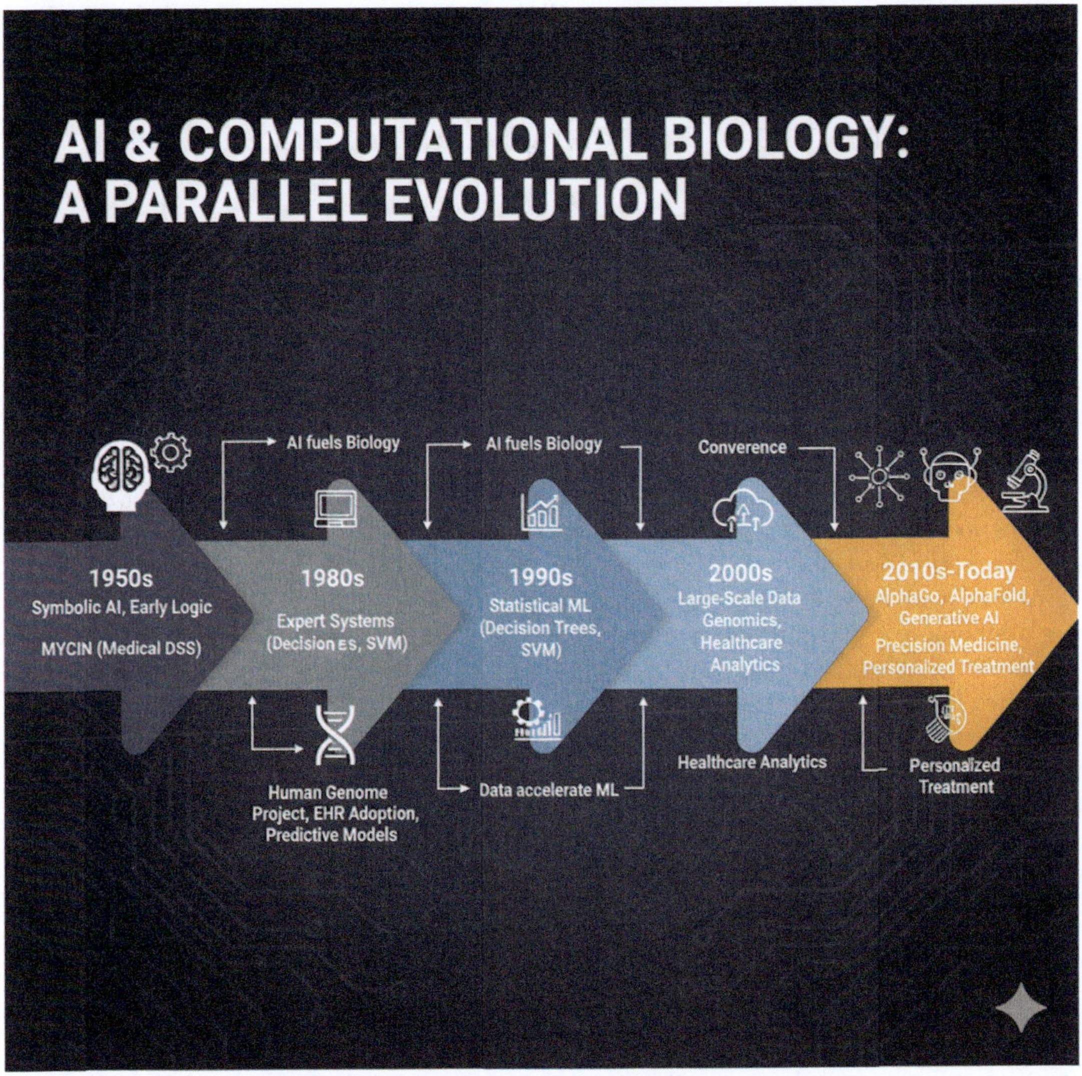

FIGURE 1.1 Parallel evolution of AI and computational biology.

with the adoption of electronic health records and predictive modeling for patient care.

From the 2000s onward, AI and ML were fueled by large-scale data and computing power, leading to applications in genomics, proteomics, imaging, and healthcare analytics. The 2010s introduced deep learning (DL), revolutionizing vision, language, and medical diagnostics, alongside breakthroughs such as AlphaGo and AlphaFold (Jumper et al., 2021; Russell and Norvig, 2021). Today, generative AI, multimodal systems, and precision medicine highlight the convergence of AI, biology, and healthcare, driving advances in personalized treatment, clinical decision support, and biomedical discovery.

1.2 Fundamentals of artificial intelligence and machine learning

1.2.1 Definition of AI and ML

The term *artificial intelligence* was coined by John McCarthy, who was an American computer scientist and is regarded as the "Father of Artificial Intelligence" (McCarthy et al., 2006) in 1956 at the Darthmouth Conference. Even though McCarthy was credited with the term, it was Alan Turing who laid the theoretical groundwork for artificial intelligence with his paper "Computing Machinery and Intelligence" in 1950 (TURING, 1950).

According to Wikipedia, *artificial intelligence (AI) is the capability of computational systems to perform tasks typically associated with human intelligence, such as learning, reasoning, problem-solving, perception, and decision-making.*

AI has evolved over the years from science fiction to reality. What was considered impossible a few decades ago is now a reality. Research studies already suggest that AI can perform as well as or even better than humans at healthcare intervention, such as surgeries, diagnosing disease, etc.

AI in an umbrella term that encompasses subfields such as ML, DL, natural language processing, robotics, etc. Deloitte survey of 1100 US managers whose organizations were already pursuing AI, 63% of companies surveyed were employing ML in their businesses. It is a broad technique at the core of many approaches to AI and there are many versions of it.

We can define ML as the field of study in AI that deals with building statistical models that can predict outcome from a set of input data and "learn" to generalize to unseen data and can automate tasks without explicit instructions or human intervention (Bishop, 2006).

A well-known example of ML software in healthcare is IBM's Watson Health. Watson Health has been trained on everything ever written in any language at any time related to cancer diagnostic and treatment. Apart from that, it also continues to ingest new data as it is published. When it is presented with a specific patient with cancer, it can recommend the treatment trial most like to treat that individual's cancer by considering their genome history, imaging, and pathology, coupled with all the information known about the treatment of cancer.

1.2.2 Types of machine learning

AI has many subsets. ML is a subset of AI (as shown in Fig. 1.2) and can be broadly divided into four types—supervised, unsupervised, semisupervised, and reinforcement

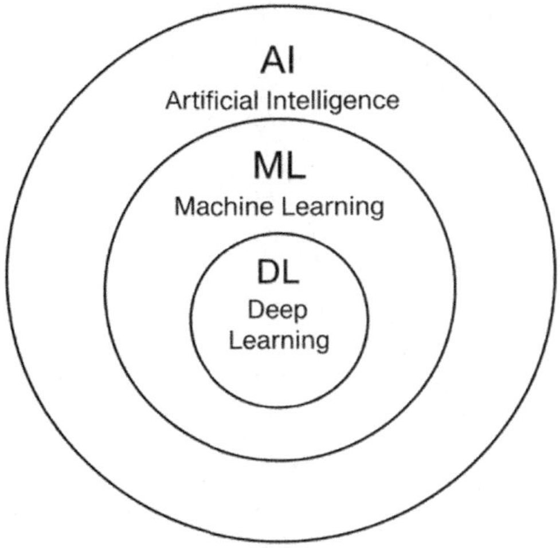

FIGURE 1.2 AI and its subsets.

learning. When model is trained on labeled data to learn the patterns between input attributes and label attribute, it is referred to as *supervised learning*. It can be further broken down into classification—classifying into a particular class—and regression tasks—predicting a value based on the data. Classifying emails into spams, sentiment analysis, prediction of house prices, and cancer detection are a few examples. *Unsupervised learning* refers to training the model on unlabeled data to learn hidden patterns in the dataset by grouping similar data points together. It can be further divided into clustering and dimensionality reduction tasks. Customer segmentation, PCA, and anomaly detection are some examples. *Semisupervised learning* refers to training your model on a small chunk of labeled data and a big chunk of unlabeled data. This is done when labeling is expensive and time-consuming. *Reinforcement learning* is the process through which a model interacts with its environment and a human user provides a certain reward or penalty depending on its performance.

1.2.3 Regression and classification

The most explored topics of ML happen to be regression and classification tasks. These are supervised learning tasks where the output or target attribute contains labels. Classification tasks involve categorizing data points into classes, say spam classification, diabetes detection, sentiment analysis, etc. based on patterns learned by a model on the given dataset. Some models that are commonly used are decision trees (splits data into branches based on feature values), random forests (an ensemble of decision trees), logistic regression (statistical model that uses probability estimation), and SVMs (finds a hyperplane to separate different classes in the data).

Regression tasks involve predicting a continuous numerical value from input data. Predicting stock prices, temperature, retail prices, etc. are good examples of the same.

Unlike classification where the output is a categorical value (class), regression outputs are continuous values. Some models that are commonly used are linear regression (models the relationship between inputs and output as linear), polynomial regression (extends linear regression by fitting a curve using polynomial terms), and SVRs (uses the principles of SVM to fit data within a margin of tolerance).

1.2.4 Clustering

Clustering is an unsupervised ML technique that groups similar data points together based on patterns or similarities without the use of predefined labels. It is widely used to uncover hidden structures in datasets, such as grouping customers by purchasing behavior, detecting anomalies such as fraudulent transactions, or identifying topics in large collections of text. Popular algorithms include K-means, which partitions data into a fixed number of clusters; hierarchical clustering, which builds nested clusters in a tree-like structure; and DBSCAN, which forms clusters based on dense regions and can handle noise effectively. Clustering is a powerful tool for data exploration and dimensionality reduction, but it also presents challenges such as determining the optimal number of clusters and dealing with high-dimensional or noisy data.

1.2.5 Deep learning

A more complex form of ML is neural network—a term that has been around since the 1960s. This term was taken from the neurons in the human brain and is supposedly built to mimic the working of the brain (Lecun et al., 2015).

Under the shade of ML lies *DL* that involves models that have neurons that are linked together to form neural networks. Just like the neurons in human brain, these neurons in the neural network fire whenever a certain threshold is crossed.

In neural network, we have inputs that are passed to the hidden layers (also called neural network layer) that comprise the neurons stacked together in a single layer. There may be many hidden layers in the model and thus the term DL, because the model is deep in structure due to many layers.

DL has vast and very significant role in healthcare that has proved to be beneficial and achieve things that humans find impossible (check for latest DL development). For example, the Food and Drug Administration now allows pharmaceutical companies to model drug interactions using AI algorithms that accurately identify drug toxicity without the usual decade of animal testing. In fact, they encourage it. Cancer detection in pathology slides is getting very close to outperforming trained pathologists, and similar pattern recognition software is getting quite accurate at reading radiographs.

1.3 What is computational biology?

The amalgamation of computer science, biology, and statistics gives birth to the interdisciplinary field of computational biology, where we analyze and interpret large-scale biological data, such as DNA and protein sequences, gene expression data, and molecular structures. You can think of it as a powerful tool to perform experiments that would take too much in a lab otherwise (Larrañaga et al., 2006).

1.3.1 Disciplines of computational biology

The interdisciplinary field of computational biology encompasses several key areas: Genomics and bioinformatics focus on analyzing genetic information, like using algorithms to find a disease-causing gene by comparing thousands of genomes. Systems biology builds computer models of entire biological systems, such as a cell, to predict how it will behave when exposed to a new drug. Structural biology and drug discovery predict the 3D shapes of molecules to design new medicines, as seen with the rapid development of COVID-19 vaccines after researchers computationally predicted the structure of the SARS-CoV-2 spike protein. Finally, computational neuroscience creates models of the brain to understand how it processes information, which helps in treating neurological disorders. These disciplines showcase how computational power is now fundamental to modern biological research (TURING, 1950).

1.3.2 Key computational methods

Computational biology is an interdisciplinary field that leverages computers to solve complex biological problems. Instead of relying solely on physical lab work, it uses mathematics, statistics, and computer science to analyze and interpret vast amounts of biological data. For instance, a computational biologist might use algorithms to identify the genetic cause of a disease by comparing thousands of DNA sequences. This field is composed of several key areas, including genomics and bioinformatics, which focuses on analyzing genetic information; systems biology, which creates computer models of entire biological systems to predict their behavior; structural biology and drug discovery, which uses computers to predict the 3D shapes of molecules to design new medicines; and computational neuroscience, which models the brain to understand its functions. In essence, these disciplines demonstrate how computational power has become a basis for modern biological research.

1.4 Introduction of medical informatics

1.4.1 Definition and scope

> *"Medical informatics is the interdisciplinary field that studies and pursues the effective use of biomedical data, information, and knowledge for scientific*

inquiry, problem solving, and decision making driven by efforts to improve human health."

Medical informatics lends the data infrastructure, algorithms, databases, and AI tools to handle complex, massive datasets from fields such as genomics, proteomics, transcriptomics, epigenetics, drug discovery, etc.

To provide solution for problem related to data, we need medical informatics to suggest appropriate methods and tools. The scope of medical informatics encompasses application of information science, data analytics, computational tools to all aspects of healthcare delivery, research, and management. Medical informatics finds applications in various fields such as follows:

- Data management
- Clinical decision support
- Telemedicine
- Medical imaging informatics
- Bioinformatics integration
- Public health applications
- Patient engagement
- Healthcare administrations

1.4.2 Health data types (EHRs, imaging, genomics, wearable device data)

A key element of medical informatics is its handling of diverse health data types. Modern healthcare generated enormous volumes of diverse data types, each with unique characteristics and applications:

i. EHRs: Electronic health records (EHRs), often also referred as electronic medical records (EMRs), are major source of clinical data. EHRs are a digital record of health information of the patient. EHRs not only contain data but also [deals with] how data is shared efficiently with clinicians and organizations involved in caring for patients, such as labs, specialists, imaging facilities, pharmacies, emergency facilities, workplace clinics, etc. [8].

ii. Medical imaging: Refers to the techniques and processes used to create images of human body or specific parts for various clinical purposes such as medical procedures and diagnosis or medical science including study of normal anatomy and function. In the wider sense, it is a part of biological imaging and incorporates radiology, endoscope, thermograph, medical photography, and microscopy. Measurement and recording techniques such as electroencephalography (EEG) and magnetoencephalography (MEG) are not primarily designed to produce images but, which produce data susceptible to be represented as maps, can be seen as forms of medical imaging.

iii. Genomics: Genomics involves DNA sequencing exploration of the function and information structure of genes, and it leads researchers to understand the creation of protein sequences and the association between genotype and phenotype. Human genomic data contains vast amounts of data. Analysis of genes or alleles identification could help in the diagnosis of disease and in the design of targeted therapies. Predicting the genetic datasets is extremely high dimensional, heterogenous, and unbalanced. Therefore, preprocessing and feature extraction were often needed by domain experts, and recently, ML and DL approaches were tried to solve the issue.

iv. Wearables: Biosensors are wearable, implantable, and ambient devices that convert biological responses into electrooptical signals and make continuous monitoring of health and wellbeing possible, even with a variety of mobile apps. Since EHRs often lack patients' self-reported experiences, human activities, and vital signs outside of clinical settings, tracking those continuously is expected to improve treatment outcomes by closely analyzing patient's condition.

1.4.3 Standards and interoperability

To enable seamless integration, sharing, and interoperability of complex data, medical informatics relies on standardized frameworks and protocols. Conformance to standards facilitates exchange and substitutability, creates independence of source from its origin, eases competition, and generates interoperability as shown in Table 1.1.

Table 1.1 Data interoperability framework.

Body	Focus area	Examples/standards
HL7 (Health Level Seven International)	Health data exchange, integration, and retrieval	HL7 v2, HL7 v3, CDA, FHIR
ISO/TC 215 (International Organization for Standardization—Technical Committee 215)	Health informatics, EHR architecture, safety, and security	ISO 13606 (EHR communication), ISO 27799 (health information security)
CEN/TC 251 (European Committee for Standardization—Technical Committee 251)	European health informatics standards, interoperability	European EHR standards, patient summaries
IEEEs (Institute of Electrical and Electronics Engineers)	Medical device communication and interoperability	IEEE 11073 (medical device data standards)
DICOM (Digital Imaging and Communications in Medicine)	Medical imaging standards and data sharing	DICOM imaging formats for X-ray, MRI, CT
IHE (Integrating the Healthcare Enterprise)	Practical integration of existing standards into workflows	Profiles combining HL7 + DICOM for interoperability
WHO (World Health Organization)	Global health informatics guidelines, especially for developing nations	ICDs (international classification of diseases), global health reporting standards

Wherever statistical analyses or case-based reimbursement are needed, data has to be in a structured form, with a trade-off regarding scope and granularity. Where communication between health professionals is paramount, poorly structured narratives tend to prevail over structured and coded data, because text is richer in detail and faster to create.

The FAIR guiding principles address the [deficiency] of data reusability for clinical data, which were primarily for scientific data management. FAIR stands for findability, accessibility, interoperability, and reusability.

Few standardization development bodies are established that create and maintain standards that ensures interoperability, data quality, and security in healthcare and medical informatics systems such as Health Level Seven International (HL7), International Organization for Standardization—Technical Committee 215 (ISO/TC 215), European Committee for Standardization—Technical Committee 251 (CEN/TC 251), etc. Figure shows several standardization bodies, their focus area, and standards. Fig. 1.1 shows some standardization development bodies with areas they focus on along with some examples/standards.

1.5 Integration of AI/ML, computational biology, and medical informatics

1.5.1 AI in computational biology

What is a train without its engine? AI plays an imperative role in computational biology allowing analysis, modeling, and prediction of complex data. DL algorithms such as convolutional neural networks (CNNs) are applied on image data (MRI, CT scans, microscopy images), recurrent neural networks (RNNs), and long short-term memory (LSTM) networks are used for DNA/RNA sequence analysis; generative adversarial networks (GANs) are used for generating and augmenting synthetic data to enrich training data; and so on. AI empowers computational biology to surpass traditional methods by leveraging ML and DL techniques to process large-scale biological data, speeding up advancements in genomics, structural biology, systems biology, and drug development (Esteva et al., 2019).

1.5.2 Impact of AI on medical informatics

The impact of AI on medical informatics has been transformative, fundamentally changing how healthcare data is processed, analyzed, and applied to patient care. From enhancing clinical decision support and improving diagnostic accuracy to enabling precision medicine and optimizing healthcare operations, AI has expanded the capabilities and potential of medical informatics systems. AI is a powerful tool that can enhance diagnostics, treatment plans, patient care, and administrative processes (Topol, 2019).

AI has impacted several areas in the medical informatics field:

i. Clinical decision support systems (CDSS)—AI has revolutionized the operations of CDSS, enabling it to process and interpret vast amount of healthcare data with unprecedented speed and accuracy. ML helps CDSS to discern patterns, recognize correlation, and derive insights from complex datasets. It can offer evidence-based personalized recommendations tailored to individual patient needs. DL models such as CNN-based CDSS can extract complex patters from medical images providing in-depth analysis.
ii. EHR enhancements—Medical data is overwhelmingly voluminous in nature, which continues to grow increasingly. Advance ML and natural language techniques offer better ways to manage and analyze EHR data more efficiently. AI-powered systems are vastly capable of scanning clinical notes, predicting disease progression, and even estimating patient morality rated using models such as CNN and graph-based learning methods.
iii. Medical imaging—The advancements in ML and DL algorithms have paved the way to the development of sophisticated AI-powered solutions capable of analyzing imaging data with unprecedented speed and accuracy. These systems can discern subtle patters, detect anomalies, and extract clinically relevant information from medical imaging data, to provide timely and accurate diagnoses. The hierarchical architecture of CNNs enables capturing of abstract features in the medical images at multiple layers of the network, such as in case of tumor detection they can identify low-level features such as edges and textures, as well as higher-level feature namely, tumor shape and spatial relationships with the surroundings. CNN has the capability of image classification, object detection, image reconstruction, abnormality detection, and generation of clinically relevant insights [10].
iv. Precision medicine and personalized healthcare—Precision medicine focuses on identifying patient phenotypes with poor clinical response of specific treatment needs. AI-powered algorithms can analyze genomics sequence to identify genetic variations associated with specific disease, which can help healthcare providers to perform targeted treatments. AI treatment plans can be customized based on patients' medical history, genetic profile, and response to various treatments. AI-driven platforms assist oncologists in selecting the most effective chemotherapy drugs based on individual patient profiles, thereby improving treatment efficacy while minimizing adverse effects.

Fig. 1.3 depicts a summarized version of contribution of AI and its impact on the various domains of medical informatics (Table 1.2).

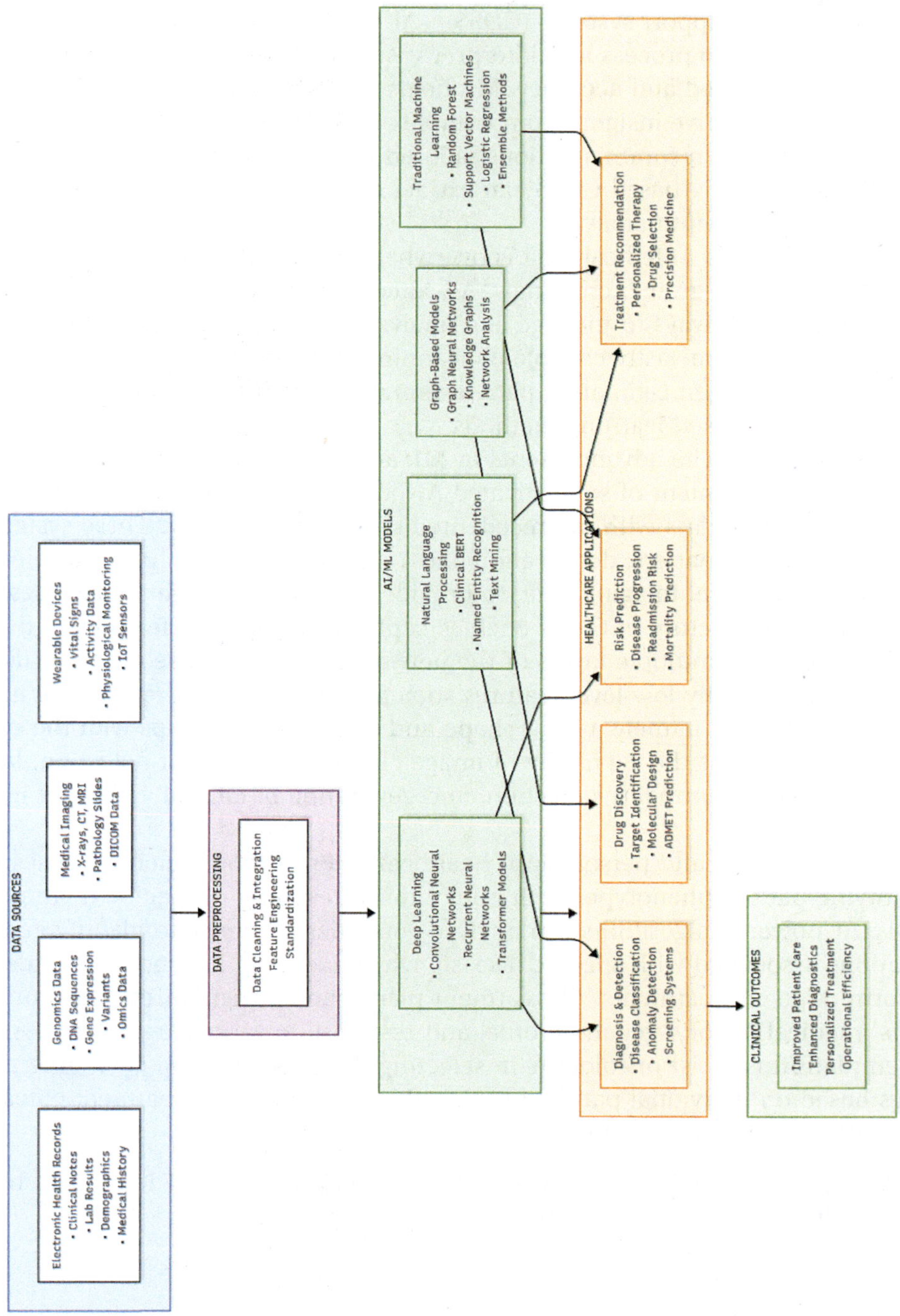

FIGURE 1.3 Intervention of AI in healthcare.

Table 1.2 Contribution and impact of AI in various domains.

Domain	AI contribution	Impact
EHR management	NLP, predictive modeling	Better record structuring, error detection
Clinical decision support	Deep learning, expert systems	Faster, evidence-based decisions
Medical imaging	CNNs, transformers	Early diagnosis of cancer, heart disease
Public health	ML forecasting	Epidemic prediction and control

1.6 Application in healthcare and life sciences

Healthcare is a multilayered system developed specifically for preventing, diagnosing, and treating diseases. In healthcare, ML has gained significant attention in fields of genomics and precision medicine. AI algorithms can analyze genomic sequences accurately to identify genetic variations associated with specific disease. AI-driven platforms such as IBM Watson Genomics and Google Deepvariant have revolutionized the field by offering accurate and rapid genetic analysis (Shortliffe and Cimino, 2014).

AI is also accelerating the drug discovery and development process by analyzing biochemical data, predicting drug interactions, and identifying potential therapeutic candidates.

Medical images, including X-rays, MRIs, and CT scans, are prominently being scanned using AI algorithms to detect disease at an early age. Google's DeepMind for retinal disease detection and IBM Watson for oncology have demonstrated high accuracy in identifying abnormalities, enabling timely interventions.

1.7 Challenges

The use of AI in healthcare presents several challenges, including data privacy and security, ethical considerations, interoperability and integration, algorithmic bias, lack of standardization and regulation, explanation and trust, integration into clinical workflow, resource constraints, resistance to change, data quality and bias in datasets, regulatory compliance, patient acceptance and engagement, and continuous learning and adaptation (Obermeyer and Emanuel, 2016). Data privacy and security are crucial for preventing unauthorized access or breaches, while ethical considerations involve transparency, accountability, and potential bias in decision-making. Interoperability and integration (*World Health Organization, International Classification of Diseases (ICD-11).' WHO Press*, 2022), (*HL7 International, "HL7 Standards*, 2023) across diverse healthcare platforms remain a persistent challenge due to the use of different standards and formats for data storage.

Algorithmic bias and fairness are also significant issues, as AI algorithms can perpetuate biases in training data, leading to disparities in healthcare outcomes. The absence of clear guidelines can lead to uncertainties in evaluating, validating, and deploying AI applications. Transparent and interpretable AI systems are essential for gaining the trust of healthcare professionals and patients. Integrating AI tools seamlessly into clinical workflows poses challenges, as healthcare professionals may face disruptions and require training to effectively use new technologies. Resource constraints, such as financial investments, skilled personnel, and infrastructure, can be barriers for smaller healthcare facilities. Resistance to change, data quality, and regulatory compliance add complexity to AI implementation. Patient acceptance and engagement with AI technologies are ongoing challenges, as they may be hesitant to fully trust AI-driven recommendations. Continuous learning and adaptation are also essential for AI systems to stay current with the latest advancements in healthcare.

1.8 Conclusion

This chapter has explored the transformative convergence of AI, ML, computational biology, and medical informatics, demonstrating how these interconnected disciplines are revolutionizing healthcare delivery, biomedical research, and life sciences applications. The integration of these fields represents a paradigm shift from traditional, siloed approaches to a unified, data-driven ecosystem that leverages computational intelligence to address complex biological and medical.

Take-home messages

- **AI and ML integration in healthcare** has become an essential and powerful tool that can handle complex and huge data generated in the field.
- **Healthcare and life sciences are being transformed** by the support that is provided with the use of AI such as in genomics, medical imaging, precision medicine, and clinical decision support.
- **Improvement scope in the challenges**—as AI is still possessing several risks to the sensitive data, ethical risks, and biasness.
- **Collaboration across domains**—computer science, biology, medicine, and informatics—is key to unlocking the full potential of these innovations.
- **Paradigm shift** as seen by the integration of AI/ML with biomedical sciences, toward personalized, patient-centric, and efficient healthcare.

References

Bishop, C.M., Pattern recognition and machine learning. (2006),

Esteva, A., Robicquet, A., Ramsundar, B., Kuleshov, V., DePristo, M., Chou, K., Cui, C., Corrado, G., Thrun, S., Dean, J., 2019. A guide to deep learning in healthcare. Nat. Med. 25 (1), 24–29. https://doi.org/10.1038/s41591-018-0316-z.

HL7 International, "HL7 standards. (2023),

Jumper, J., Evans, R., Pritzel, A., Green, T., Figurnov, M., Ronneberger, O., Tunyasuvunakool, K., Bates, R., Žídek, A., Potapenko, A., Bridgland, A., Meyer, C., Kohl, S.A.A., Ballard, A.J., Cowie, A., Romera-Paredes, B., Nikolov, S., Jain, R., Adler, J., Back, T., Petersen, S., Reiman, D., Clancy, E.,

Zielinski, M., Steinegger, M., Pacholska, M., Berghammer, T., Bodenstein, S., Silver, D., Vinyals, O., Senior, A.W., Kavukcuoglu, K., Kohli, P., Hassabis, D., 2021. Highly accurate protein structure prediction with AlphaFold. Nature 596 (7873), 583–589. https://doi.org/10.1038/s41586-021-03819-2.

Larrañaga, P., Calvo, B., Santana, R., Bielza, C., Galdiano, J., Inza, I., Lozano, J.A., Armañanzas, R., Santafé, G., Pérez, A., Robles, V., 2006. Machine learning in bioinformatics. Briefings Bioinf. 7 (1), 86–112. https://doi.org/10.1093/bib/bbk007.

Lecun, Y., Bengio, Y., Hinton, G., 2015. Deep learning. Nature 521 (7553), 436–444. https://doi.org/10.1038/nature14539.

McCarthy, J., Minsky, M.L., Rochester, N., Shannon, C.E., 2006. A proposal for the dartmouth summer research project on artificial intelligence. AI Mag. 27 (4), 12–14.

Obermeyer, Z., Emanuel, E.J., 2016. Predicting the future-big data, machine learning, and clinical medicine. N. Engl. J. Med. 375 (13), 1216–1219. https://doi.org/10.1056/NEJMp1606181.

Russell, S., Norvig, P. Artificial intelligence: a modern approach. (2021),

Shortliffe, E.H., Cimino, J.J., 2014. Biomedical Informatics: Computer Applications in Health Care and Biomedicine: Fourth Edition. Springer London, United States, pp. 1–965. https://doi.org/10.1007/978-1-4471-4474-8.

Topol, E.J. Deep medicine: how artificial intelligence can make healthcare human again. (2019),

Turing, A.M., 1950. I.—computing machinery and intelligence. Mind LIX (236), 433–460. https://doi.org/10.1093/mind/lix.236.433.

World Health Organization, 2022. International Classification of Diseases (ICD-11). WHO Press.

2

AI/ML applications in medical imaging

Yukti Sabikhi[1], Anshika Singh[1], Sameen Masroor[1], Chhavi Dudeja[1], Syed Sibtain Khalid[2], Aafia Tasneem[3], Hammad Mansoor[4,*], Emad Zeni[5], Shamim Ahmed Shamim[6] and Mohammad "Sufian" Badar[7,8,9,10]

[1]JAMIA HAMDARD UNIVERSITY, SCHOOL OF ENGINEERING SCIENCES & TECHNOLOGY, NEW DELHI, INDIA; [2]DEPARTMENT OF COMPUTER SCIENCE AND ENGINEERING, SCHOOL OF ENGINEERING SCIENCES AND TECHNOLOGY (SEST), JAMIA HAMDARD UNIVERSITY, NEW DELHI, INDIA; [3]BADAR MEDICAL CENTER (BMC), NEW DELHI, INDIA; [4]TELUS DIGITAL, CANDOR TECH SPACE, SECTOR 135, NOIDA, UTTAR PRADESH, INDIA; [5]LAMAR UNIVERSITY, BEAUMONT, TX, UNITED STATES; [6]DEPARTMENT OF NUCLEAR MEDICINE, AIIMS, NEW DELHI, INDIA; [7]DEPARTMENT OF BIOENGINEERING, UNIVERSITY OF CALIFORNIA, RIVERSIDE, CA, UNITED STATES; [8]UNIVERSAL SCIENTIFIC EDUCATION AND RESEARCH NETWORK (USERN), TEHRAN, IRAN; [9]SPI DARBHANGA, DARBHANGA, BIHAR, INDIA; [10]DEPARTMENT OF COMPUTER SCIENCE AND ENGINEERING (BIOINFORMATICS), SCHOOL OF ENGINEERING SCIENCES AND TECHNOLOGY (SEST), JAMIA HAMDARD UNIVERSITY, NEW DELHI, INDIA

*E-MAIL ADDRESS: HAMMADMANSOOR20@GMAIL.COM.

2.1 Introduction

Medical imaging is the process by which an image of the structure and function is created for various kinds of human tissue and organs for clinical purposes or to enable medical science to study thoroughly the normal and abnormal anatomy and physiology of the body (Abhisheka et al., 2023). These imaging techniques become an alternative to exposing internal structures to anomalies and treatment beneath the skin and bones (Obuchowicz et al., 2024a). Radiology, using imaging techniques such as X-ray radiography (Irede et al., 2026), X-ray computed tomography (CT) (Bellens et al., 2024), endoscopy, magnetic resonance imaging (MRI), magnetic resonance spectroscopy (MRS) (Saleh et al., 2025), positron emission tomography (PET) (Sammartino et al., 2024), thermography (Abdulkareem and Hashim, 2024; Shi et al., 2024); medical photography (Singh et al., 2022); electrical source imaging (ESI); digital mammography; tactile imaging; magnetic source imaging (MSI) (Fu et al., 2025); medical optical imaging;

Applying Artificial Intelligence to Computational Biology and Medical Informatics. https://doi.org/10.1016/B978-0-443-44915-4.00002-2

single-photon emission computed tomography (SPECT) (Massari and Mok, 2023); ultrasonic and electrical impedance tomography (EIT) (Biernacki et al., 2022), are included in this adjunct.

Artificial Intelligence (AI), in particular, the deep learning models, has shown an unparalleled ability to identify useful information that is buried in the medical images. The models, after being trained on large datasets, have the ability to identify complex patterns and subtle details (Pinto-Coelho, 2023a). This ability, in addition to enhancing the accuracy of diagnosis, brings new information to the image features most important to making decisions regarding patients' treatment. In regard to the speed of detecting abnormalities and the body's anatomical structures, as well as disease course predictions, AI aids healthcare professionals (Ennab and Mcheick, 2024). The prompt and accurate reading of medical images is especially important for spotting diseases in their early stages that other techniques cannot detect. They can be greatly beneficial in both early diagnosis and treatment, which leads to better results for patients (Hussain et al., 2022).

Moreover, advances in ML-based image segmentation have also allowed us to precisely quantify these critical structures, such as tumors and blood vessels. This precision can be an important dimension for the plan of treatment, thereby enabling therapists to intervene optimally and accordingly tailor therapies to become even more individualized with regard to the patient (Xu et al., 2025). In conclusion, AI-based applications in the field of medical imaging have huge potential for the future in enhancing diagnosis and predicting disease properties.

This chapter emphasizes the pivotal position of medical imaging in contemporary healthcare, reflecting a broad range of methods employed to image and analyze the internal structures of the human body. It also points out the revolutionary role of Artificial Intelligence (AI), specifically deep learning, to improve image interpretation, raise diagnostic accuracy, and facilitate early detection of disease.

2.2 Overview of AI/ML in medical imaging

AI encompasses computational techniques that aim to replicate human cognitive abilities such as reasoning, learning, and decision-making. Rather than being a single technology, AI represents a broad, multidisciplinary domain that integrates ML, natural language processing (NLP), computer vision, robotics, and advanced decision-support systems (Khan et al., 2025). Its rapid growth in healthcare has been facilitated by the availability of large-scale biomedical data, increased computational capacity, and the development of sophisticated algorithms.

Within this ecosystem, ML constitutes the most widely applied subset of AI, as it enables systems to automatically identify patterns and generate predictions from data without explicit programming(Sa(Ke et al., 2022). ML approaches are generally categorized into supervised learning, which relies on labelled datasets for tasks such as tumor classification or disease risk prediction; unsupervised learning, which uncovers hidden structures in unlabelled data and is particularly useful in clustering medical images or genomic profiles(Trezza et al., 2024); semisupervised learning, which leverages both labelled and unlabelled datasets to enhance performance when annotations are limited;

and reinforcement learning, which optimizes sequential decision-making through feedback mechanisms and is increasingly explored in treatment planning and robotics.

A transformative advancement within ML is deep learning (DL), which utilizes multi-layered neural networks to automatically extract and learn hierarchical features. Architectures such as convolutional neural networks (CNNs) have demonstrated exceptional performance in image-intensive domains, including radiology, histopathology, and ophthalmology, often surpassing traditional ML techniques (Luo et al., 2025). Alongside DL, NLP techniques are reshaping healthcare by extracting clinically meaningful insights from unstructured sources such as electronic health records (EHRs), physician notes, and medical literature (Derek and Collings, 2025). Similarly, AI-enabled voice systems and digital assistants are reducing the documentation burden, enhancing patient engagement, and supporting decision-making in real time (Alshami et al., 2025). Medical robotics represents another frontier, offering precision in surgical interventions, rehabilitation, and assistive care.

These advances highlight a paradigm shift from “artificial intelligence” to “augmented intelligence”, emphasizing AI as a partner that strengthens, rather than replaces, human expertise (Mohamed, 2025). The healthcare AI ecosystem is currently shaped by three major stakeholders: EHR vendors embedding ML and NLP tools, technology giants such as Google, IBM, and Microsoft providing scalable platforms, and specialized start-ups driving innovations in diagnostics, chatbots, and telemedicine. On a global scale, adoption is accelerated through cross-sector collaborations, with the United States leading industry-driven applications, the UK and Canada fostering public–private partnerships, and China rapidly advancing in large-scale AI integration (Ávila et al., 2025). Collectively, the fundamentals of AI and ML form the conceptual and technological basis for the transformative role of medical imaging in diagnostics and healthcare innovation (Pinto-Coelho, 2023b).

AI and ML are transforming medical imaging through the computerized detection, segmentation, and classification of sophisticated disease patterns (Obuchowicz et al., 2024b). Behind these applications are Convolutional Neural Networks (CNNs), which draw hierarchical features from imaging data to a much better extent than conventional handcrafted methods. Appropriate preprocessing involving normalization, augmentation, and denoising is needed to enhance the quality of the data as well as model generalization, while partitioning the dataset into training, validation, and testing sets provides reproducibility (Yaqoob, Musheer Aziz and Verma, 2023). The general workflow is illustrated in Fig. 2.1.

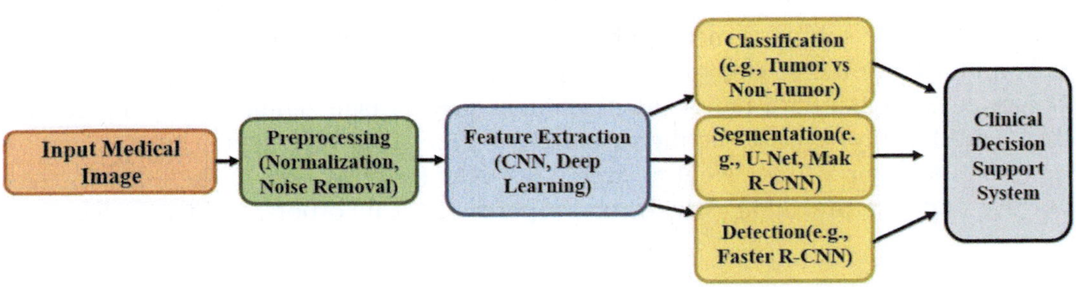

FIGURE 2.1 General workflow of AI/ML in medical imaging.

Table 2.1 Key AI/ML models and representative applications in medical imaging.

Model	Tasks
CNN	Image classification
U-Net	Image segmentation
ResNet	Feature extraction and classification
Faster R–CNN	Object detection
GANs	Image generation and augmentation

A number of CNN designs have become broadly used. The VGG-19 is characterized by good feature extraction abilities with robust accuracy and stability across datasets. The ResNet-50, through skip connections, attains good training accuracy but overfits easily. LeNet-5, though very light, is characterized by consistent and dependable performance, thus suitable in resource-limited clinical environments (Xu et al., 2024). DenseNet promotes sharing of features, as well as gradient vanishing alleviation, but sometimes underperforms compared to VGG-19 or LeNet. Comparative studies indicate that the most robust are usually VGG-19 and LeNet-5, while the optimal performance for ResNet-50 and DenseNet needs appropriate tuning (Vempati et al., 2024). A comparative overview of these models and their applications is summarized in Table 2.1.

Besides conventional CNNs, there are customized architectures for specific jobs. Segmentation is particularly performed very well with U-Net to enable pixel-level mapping of lesions or tumors(Azad et al., 2024). Detection is particularly done very effectively with Faster R–CNN through localization of regions of interest(Xue et al., 2021). There has been successful implementation in breast cancer imaging with the hybrid Faster R–CNN + U-Net architecture: Faster R–CNN identifies suspicious regions with localization, while U-Net is used to fine-tune segmentation of tumor boundaries (Yan et al., 2025). The integration optimizes diagnostic performance while dealing with detection as well as segmentation in one pipeline. The performance assessment of these models is usually dependent on accuracy, sensitivity, specificity, precision, recall, and F1-score(Chicco and Jurman, 2020). Each model has strengths and trade-offs in offering them. For instance, U-Net has great segmentation but is best accompanied by detection models, whereas ResNet-50 has robust training results but is prone to overfitting when not adequately regularized. Briefly, CNN-based model integration—from early designs such as LeNet to sophisticated hybrids has greatly promoted medical image analysis. Such methodologies improve diagnostic efficiency, guarantee sounder decision-making, and offer scalable solutions that can be applied across imaging modalities.

AI and ML are increasingly being integrated into the clinical radiological workflow (Najjar, 2023). A standard radiologic pipeline involves image acquisition via modalities like CT, MRI, X-ray, or PET, and then archiving in Picture Archiving and Communication Systems (PACS) or Vendor Neutral Archives (VNA) (Li et al., 2024). Radiologists then perceive and interpret images with final reports being incorporated in the Electronic

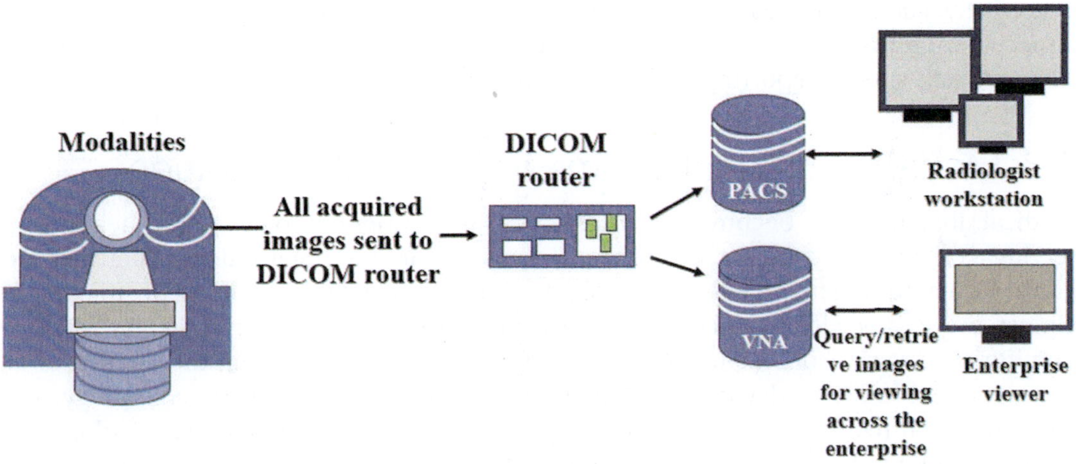

FIGURE 2.2 Standard radiology workflow with AI/ML integration points.

Medical Record (EMR) (Castillo et al., 2021). AI/ML are integrated at several points within the radiologic pipeline to aid in triage, lesion detection, segmentation, and structured reporting, as shown in Fig. 2.2. Standard workflow typically involves five steps: image acquisition, DICOM routing, archiving within PACS/VNA, radiologists' visualization, and report generation with incorporation in EMR(Inamdar, 2020). AI can be added at several points within this paradigm to enhance efficiency, improve review speed, and supply decision-making.

The incorporation of AI/ML within radiology can be characterized at varying levels of maturity. At the research phase, AI models are developed and tested post-hoc on historical PACS data with results saved outside of PACS/EMR for review. At the production phase, prequalified models are incorporated directly within PACS/EMR, facilitating real-time interpretation and automated flagging of abnormalities. At the feedback phase, radiologists' corrections and annotations are cycled back within the system, defining a closed learning loop that improves accuracy within time(Girum et al., 2021). This step-wise maturity scale both assures and facilitates adaptability within clinical practice. Real-world examples show how AI is being increasingly integrated with prevailing diagnostic workflows(Ortega-Robles and Arias-Carrión, 2025). For example, with brain metastases detection using MRI, initially CNNs were used on retrospective data and then integrated within PACS viewers to enable real-time detection before being fine-tuned by constant radiologist feedback in order to eliminate false positives and increase sensitivity (Dikici et al., 2020). Analogously, with breast cancer detection using mammography, deep learning structures like Faster R–CNN and U-Net were used for lesion association and segmentation. These were incorporated within PACS so that suspicious calcifications and masses were identified and highlighted semiautomatically by the system, and radiologists' inputs further refined detection thresholds with the end-result being better diagnostic accuracy and lowered workload(Abhisheka et al., 2023). Individually and collectively, such examples underscore the point that AI/ML is neither a substitute nor a replacement for radiologists but rather a collaborative assistant that enhances diagnostic accuracy and speeds reporting and lays the groundwork for constant system learning.

2.3 Modalities in medical imaging and AI integration

Medical imaging has become an indispensable component of modern healthcare, serving as the primary means for noninvasive visualization of anatomical structures and physiological functions. Imaging technologies have changed clinical decision-making in almost every specialty by facilitating early diagnosis, therapy planning, and disease monitoring. A variety of imaging modalities, each tailored for particular therapeutic uses, have emerged in recent decades as a result of developments in physics, technology, and computational techniques (Weissleder et al., 2016). These modalities differ in their underlying principles, ranging from ionizing radiation and magnetic resonance to sound waves and molecular tracers.

AI has emerged as a transformative force in medical imaging, extending far beyond its early applications in automated pattern recognition. It is now integrated across the entire imaging workflow from acquisition and reconstruction to image analysis, interpretation, and clinical reporting, fundamentally reshaping diagnostic practices (Najjar, 2023). Deep learning in particular has demonstrated the ability to learn complex, high-dimensional representations directly from imaging data, surpassing traditional rule-based or handcrafted feature methods. This integration has not only enhanced diagnostic accuracy but also enabled new capabilities such as image reconstruction from incomplete data, automated quality control, and multimodal data fusion (Kumar, 2022).

2.3.1 X-ray imaging and automated interpretation

Radiographic imaging remains the most widely deployed modality in routine diagnostics due to its low cost, high availability, and ability to capture rapid two-dimensional projections of anatomical structures (A(Li et al., 2024). Chest radiography is among the most frequently conducted imaging examinations worldwide, providing a first-line diagnostic tool in pulmonary, cardiac, and musculoskeletal assessment (Lewis et al., 2021). Despite their utility, the interpretation of X-rays is limited by low soft-tissue contrast, superimposition of anatomical features, and considerable interobserver variability, which collectively hinder diagnostic precision.

The emergence of AI has enabled automated interpretation strategies capable of addressing several of these limitations. Deep convolutional neural networks and transformer-based architectures have been trained on large annotated repositories such as ChestX-ray14, MIMIC-CXR, and CheXpert to detect pathologies including pneumonia, tuberculosis, pneumothorax, lung nodules, and fractures with performance approaching that of expert radiologists (Irvin et al., 2019; Rajpurkar et al., 2018). Automated systems extend their utility beyond disease classification by enabling abnormality localization through saliency mapping, bounding box generation, and heatmap overlays, thereby enhancing interpretability and supporting clinical decision-making (Yuan et al., 2024). Building on these capabilities, AI-driven approaches in X-ray imaging further contribute to workflow optimization through automated triage, real-time detection of critical pathologies, image quality assessment, and preliminary report generation (Lakhani and Sundaram, 2017). These advancements are particularly impactful in resource-limited environments, where the scarcity of subspecialty radiologists underscores the need for reliable and efficient decision-support systems.

Clinical translation of AI-driven X-ray analysis continues to face critical limitations. Distributional shifts arising from protocol variability, hardware diversity, and demographic heterogeneity undermine model robustness, whereas annotation noise, insufficient validation across external cohorts, and limited explainability further restrict adoption. Robust prospective trials, standardized multi-center benchmarking, and stringent regulatory oversight remain prerequisites for reliable integration (Kelly et al., 2019; Tang et al., 2021).

2.3.2 Computed tomography with AI-based analysis

Computed tomography (CT) offers high-resolution, cross-sectional imaging, making it indispensable in trauma assessment, oncology, and vascular imaging. The modality's ability to provide detailed anatomical information is counterbalanced by challenges such as high radiation doses and the complexity of interpreting large volumes of dataAI has been increasingly integrated into CT imaging to address these issues (Buzug, 2011). AI applications in CT encompass several areas, including image reconstruction, lesion detection, and quantitative analysis. Deep learning-based reconstruction techniques have demonstrated the potential to reduce noise and artifacts, thereby enhancing image quality and allowing for dose reduction without compromising diagnostic accuracy (Brady et al., 2021). These advancements facilitate the use of lower radiation doses in clinical practice, aligning with the principle of as low as reasonably achievable (ALARA) (Dudhe et al., 2024). In lesion detection, AI algorithms have been developed to identify and characterize abnormalities such as lung nodules, tumors, and vascular anomalies. These systems utilize convolutional neural networks (CNNs) to analyze CT images, providing automated detection that can assist radiologists in identifying clinically significant findings (Huang et al., 2019). The integration of AI in this context aims to improve diagnostic accuracy and efficiency, particularly in high-volume settings. Quantitative analysis through AI enables the extraction of detailed metrics from CT images, such as tumor volume and tissue density. These quantitative assessments can be utilized for treatment planning, monitoring therapeutic response, and prognostication (Paudyal et al., 2023). The ability to perform such analyses automatically and consistently is a significant advantage in clinical workflows.

2.3.3 Magnetic resonance imaging (MRI) with deep learning enhancement

Magnetic resonance imaging (MRI) provides high-contrast visualization of soft tissues without ionizing radiation, making it a primary modality for neuroimaging, musculoskeletal assessment, and oncologic evaluation. MRI produces multiparametric datasets including T1-, T2-, and diffusion-weighted images, functional imaging sequences, and dynamic contrast-enhanced studies. The complexity and volume of these datasets create challenges for conventional analysis, motivating the integration of AI for reconstruction, enhancement, and interpretation (Delongchamps et al., 2011). Deep learning methods have been applied to accelerate MRI acquisition by predicting fully sampled images from undersampled k-space data (Zeng et al., 2021). These approaches enable shorter scan times while maintaining image fidelity and signal-to-noise ratio, which improves patient comfort and throughput. AI-driven denoising and artifact correction have also been implemented to mitigate motion artifacts, field inhomogeneity effects, and other acquisition-related distortions (Yang et al., 2024).

In image analysis, convolutional and attention-based neural networks facilitate automated segmentation of organs, tumors, and lesions. These models provide quantitative metrics such as volumetric measurements, tissue characterization, and morphological assessment, which support diagnosis, treatment planning, and longitudinal monitoring (Havaei et al., 2017). Integration of radiomics with deep learning further enables the extraction of high-dimensional features from MRI scans, offering predictive biomarkers for disease progression, therapy response, and patient stratification (Xia et al., 2024). The implementation of AI in MRI also addresses operational efficiency. Automated scan quality assessment, protocol selection, and detection of anomalies during acquisition improve workflow and reduce the likelihood of repeated scans. AI-assisted reconstruction and segmentation decrease radiologist workload and enable standardized evaluation across large datasets (Hammernik et al., 2018).

2.3.4 Ultrasound imaging with AI-assisted detection

Ultrasound imaging is a real-time, nonionizing modality widely used in obstetrics, cardiology, abdominal imaging, and emergency medicine. It offers portability and bedside applicability but is highly operator-dependent, with image quality affected by probe angle, patient anatomy, and acquisition technique (Tenajas et al., 2023). The inherent variability and the presence of speckle noise make interpretation challenging, motivating the integration of AI to assist in detection, classification, and standardization. AI applications in ultrasound focus on automated recognition of anatomical structures, lesion detection, and measurement tasks (Akkus et al., 2019). Convolutional neural networks and deep learning models have been developed to identify fetal biometric parameters, cardiac chambers, liver lesions, and thyroid nodules with high accuracy. These systems can provide real-time feedback to the operator, enhancing the quality and consistency of image acquisition while reducing dependence on operator expertise. Automated segmentation and classification using AI also enable quantitative assessment of tissue morphology and function. For example, in echocardiography, deep learning models can calculate ejection fraction, chamber volumes, and wall motion abnormalities, supporting diagnostic and prognostic evaluation (Jiang et al., 2025). In obstetric imaging, AI facilitates standard plane detection and biometric measurements, improving efficiency and reducing measurement variability.

In ultrasound imaging, AI enhances both interpretation and workflow efficiency by identifying abnormal scans, highlighting key areas, and checking image quality. Emerging techniques involve 3D/4D reconstruction, elastography analysis, and integration with other imaging methods to improve diagnostic accuracy (Boulerbah et al., 2025).

2.3.5 Emerging imaging modalities supported by AI

Recent advances in medical imaging have introduced emerging modalities that extend beyond conventional radiography, CT, MRI, and ultrasound, offering enhanced functional, molecular, and structural information. Techniques such as photoacoustic imaging, spectral CT, hybrid PET/MRI, optical coherence tomography (OCT), and functional near-infrared spectroscopy (fNIRS) provide complementary insights into tissue physiology, perfusion, and molecular composition (Tian et al., 2023). The complexity and novelty of these modalities necessitate AI-based approaches for effective image reconstruction, analysis, and interpretation. AI has been applied to accelerate acquisition and improve image quality in these emerging modalities. In photoacoustic imaging, deep learning models reconstruct high-resolution volumetric images from limited sensor data, enabling rapid functional imaging of vasculature and tissue oxygenation (Zhu et al., 2022). In spectral CT, AI assists in material decomposition and quantitative mapping of contrast agents, improving detection of lesions and vascular abnormalities. Hybrid PET/MRI benefits from AI-driven attenuation correction, motion compensation, and lesion segmentation, enhancing molecular imaging sensitivity and reducing scan times (Hussain et al., 2024).

2.3.6 Applications of AI/ML in clinical diagnosis

These days, rapidly advancing AI in oncology is largely focused on understanding the molecular biology of cancer, processing big biomedical datasets, and making early predictions through NGS and Imaging. An AI system in bioinformatics helps in biomarker discovery, custom drug design, and targeted therapy. AI is classified as General AI, Super AI, and Narrow AI(Kumar, 2017), where Narrow AI is application-specific, such as Siri, Alexa, or a self-driving car. According to recent initiatives, such as Nvidia planning to build an AI supercomputer for medical research, this situation underpins the urgent need for oncologists to embrace AI-based skills for safe and effective clinical use (Aron, 2011).

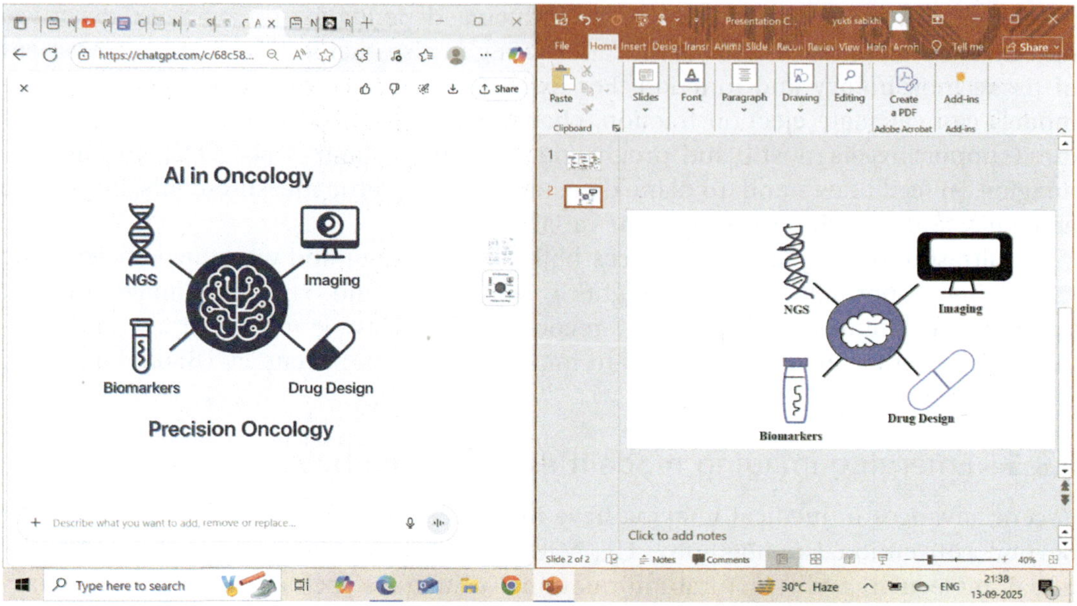

FIGURE 2.3 Applications of AI/ML in medical imaging.

With uses ranging from robotic-assisted surgery and precision oncology to drug development and clinical diagnosis (Fig. 2.3), AI has become a disruptive force in contemporary healthcare. AI systems enable computers and robots to manage massive medical datasets, simulate human intellect, and understand the cellular makeup of diseases like cancer. Clinical practice is being redefined by both the virtual and physical branches of AI: the physical branch supports robotic surgery and nano-robotics for targeted drug delivery, while the virtual branch uses deep learning (DL) algorithms to interpret electronic health records and direct precision decision-making (Hamet and Tremblay, 2017). In diagnostics, ML algorithms, such as supervised, unsupervised, and reinforcement learning, have improved clinical decision-making by revealing patterns in biomedical data sets. ML has already made strides in protein–protein interaction mapping, therapeutic target discovery, and discovery of new DNA variants linked to disease risk (Rapakoulia et al., 2014). Likewise, AI-powered "care bots" and medical devices are being used to track patients in real-time, especially in elderly care and in surgical support (Larson et al., 2014).

The worldwide use of AI for healthcare is illustrated by projects like NHS 24, the Scotland-based telephonic patient care project that uses DL-based algorithms, and Babylon Health, which applies semantic web technology to provide digital clinical assistance (McCartney, 2018). In cancer, AI-assisted platforms scrutinize immense amounts of data from next-generation sequencing (NGS), proteomics, and imaging, facilitating precision oncology strategies for biomarker discovery, tailored treatment, and early detection of cancer (Dlamini et al., 2020).

AI has also revolutionized medical imaging technologies. Deep learning–based models can detect abnormal cellular growths with higher sensitivity than conventional methods, significantly advancing radiology, MRI, and neuroradiography(Recht and Bryan, 2017). Mobile applications such as SkinVision demonstrate AI's potential in early cancer detection, achieving up to 95% accuracy in identifying malignant skin lesions (Malhi and Yiu, 2021). Likewise, AI-based computational pathology uses whole-slide imaging (WSI) and deep neural networks to create high-quality tumor images and new biomarkers, enhancing diagnostic accuracy and drug discovery pipelines (Bera et al., 2019).

In clinical practice, AI-assisted ML models assist in preoperative risk stratification, intraoperative decision support, and postoperative predictive outcomes. Random forest and support vector machine models have been shown to possess robust predictive capability in breast cancer survival prognosis and in minimizing unnecessary surgical resections (Ganggayah et al., 2019). In addition, the construction of Collective Surgical Consciousness (CSC) highlights the incorporation of AI in real-time operating room monitoring and population-level analysis(Harvey et al., 2016). Even with all these developments, ethical issues are a cause of major concern. The black box problem of deep learning models, patient overreliance on the AI systems, transparency issues, and bias issues due to population-specific genetic information are all hindrances to clinical integration with safety (Buchanan et al., 2020). In addition, improved algorithmic precision, interpretability, and clinician training are also constantly required for safe implementation of AI in healthcare.

In all, AI has proven to be an immensely powerful interdisciplinary tool with the potential to transform medicine by enhancing diagnostic accuracy, speeding up drug discovery, and tailoring treatment plans. Despite challenges, the continued integration of computational biology, medical imaging, and ML capabilities continues to move us forward toward more cost-effective, faster, and safer healthcare solutions.

2.3.7 Cardiovascular disease imaging analysis

AI, specifically ML and deep learning (DL), is now essential in cardiovascular imaging through improved diagnostic precision, prognosis estimation, and workflow optimization. These methods solve problems like observer variability, overlooked diagnoses, and time-consuming analyses, thus allowing earlier diagnoses and better patient stratification (Mesko, 2017).

Electrocardiography (ECG) is an inexpensive and commonly available modality that has been greatly aided by ML-based methods. Convolutional neural networks (CNNs) have achieved high accuracy in left ventricular dysfunction detection (AUC 0.93), prediction of atrial fibrillation under sinus rhythm (AUC 0.90), and arrhythmia classification, sometimes even exceeding board-certified cardiologists. Other applications involve age and sex prediction, screening for hyperkalemia, and myocardial infarction detection.

Yet, limitations such as data labeling mistakes and the "black-box" character of neural networks hinder clinical translation (Attia et al., 2019).

Another imaging technique is Echocardiographic interpretation, which has typically relied upon specialized interpretation, but ML has mechanized classification of morphologic and functional characteristics. CNN-based models have discriminated hypertrophic cardiomyopathy, cardiac amyloidosis, and pulmonary arterial hypertension with AUCs >0.85, whereas supervised ML models attained sensitivity as high as 96% in discriminating pathological from physiological hypertrophy. In addition, CNNs have categorized echocardiographic views with accuracies >90% and reclassified the severity of diastolic dysfunction with 99% accuracy. Even with promising outcomes, small dataset sizes and overfitting remain challenges (Sengupta et al., 2016).

Computed Tomography Coronary Angiography (CTCA)-derived ML models have enhanced diagnostic and prognostic evaluation. Ensemble models have accurately predicted 5-year all-cause mortality with greater precision than traditional risk scores (AUC 0.79 vs. 0.61–0.63). ML also better detects lesion-specific ischemia and MACE prediction using CT-FFR, reaching AUCs of up to 0.94. However, external verification and larger populations are needed before general adoption (von Knebel Doeberitz et al., 2019).

Cardiac Nuclear Imaging in which ML has enhanced interpretation of myocardial perfusion single-photon emission computed tomography (SPECT), is a prevalent noninvasive stress imaging modality. CNN models have surpassed standard total perfusion deficit scoring in the detection of coronary artery disease, with higher sensitivity and AUCs. Other ML algorithms that combine perfusion and functional parameters have attained diagnostic accuracies above 85%, outperforming expert interpretation (Arsanjani et al., 2013).

Cardiac Magnetic Resonance (CMR) offers high-resolution functional and structural information, for which ML has been useful in automated segmentation, functional evaluation, and pathology classification. Deep learning algorithms on large annotated databases (e.g., ACDC challenge) delivered >95% accuracy in diagnosis and clinical index extraction. Further, supervised ML used on right ventricular motion evaluation has been able to predict survival in pulmonary hypertension with greater accuracy than standard imaging markers. These benefits notwithstanding, computational cost and interpretability continue to be challenges (Bernard et al., 2018).

Electronic Health Record (EHR)-derived features integrated with imaging data provide an added level for predictive modeling. Large patient cohort deep feature learning has provided prediction of nonhypertensive heart failure (AUC 0.87) and myocardial infarction (AUC 0.85). Nonetheless, heterogeneity in healthcare systems, limited generalizability, and the "black-box" problem limit strong clinical uptake (Miotto et al., 2016).

In general, AI/ML applications in cardiovascular imaging show impressive promise to automate image interpretation, enhance diagnostic accuracy, and foretell clinical outcomes. ECG, echocardiography, CTCA, nuclear imaging, and CMR have all yielded promising results, most often with performance outperforming conventional approaches and even experienced clinicians. In spite of limitations like data quality, generalizability, and interpretability of the algorithm, continuous improvements in computational techniques and incorporation with multimodal datasets are leading the way toward precision cardiology.

2.3.8 Ophthalmic imaging and retinal disease detection

AI and ML have surfaced as revolutionary resources in eye imaging, especially for identifying and treating retinal conditions like diabetic retinopathy (DR), age-related macular degeneration (AMD), and glaucoma. Deep learning (DL) frameworks, especially convolutional neural networks (CNNs), have shown exceptional capability to detect subtle pathological traits in fundus images and optical coherence tomography (OCT) scans that might not be easily visible to human specialists. Recent extensive screening initiatives in India, like the AIDRSS project in Kolkata, demonstrated over 92% sensitivity and 88% specificity for DR detection from more than 10,000 retinal images, highlighting the practicality of AI-supported mass screening in resource-constrained environments (Dey et al., 2025). Additional support from systematic reviews and meta-analyses consistently shows that AI-based DR screening matches, and in certain instances exceeds, manual grading, with combined sensitivity and specificity reliably higher than 0.90 in high-quality imaging datasets (Tahir et al., 2025).

It seems there may have been an error in the input, as there is no text provided for paraphrasing. Please provide the text you would like me to rephrase. Beyond diabetic retinopathy, AI algorithms are being increasingly utilized for age-related macular degeneration and other retinal conditions: OCT-based deep learning systems have demonstrated high accuracy in distinguishing between neovascular and nonneovascular AMD, as well as in forecasting disease progression. It seems you have provided only a closing parenthesis without any text to paraphrase. Please provide the text you would like me to rephrase. New hybrid and "foundation" models developed on multimodal retinal datasets now provide adaptable frameworks for various downstream tasks, such as chorioretinal pathology classification and interpretable disease staging, enhancing their practicality for real-world applications (Driban et al., 2024; Shi et al., 2024)

2.3.9 Neurological disorder diagnosis through imaging

AI and ML have developed into revolutionary resources for diagnosing neurological disorders via imaging, providing enhanced precision, effectiveness, and early identification relative to traditional methods. Neuroimaging methods including magnetic resonance imaging (MRI), functional MRI (fMRI), computed tomography (CT), and positron emission tomography (PET) produce large volumes of intricate data that can be difficult for human specialists to examine thoroughly. AI/ML algorithms—especially deep learning and convolutional neural networks (CNNs)—can automatically identify nuanced imaging features that might go unnoticed by humans, facilitating earlier and more accurate detection of conditions such as Alzheimer's disease, Parkinson's disease, epilepsy, multiple sclerosis, and brain tumors. For example, classifiers based on ML have demonstrated significant ability in distinguishing healthy individuals from those with Alzheimer's by detecting patterns of hippocampal atrophy or cortical thinning from MRI images (Wen et al., 2020). Likewise, deep learning techniques utilized for PET and fMRI data can identify irregular connectivity patterns in Parkinson's disease and other movement disorders, assisting with differential diagnosis and treatment strategies (Litjens et al., 2017).

In epilepsy, the use of ML to automate seizure localization from fused fMRI and EEG imaging data diminishes reliance on invasive techniques (Fahoum et al., 2013). Additionally, AI-based imaging methods are enhancing precision medicine by forecasting disease development and individualized treatment responses, exemplified in multiple sclerosis, where ML models monitor lesion changes and cognitive deterioration (Biernacki et al., 2022).

In addition to diagnostic precision, AI/ML tools also improve reproducibility, lessen variations between observers, and assist radiologists in managing large imaging workloads. Even with these advancements, issues like data diversity, algorithm transparency, and integration into clinical practices persist. Nonetheless, the combination of AI/ML with neuroimaging is leading to individualized, data-informed treatment of neurological conditions.

2.3.10 Predictive analytics for treatment planning

AI and ML are crucial in predictive analytics for treatment strategies, facilitating personalized and data-informed methods for patient care. Utilizing extensive datasets from electronic health records (EHRs), genomic data, imaging, and actual clinical outcomes, AI/ML algorithms can uncover concealed patterns that guide treatment choices and forecast patient reactions. In oncology, for example, ML models evaluate tumor genomics and medical history to predict patient responses to chemotherapy, immunotherapy, or targeted treatments, thus reducing trial-and-error tactics and minimizing side effects (Kourou et al., 2015).

In neurology, predictive ML models utilized on imaging and clinical data can forecast disease progression in disorders such as Alzheimer's disease or multiple sclerosis, aiding early intervention approaches (Pellegrini et al., 2018). Likewise, in cardiology, predictive models enhanced by AI utilize imaging, blood biomarkers, and patient demographics to refine treatment strategies for heart failure and arrhythmias (Johnson et al., 2018). Apart from disease-specific scenarios, reinforcement learning algorithms are being designed to suggest adaptive treatment plans that change alongside patient development, like enhancing insulin dosing in diabetes care (Yu et al., 2023). These forecasting systems aid clinicians in making evidence-driven choices and enhance precision medicine by customizing treatments to personal patient profiles. Nevertheless, obstacles persist, such as data diversity, model clarity, and the incorporation of predictive instruments into clinical processes. Despite these challenges, AI/ML-based predictive analytics is increasingly acknowledged as a fundamental element of contemporary treatment planning, improving outcomes while lowering healthcare expenses.

2.4 Challenges and limitations

The use of AI and ML in medical imaging introduces significant challenges that hinder their incorporation into standard clinical practice. The primary issues involve data management, algorithm reliability, performance evaluation, and regulatory supervision. Data governance necessitates secure storage, consistent annotation, and upkeep of superior quality imaging datasets, whereas algorithm resilience is compromised by bias arising from insufficient diversity in training groups, disease categories, imaging modalities, and clinical settings. This prompts worries regarding fairness, generalizability, and transparency in practical applications (Saw and Ng, 2022). Evaluating performance presents extra challenges, since choosing suitable assessment metrics, avoiding test data leakage, and defining trustworthy reference standards continue to be unresolved problems (Varoquaux and Cheplygina, 2022). The oversight of adaptive, continuously learning models adds complexity to the scenario, as guaranteeing their safety, accountability, and clinical dependability remains a challenge that has not been completely solved. In reaction, organizations like the U.S. Food and Drug Administration (FDA) have established frameworks such as the Predetermined Change Control Plan (PCCP) and encouraged the creation of Good Machine Learning Practices (GMLPs) to tackle these regulatory challenges (U.S. Food and Drug Administration, 2018).

In addition to these, transparency and data constraints continue to be key barriers to advancement. Numerous medical devices powered by AI/ML do not possess adequate peer-reviewed validation, open-source code, or reproducible datasets, limiting independent verification and eroding stakeholder trust(van Leeuwen et al., 2021).

Clinical datasets are often limited in size, requiring innovative methods like data augmentation through synthetic images, weak labeling, or phantom-based techniques to enhance training resilience (Maynord et al., 2023). Furthermore, adaptive AI/ML systems pose issues related to ongoing updates, as a decline in performance can arise if the recalibration techniques are insufficient(Feng et al., 2022). Validation in real-world settings presents difficulties, as the effectiveness of AI/ML tools, like computer-aided triage systems, is significantly influenced by diverse clinical contexts and their integration into workflows (Thompson et al., 2022).

To tackle these challenges, the FDA and research organizations are funding regulatory science tools (RSTs), cooperative projects, and structured methods for robustness testing, fairness assessment, and failure mode examination (U.S. Food and Drug Administration, 2018). Together, these challenges highlight the importance of collaboration among multiple stakeholders to guarantee that medical imaging AI/ML systems attain not just technical superiority but also clinical dependability, safety, and ethical endorsement.

The emergence of AI in medical imaging has sparked both enthusiasm and concern, as some contemplate the possible replacement of radiologists; yet, its implementation in clinical settings remains limited by various ongoing challenges. Historically, advancement was constrained by minimal datasets, insufficient computational power, and disjointed research initiatives, and while deep learning breakthroughs in 2012 and later improvements in computational capacity and extensive datasets sped up progress, significant obstacles persist. Present AI/ML systems face challenges with data constraints, algorithmic bias, inadequate robustness among varied patient populations, inadequate transparency and reproducibility, as well as ongoing regulatory and legal issues concerning adaptive models. These problems not only erode clinical trust but also emphasize that AI is still distant from independent operation, necessitating radiologists' expertise for supervision, contextual analysis, and decisions centered on patient care. Consequently, the future of AI in medical imaging is characterized less by the obsolescence of professionals and more by tackling these obstacles to facilitate safe, equitable, and efficient incorporation into healthcare (Greenspan et al., 2016).

2.5 Future scope

AI and ML are poised to transform the future of medical imaging by enhancing diagnostic precision and clinical decision-making. Given the large volumes of imaging data produced by healthcare systems, implementing federated learning will enable collaborative model training across various institutions while ensuring patient privacy is preserved. This will improve the reliability, applicability, and widespread use of AI-driven imaging solutions globally.

An essential focus is the advancement of explainable AI, which enhances transparency and interpretability. Clinicians will obtain precise predictions and also comprehend the reasoning behind them, thus boosting trust and acceptance of AI in clinical settings. Furthermore, real-time AI-driven imaging technologies are anticipated to assist radiologists in crucial situations like surgeries and emergency procedures, where prompt analysis is essential.

The incorporation of multi-modal data—comprising imaging, genomics, electronic health records, and molecular biomarkers—signifies a significant advancement toward comprehensive disease profiling and individualized medicine. This convergence can facilitate personalized treatment plans, enhanced prognostic precision, and improved patient results. Moreover, analysis of imaging biomarkers powered by AI could speed up drug discovery and therapeutic advancements, connecting diagnostics with the creation of treatments.

Over time, AI/ML will transition from a supportive tool to a proactive partner in clinical processes. In addition to analyzing images, it will assist in forecasting disease advancement, tracking treatment outcomes, and even aiding preventive healthcare measures. This shift establishes AI/ML not just as a diagnostic tool but as a fundamental element of precision medicine, influencing the future of healthcare delivery.

2.6 Conclusion

AI and ML are changing medical imaging at an extraordinary pace, providing opportunities to enhance diagnosis, prognosis, and clinical decision-making that have never been seen before. AI-based tools are reinventing workflows, from image reconstruction and segmentation to predictive modeling and precision medicine, allowing the practice of providing care to be accomplished in more accurate, efficient, and personalized ways. However, hurdles to future work still remain, particularly relating to standardization of data, the generalizability of AI models, interpretations of ML algorithms, and ethical questions around patient privacy and regulatory approval. Overcoming these challenges will require collaboration among computer scientists, clinicians, regulators, and those involved in the healthcare industry to establish AI systems that are scientifically sound and have clinical meaning.

In the future, the potential to use AI/ML and the promise of precision healthcare will be further enhanced by a convergence of AI/ML tools and integration with other technologies like multimodal imaging, digital pathology, and genomics. Future directions must concentrate on AI that is explainable, implementations of federated learning, and continuous evaluation of models in real-world clinical settings. It is the hope that the continued development of AI that is transparent, scalable, and equitable will result in medical imaging evolving into a more intelligent ecosystem that not only enhances the accuracy of diagnosis but also helps to address the inequities associated with access to healthcare. Ultimately, with regard to medical imaging, we as a field are ensuring that AI/ML advances the integration of technology with end-user clinical needs, ethical value, and patient-centric outcomes.

References

Abdulkareem A.F. Hashim A.Q. 2024 Infrared medical thermography, medical applications, and its basic principles: a review. BIO Web of Conferences 97 21174458 EDP Sciences Iraq https://doi.org/10.1051/bioconf/20249700140.

Abhisheka, B., Biswas, S.K., Purkayastha, B., 2023. A comprehensive review on breast cancer detection, classification and segmentation using deep learning. Arch. Comput. Methods Eng. 30 (8), 5023–5052. https://doi.org/10.1007/s11831-023-09968-z.

Akkus, Z., Cai, J., Boonrod, A., Zeinoddini, A., Weston, A.D., Philbrick, K.A., Erickson, B.J., 2019. A survey of deep-learning applications in ultrasound: artificial intelligence–powered ultrasound for improving clinical workflow. J. Am. Coll. Radiol. 16 (9), 1318–1328. https://doi.org/10.1016/j.jacr.2019.06.004.

Alshami, A., Nashwan, A., AlDardour, A., Qusini, A., 2025. Artificial Intelligence in rehabilitation: a narrative review on advancing patient care. Rehabilitacion 59 (2), 100911. https://doi.org/10.1016/j.rh.2025.100911.

Aron, J., 2011. How innovative is Apple's new voice assistant, Siri? New Scientist 212 (2836), 24. https://doi.org/10.1016/s0262-4079(11)62647-x.

Arsanjani, R., Xu, Y., Dey, D., Fish, M., Dorbala, S., Hayes, S., Berman, D., Germano, G., Slomka, P., 2013. Improved accuracy of myocardial perfusion SPECT for the detection of coronary artery disease using a support vector machine algorithm. J. Nucl. Med. 54 (4), 549–555. https://doi.org/10.2967/jnumed.112.111542.

Attia, Z.I., Friedman, P.A., Noseworthy, P.A., Lopez-Jimenez, F., Ladewig, D.J., Satam, G., Pellikka, P.A., Munger, T.M., Asirvatham, S.J., Scott, C.G., Carter, R.E., Kapa, S., 2019. Age and sex estimation using artificial intelligence from standard 12-lead ECGs. Circulation: Arrhythmia and Electrophysiology 12 (9). https://doi.org/10.1161/CIRCEP.119.007284.

Ávila, L.V., Shulla, K., Favarin, R., Filho, W.L., Trevisan, M., Salvia, A.L., Giazzon, L., Santini, É., 2025. The role of science and technology parks in meeting the sustainable development goals (importance of sustainability for the STPS). Discover Sustainability 6 (1). https://doi.org/10.1007/s43621-025-01655-2.

Azad, R., Aghdam, E.K., Rauland, A., Jia, Y., Avval, A.H., Bozorgpour, A., Karimijafarbigloo, S., Cohen, J.P., Adeli, E., Merhof, D., 2024. Medical image segmentation review: the success of U-Net. IEEE Trans. Pattern Anal. Mach. Intell. 46 (12), 10076–10095. https://doi.org/10.1109/TPAMI.2024.3435571.

Bellens, S., Guerrero, P., Vandewalle, P., Dewulf, W., 2024. Machine learning in industrial X-ray computed tomography – a review. CIRP J. Manuf. Sci. Technol. 51, 324–341. https://doi.org/10.1016/j.cirpj.2024.05.004.

Bera, K., Schalper, K.A., Rimm, D.L., Velcheti, V., Madabhushi, A., 2019. Artificial intelligence in digital pathology — new tools for diagnosis and precision oncology. Nat. Rev. Clin. Oncol. 16 (11), 703–715. https://doi.org/10.1038/s41571-019-0252-y.

Bernard, O., Lalande, A., Zotti, C., Cervenansky, F., Yang, X., Heng, P.A., Cetin, I., Lekadir, K., Camara, O., Gonzalez Ballester, M.A., Sanroma, G., Napel, S., Petersen, S., Tziritas, G., Grinias, E., Khened, M., Kollerathu, V.A., Krishnamurthi, G., Rohe, M.M., Pennec, X., Sermesant, M., Isensee, F., Jager, P., Maier-Hein, K.H., Full, P.M., Wolf, I., Engelhardt, S., Baumgartner, C.F., Koch, L.M., Wolterink, J.M., Isgum, I., Jang, Y., Hong, Y., Patravali, J., Jain, S., Humbert, O., Jodoin, P.M., 2018. Deep learning techniques for automatic MRI cardiac multi-structures segmentation and diagnosis: is the problem solved? IEEE Trans. Med. Imag. 37 (11), 2514–2525. https://doi.org/10.1109/TMI.2018.2837502.

Biernacki, T., Kokas, Z., Sandi, D., Füvesi, J., Fricska-Nagy, Z., Faragó, P., Zsigmond Kincses, T., Klivényi, P., Bencsik, K., Vécsei, L., 2022. Emerging biomarkers of multiple sclerosis in the blood and the CSF: a focus on Neurofilaments and therapeutic Considerations. Int. J. Mol. Sci. 23 (6), 3383. https://doi.org/10.3390/ijms23063383.

Boulerbah, R., Chaabane, A., Al-Naib, I., Alqadami, A.S.M., Aissaoui, D., Attia, H., 2025. Recent advancements in breast cancer detection: a Holistic review of Microwaves, ultrasound, and Photo-Acoustic imaging techniques. IEEE Journal of Microwaves. 5 (4), 776–792. https://doi.org/10.1109/JMW.2025.3580503.

Brady, S.L., Trout, A.T., Somasundaram, E., Anton, C.G., Li, Y., Dillman, J.R., 2021. Improving image quality and reducing radiation dose for pediatric CT by using deep learning reconstruction. Radiology 298 (1), 180–188. https://doi.org/10.1148/RADIOL.2020202317.

Buchanan, C., Howitt, M.L., Wilson, R., Booth, R.G., Risling, T., Bamford, M., 2020. Predicted Influences of artificial intelligence on the domains of Nursing: Scoping review. JMIR Nursing 3 (1). https://doi.org/10.2196/23939.

Buzug, T.M., 2011. Computed Tomography. Springer Nature, pp. 311–342. https://doi.org/10.1007/978-3-540-74658-4_16.

Castillo, C., Steffens, T., Sim, L., Caffery, L., 2021. The effect of clinical information on radiology reporting: a systematic review. J. Med. Radiat. Sci. 68 (1), 60–74. https://doi.org/10.1002/jmrs.424.

Chicco, D., Jurman, G., 2020. The advantages of the Matthews correlation coefficient (MCC) over F1 score and accuracy in binary classification evaluation. BMC Genom. 21 (1). https://doi.org/10.1186/s12864-019-6413-7.

Delongchamps, N.B., Rouanne, M., Flam, T., Beuvon, F., Liberatore, M., Zerbib, M., Cornud, F., 2011. Multiparametric magnetic resonance imaging for the detection and localization of prostate cancer: combination of T2-weighted, dynamic contrast-enhanced and diffusion-weighted imaging. BJU Int. 107 (9), 1411–1418. https://doi.org/10.1111/j.1464-410X.2010.09808.x.

Derek, ictor, Collings, P., 2025. Natural Language Processing (NLP) in Healthcare AI: Enhancing Clinical Insight Extraction from Unstructured Patient Data. Authorea. https://doi.org/10.22541/au.174845034.41890518/v1.

Dey, A.K., Walia, P., Somvanshi, G., Ali, A., Das, S., Paul, P., Ghosh, M., 2025. AI-driven diabetic retinopathy screening: Multicentric validation of AIDRSS in India. arXiv. https://doi.org/10.48550/arXiv.2501.05826.

Dikici, E., Bigelow, M., Prevedello, L.M., White, R.D., Erdal, B.S., 2020. Integrating AI into radiology workflow: levels of research, production, and feedback maturity. J. Med. Imaging 7 (1). https://doi.org/10.1117/1.JMI.7.1.016502.

Dlamini, Z., Francies, F.Z., Hull, R., Marima, R., 2020. Artificial intelligence (AI) and big data in cancer and precision oncology. Comput. Struct. Biotechnol. J. 18, 2300–2311. https://doi.org/10.1016/j.csbj.2020.08.019.

Driban, M., Yan, A., Selvam, A., Ong, J., Vupparaboina, K.K., Chhablani, J., 2024. Artificial intelligence in chorioretinal pathology through fundoscopy: a comprehensive review. International Journal of Retina and Vitreous 10 (1). https://doi.org/10.1186/s40942-024-00554-4.

Dudhe, S.S., Miswhra, G., Parihar, P., Nimodia, D., Kumari, A., 2024. Radiation dose optimization in radiology: a comprehensive review of Safeguarding patients and Preserving image fidelity. Cureus 16 (5). https://doi.org/10.7759/cureus.60846.

Ennab, M., Mcheick, H., 2024. Enhancing interpretability and accuracy of AI models in healthcare: a comprehensive review on challenges and future directions. Frontiers in Robotics and AI 11. https://doi.org/10.3389/frobt.2024.1444763.

Fahoum, F., Zelmann, R., Tyvaert, L., Dubeau, F., Gotman, J., Ward, L.M., 2013. Epileptic Discharges Affect the Default mode network – fMRI and Intracerebral EEG evidence. PLoS One 8 (6), e68038. https://doi.org/10.1371/journal.pone.0068038.

Feng, J., Gossmann, A., Sahiner, B., Pirracchio, R., 2022. Bayesian logistic regression for online recalibration and revision of risk prediction models with performance guarantees. J. Am. Med. Inform. Assoc. 29 (5), 841–852. https://doi.org/10.1093/jamia/ocab280.

Fu, X., Yang, Y., Cao, Y., Han, Q., Guo, X., Xu, Y., Ning, X., 2025. Magnetic source imaging registration based on self-supervised learning and multi-view differentiable rendering. Inf. Fusion 122, 103161. https://doi.org/10.1016/j.inffus.2025.103161.

Ganggayah, M.D., Taib, N.A., Har, Y.C., Lio, P., Dhillon, S.K., 2019. Predicting factors for survival of breast cancer patients using machine learning techniques. BMC Med. Inf. Decis. Making 19 (1). https://doi.org/10.1186/s12911-019-0801-4.

Girum, K.B., Crehange, G., Lalande, A., 2021. Learning with context feedback loop for robust medical image segmentation. IEEE Trans. Med. Imag. 40 (6), 1542–1554. https://doi.org/10.1109/TMI.2021.3060497.

Greenspan, H., Van Ginneken, B., Summers, R.M., 2016. Guest Editorial deep learning in medical imaging: overview and future promise of an Exciting new technique. IEEE Trans. Med. Imag. 35 (5), 1153–1159. https://doi.org/10.1109/TMI.2016.2553401.

Hamet, P., Tremblay, J., 2017. Artificial intelligence in medicine. Metabolism 69, S36. https://doi.org/10.1016/j.metabol.2017.01.011.

Hammernik, K., Klatzer, T., Kobler, E., Recht, M.P., Sodickson, D.K., Pock, T., Knoll, F., 2018. Learning a variational network for reconstruction of accelerated MRI data. Magn. Reson. Med. 79 (6), 3055–3071. https://doi.org/10.1002/mrm.26977.

Harvey, C., Koubek, R., Bégat, V., Jacob, S., 2016. Usability evaluation of a blood glucose monitoring system with a spill-resistant vial, easier strip handling, and connectivity to a mobile app: Improvement of patient convenience and satisfaction. J. Diabetes Sci. Technol. 10 (5), 1136–1141. https://doi.org/10.1177/1932296816658058.

Havaei, M., Davy, A., Warde-Farley, D., Biard, A., Courville, A., Bengio, Y., Pal, C., Jodoin, P.M., Larochelle, H., 2017. Brain tumor segmentation with deep neural networks. Med. Image Anal. 35, 18–31. https://doi.org/10.1016/j.media.2016.05.004.

Huang, X., Sun, W., Tseng, T.L., Li, C., Qian, W., 2019. Fast and fully-automated detection and segmentation of pulmonary nodules in thoracic CT scans using deep convolutional neural networks. Comput. Med. Imag. Graph. 74, 25–36. https://doi.org/10.1016/j.compmedimag.2019.02.003.

Hussain, S., Mubeen, I., Ullah, N., Shah, S.S.U.D., Khan, B.A., Zahoor, M., Ullah, R., Khan, F.A., Sultan, M. A., 2022. Modern diagnostic imaging technique applications and risk factors in the medical field: a review. BioMed Res. Int. 2022. https://doi.org/10.1155/2022/5164970.

Hussain, D., Abbas, N., Khan, J., 2024. Recent breakthroughs in PET-CT Multimodality imaging: innovations and clinical impact. Bioengineering 11 (12), 1213. https://doi.org/10.3390/bioengineering11121213.

Inamdar, R., 2020. Chapter 86 - Enterprise image management and medical informatics. In: Ernesto Iadanza,Clinical Engineering Handbook (Second Edition). Academic Press, pp. 616–637. https://doi.org/10.1016/B978-0-12-813467-2.00087-0.

Irede, E.L., Aworinde, O.R., Lekan, O.K., Amienghemhen, O.D., Okonkwo, T.P., Paul Onivefu, A., Ifijen, I. H., 2026. Medical imaging: a critical review on X-ray imaging for the detection of Infection. Biomedical Materials & Devices 4 (1), 1–45. https://doi.org/10.1007/s44174-024-00212-1.

Irvin, J., Rajpurkar, P., Ko, M., Yu, Y., Ciurea-Ilcus, S., Chute, C., Marklund, H., Haghgoo, B., Ball, R., Shpanskaya, K., Seekins, J., Mong, D.A., Halabi, S.S., Sandberg, J.K., Jones, R., Larson, D.B., Langlotz, C.P., Patel, B.N., Lungren, M.P., Ng, A.Y., 2019. CheXpert: a large chest radiograph dataset with Uncertainty Labels and expert comparison. Proc. AAAI Conf. Artif. Intell. 33 (01), 590–597. https://doi.org/10.1609/aaai.v33i01.3301590.

Jiang, L., Zuo, H.J., Chen, C., 2025. Artificial intelligence in echocardiography: applications and future directions. Fundam. Res.. https://doi.org/10.1016/j.fmre.2025.01.020.

Johnson, K.W., Torres Soto, J., Glicksberg, B.S., Shameer, K., Miotto, R., Ali, M., Ashley, E., Dudley, J.T., 2018. Artificial intelligence in cardiology. J. Am. Coll. Cardiol. 71 (23), 2668–2679. https://doi.org/10.1016/j.jacc.2018.03.521.

Ke, X.Y., Hou, W., Huang, Q., Hou, X., Bao, X.Y., Kong, W.X., Li, C.X., Qiu, Y.Q., Hu, S.Y., Dong, L.H., 2022. Advances in electrical impedance tomography-based brain imaging. Milit. Med. Res. 9 (1). https://doi.org/10.1186/s40779-022-00370-7.

Kelly, C.J., Karthikesalingam, A., Suleyman, M., Corrado, G., King, D., 2019. Key challenges for delivering clinical impact with artificial intelligence. BMC Med. 17 (1). https://doi.org/10.1186/s12916-019-1426-2.

Khan, M.A., Shafiq, U., Hamza, A., Mirza, A.M., Baili, J., AlHammadi, D.A., Cho, H.-C., Chang, B., 2025. A novel network-level fused deep learning architecture with shallow neural network classifier for gastrointestinal cancer classification from wireless capsule endoscopy images. BMC Med. Inform. Decis. Mak. 25, 150. https://doi.org/10.1186/s12911-025-02966-0.

Kourou, K., Exarchos, T.P., Exarchos, K.P., Karamouzis, M.V., Fotiadis, D.I., 2015. Machine learning applications in cancer prognosis and prediction. Comput. Struct. Biotechnol. J. 13, 8–17. https://doi.org/10.1016/j.csbj.2014.11.005.

Kumar, R., 2017. Artificial Intelligence—Basics. Springer Nature, pp. 33–49. https://doi.org/10.1007/978-1-4842-3069-5_3.

Kumar, A., 2022. Deep learning for multi-modal medical imaging fusion: enhancing diagnostic accuracy in complex disease detection. Int J Eng Technol Res Manag 6 (11), 183. https://doi.org/10.5281/zenodo.15033792.

Lakhani, P., Sundaram, B., 2017. Deep learning at chest radiography: automated classification of pulmonary tuberculosis by using convolutional neural networks. Radiology 284 (2), 574–582. https://doi.org/10.1148/radiol.2017162326.

Larson, J.A., Johnson, M.H., Bhayani, S.B., 2014. Application of surgical safety standards to robotic surgery: five principles of ethics for nonmaleficence. J. Am. Coll. Surg. 218 (2), 290–293. https://doi.org/10.1016/j.jamcollsurg.2013.11.006.

Lewis, B.T., Herr, K.D., Hamlin, S.A., Henry, T., Little, B.P., Naeger, D.M., Hanna, T.N., 2021. Imaging manifestations of chest trauma. Radiographics 41 (5), 1321–1334. https://doi.org/10.1148/rg.2021210042.

Li, J., Tian, Y., Zhou, T., 2024. Picture Archiving and Communication Systems. Springer Science and Business Media LLC, pp. 97–127. https://doi.org/10.1007/978-981-97-9551-2_4.

Litjens, G., Kooi, T., Bejnordi, B.E., Setio, A.A.A., Ciompi, F., Ghafoorian, M., van der Laak, J.A.W.M., van Ginneken, B., Sánchez, C.I., 2017. A Survey on Deep Learning in Medical Image Analysis, 42. Medical Image Analysis, pp. 60–88. https://doi.org/10.1016/j.media.2017.07.005.

Luo, X., Wang, S., Liu, J., Dong, X., He, P., Yang, Q., Chen, X., Zhou, F., Zhang, T., Feng, S., Han, P., Zhou, Z., Xiang, M., Qian, J., Ma, H., Zhou, S., Lu, L., Zuo, C., Geng, Z., Yi, W., Liu, F., 2025. Revolutionizing optical imaging: computational imaging via deep learning. Photonics Insights 4 (2), R03. https://doi.org/10.3788/pi.2025.r03.

Malhi, I.S., Yiu, Z.Z.N., 2021. Algorithm-based smartphone apps to assess risk of skin cancer in adults: critical appraisal of a systematic review. Br. J. Dermatol. 184 (4), 638–639. https://doi.org/10.1111/bjd.19502.

Massari, R., Mok, G.S.P., 2023. Editorial: new trends in single photon emission computed tomography (SPECT). Front. Med. 10. https://doi.org/10.3389/fmed.2023.1349877.

Maynord, M., Farhangi, M.M., Fermüller, C., Aloimonos, Y., Levine, G., Petrick, N., Sahiner, B., Pezeshk, A., 2023. Semi-supervised training using cooperative labeling of weakly annotated data for nodule detection in chest CT. Med. Phys. 50 (7), 4255–4268. https://doi.org/10.1002/mp.16219.

McCartney, M., 2018. Margaret McCartney: AI in medicine must be rigorously tested. BMJ 361, k1752. https://doi.org/10.1136/bmj.k1752.

Mesko, B., 2017. The role of artificial intelligence in precision medicine. Expert Review of Precision Medicine and Drug Development 2 (5), 239–241. https://doi.org/10.1080/23808993.2017.1380516.

Miotto, R., Li, L., Kidd, B.A., Dudley, J.T., 2016. Deep patient: an unsupervised representation to predict the future of patients from the electronic health records. Sci. Rep. 6. https://doi.org/10.1038/srep26094.

Mohamed, N., 2025. Augmented intelligence: a comprehensive review of the Flexibility between human and artificial intelligence. J. Inst. Eng.: Series B. https://doi.org/10.1007/s40031-025-01213-4.

Najjar, R., 2023. Redefining radiology: a review of artificial intelligence integration in medical imaging. Diagnostics 13 (17), 2760. https://doi.org/10.3390/diagnostics13172760.

Obuchowicz, R., Strzelecki, M., Piórkowski, A., 2024. Clinical applications of artificial intelligence in medical imaging and image processing—a review. Cancers 16 (10), 1870. https://doi.org/10.3390/cancers16101870.

Ortega-Robles, E., Arias-Carrión, O., 2025. Tremor: clinical frameworks, network dysfunction and therapeutics. Brain Sci. 15 (8), 799. https://doi.org/10.3390/brainsci15080799.

Paudyal, R., Shah, A.D., Akin, O., Do, R.K.G., Konar, A.S., Hatzoglou, V., Mahmood, U., Lee, N., Wong, R.J., Banerjee, S., Shin, J., Veeraraghavan, H., Shukla-Dave, A., 2023. Artificial intelligence in CT and MR imaging for oncological applications. Cancers 15 (9). https://doi.org/10.3390/cancers15092573.

Pellegrini, E., Ballerini, L., Hernandez, M.d.C.V., Chappell, F.M., González-Castro, V., Anblagan, D., Danso, S., Muñoz-Maniega, S., Job, D., Pernet, C., Mair, G., MacGillivray, T.J., Trucco, E., Wardlaw, J. M., 2018. Machine learning of neuroimaging for assisted diagnosis of cognitive impairment and dementia: a systematic review. Alzheimer's Dementia: Diagnosis, Assessment and Disease Monitoring 10, 519–535. https://doi.org/10.1016/j.dadm.2018.07.004.

Pinto-Coelho, L., 2023. How artificial intelligence is shaping medical imaging technology: a survey of innovations and applications. Bioengineering 10 (12), 1435. https://doi.org/10.3390/bioengineering10121435.

Rajpurkar, P., Irvin, J., Ball, R.L., Zhu, K., Yang, B., Mehta, H., Duan, T., Ding, D., Bagul, A., Langlotz, C.P., Patel, B.N., Yeom, K.W., Shpanskaya, K., Blankenberg, F.G., Seekins, J., Amrhein, T.J., Mong, D.A., Halabi, S.S., Zucker, E.J., Ng, A.Y., Lungren, M.P., 2018. Deep learning for chest radiograph diagnosis: a retrospective comparison of the CheXNeXt algorithm to practicing radiologists. PLoS Med. 15 (11). https://doi.org/10.1371/journal.pmed.1002686.

Rapakoulia, T., Theofilatos, K., Kleftogiannis, D., Likothanasis, S., Tsakalidis, A., Mavroudi, S., 2014. EnsembleGASVR: a novel ensemble method for classifying missense single nucleotide polymorphisms. Bioinformatics 30 (16), 2324–2333. https://doi.org/10.1093/bioinformatics/btu297.

Recht, M., Bryan, R.N., 2017. Artificial intelligence: Threat or Boon to radiologists? J. Am. Coll. Radiol. 14 (11), 1476–1480. https://doi.org/10.1016/j.jacr.2017.07.007.

Saleh, S.R., Yousif, S.A., Ahmed, I.Q., 2025. Brain lesion MRI and co-related MRS spectroscopy dataset. Data Brief 59. https://doi.org/10.1016/j.dib.2025.111288.

Sammartino, A.M., Bonfioli, G.B., Dondi, F., Riccardi, M., Bertagna, F., Metra, M., Vizzardi, E., 2024. Contemporary role of positron emission tomography (PET) in Endocarditis: a narrative review. J. Clin. Med. 13 (14). https://doi.org/10.3390/jcm13144124.

Saw, S.N., Ng, K.H., 2022. Current challenges of implementing artificial intelligence in medical imaging. Phys. Med. 100, 12–17. https://doi.org/10.1016/j.ejmp.2022.06.003.

Sengupta, P.P., Huang, Y.M., Bansal, M., Ashrafi, A., Fisher, M., Shameer, K., Gall, W., Dudley, J.T., 2016. Cognitive machine-learning algorithm for cardiac imaging; A Pilot study for Differentiating Constrictive Pericarditis from restrictive cardiomyopathy. Circ., Cardiovasc. Imaging 9 (6). https://doi.org/10.1161/CIRCIMAGING.115.004330.

Shi, D., Zhang, W., Chen, X., Liu, Y., Yang, J., Huang, S., Tham, Y.C., Zheng, Y., He, M., 2024. EyeFound: a multimodal generalist foundation model for ophthalmic imaging. arXiv. https://doi.org/10.48550/arXiv.2405.11338.

Singh, J., Ebersole, J.S., Brinkmann, B.H., 2022. From theory to practical fundamentals of electroencephalographic source imaging in localizing the epileptogenic zone. Epilepsia 63 (10), 2476–2490. https://doi.org/10.1111/epi.17361.

Tahir, H.N., Ullah, N., Tahir, M., Domnic, I.S., Prabhakar, R., Meerasa, S.S., AbdElneam, A.I., Tahir, S., Ali, Y., 2025. Artificial intelligence versus manual screening for the detection of diabetic retinopathy: a comparative systematic review and meta-analysis. Front. Med. 12. https://doi.org/10.3389/fmed.2025.1519768.

Tang, Z., Zhao, W., Xie, X., Zhong, Z., Shi, F., Ma, T., Liu, J., Shen, D., 2021. Severity assessment of COVID-19 using CT image features and laboratory indices. Phys. Med. Biol. 66 (3), 035015. https://doi.org/10.1088/1361-6560/abbf9e.

Tenajas, R., Miraut, D., Illana, C.I., Alonso-Gonzalez, R., Arias-Valcayo, F., Herraiz, J.L., 2023. Recent advances in artificial intelligence-assisted ultrasound Scanning. Appl. Sci. 13 (6), 3693. https://doi.org/10.3390/app13063693.

Thompson Y.L.E. Levine G. Chen W. Sahiner B. Li Q. Petrick N. Samuelson F. 2022 Wait-Time-Saving Analysis and Clinical Effectiveness of Computer-Aided Triage and Notification (CADt) Devices Based on Queueing Theory. Progress in Biomedical Optics and Imaging - Proceedings of SPIE SPIE United States 12035 doi: 10.1117/12.2603184.

Tian, Y., Xu, Z., Ma, Y., Ding, W., Wang, R., Gao, Z., Cheng, G., He, L., Zhao, X., 2023. Survey on deep learning in multimodal medical imaging for cancer detection. Neural Comput. Appl. 1–16. https://doi.org/10.1007/s00521-023-09214-4.

Trezza, A., Visibelli, A., Roncaglia, B., Spiga, O., Santucci, A., 2024. Unsupervised learning in precision medicine: Unlocking personalized healthcare through AI. Appl. Sci. 14 (20), 9305. https://doi.org/10.3390/app14209305.

U.S. Food and Drug Administration, 2018. Medical devices: general controls. https://www.fda.gov/medical-devices/regulatory-controls/general-controls-medical-devices.

van Leeuwen, K.G., Schalekamp, S., Rutten, M.J.C.M., van Ginneken, B., de Rooij, M., 2021. Artificial intelligence in radiology: 100 commercially available products and their scientific evidence. Eur. Radiol. 31 (6), 3797–3804. https://doi.org/10.1007/s00330-021-07892-z.

Varoquaux, G., Cheplygina, V., 2022. Machine learning for medical imaging: methodological failures and recommendations for the future. Npj Digital Medicine 5 (1). https://doi.org/10.1038/s41746-022-00592-y.

Vempati, G., Paladugu, T., Sama, J., Kamineni, R.T., Kamepalli, S., 2024. Evaluating CNN and deep learning models for bone fracture detection: a comparative study of VGG19, ResNet-50, LeNet and DenseNet. In: 1st International Conference on Innovations in Communications, Electrical and Computer Engineering, ICICEC 2024. Institute of Electrical and Electronics Engineers Inc., India. https://doi.org/10.1109/ICICEC62498.2024.10808468.

von Knebel Doeberitz, P.L., De Cecco, C.N., Schoepf, U.J., Duguay, T.M., Albrecht, M.H., van Assen, M., Bauer, M.J., Savage, R.H., Pannell, J.T., De Santis, D., Johnson, A.A., Varga-Szemes, A., Bayer, R.R., Schönberg, S.O., Nance, J.W., Tesche, C., 2019. Coronary CT angiography–derived plaque quantification with artificial intelligence CT fractional flow reserve for the identification of lesion-specific ischemia. Eur. Radiol. 29 (5), 2378–2387. https://doi.org/10.1007/s00330-018-5834-z.

Weissleder, R., Schwaiger, M.C., Gambhir, S.S., Hricak, H., 2016. Imaging approaches to optimize molecular therapies. Sci. Transl. Med. 8 (355). https://doi.org/10.1126/scitranslmed.aaf3936.

Wen, J., Thibeau-Sutre, E., Diaz-Melo, M., Samper-González, J., Routier, A., Bottani, S., Dormont, D., Durrleman, S., Burgos, N., Colliot, O., Alzheimer's Disease Neuroimaging Initiative, 2020. Convolutional neural networks for classification of Alzheimer's disease: overview and reproducible evaluation. Med. Image Anal. 63, 101694. https://doi.org/10.1016/j.media.2020.101694.

Xia, T., Zhao, B., Li, B., Lei, Y., Song, Y., Wang, Y., Tang, T., Ju, S., 2024. MRI-based radiomics and deep learning in biological characteristics and prognosis of hepatocellular carcinoma: opportunities and challenges. J. Magn. Reson. Imag. 59 (3), 767–783. https://doi.org/10.1002/jmri.28982.

Xu, J., Lu, C., Jin, S., Meng, Y., Fu, X., Zeng, X., Nussinov, R., Cheng, F., 2025. Deep learning-based cell-specific gene regulatory networks inferred from single-cell multiome data. Nucleic Acids Res. 53 (5), gkaf138. https://doi.org/10.1093/nar/gkaf138.

Xu, Y., Quan, R., Xu, W., Huang, Y., Chen, X., Liu, F., October 16, 2024. Advances in medical image segmentation: a comprehensive review of traditional, deep learning and hybrid approaches. Bioengineering (Basel) 11 (10), 1034. https://doi.org/10.3390/bioengineering11101034. PMID: 39451409; PMCID: PMC11505408.

Xue, L., Yan, W., Luo, P., Zhang, X., Chaikovska, T., Liu, K., Gao, W., Yang, K., 2021. Detection and localization of hand fractures based on GA_Faster R-CNN. Alex. Eng. J. 60 (5), 4555–4562. https://doi.org/10.1016/j.aej.2021.03.005.

Yan, F., Du, Z., Huang, Y.A., 2025. DAGFormer: a graph-based domain adaptation approach for single-cell cancer drug response prediction. PLoS Comput. Biol. 21 (12), e1013832. https://doi.org/10.1371/journal.pcbi.1013832.

Yang, H., Wang, G., Li, Z., Li, H., Zheng, J., Hu, Y., Cao, X., Liao, C., Ye, H., Tian, Q., 2024. Artificial intelligence for neuro MRI acquisition: a review. Magn. Reson. Mater. Phys., Biol. Med. 37 (3), 383–396. https://doi.org/10.1007/s10334-024-01182-7.

Yaqoob, A., Aziz, R.M., verma, N.K., 2023. Applications and techniques of machine learning in cancer classification: a systematic review. Human-Centric Intelligent Systems 3 (4), 588–615. https://doi.org/10.1007/s44230-023-00041-3.

Yu, Y.-C., Zhang, W., O'Gara, D., Li, J.-S., Chang, S.-H., 2023. A moment kernel machine for clinical data mining to inform medical decision making. Sci. Rep. 13, 10459. https://doi.org/10.1038/s41598-023-36752-7.

Yuan, H., Yu, K., Xie, F., Liu, M., Sun, S., 2024. Automated machine learning with interpretation: a systematic review of methodologies and applications in healthcare. Medicine Advances 2 (3), 205–237. https://doi.org/10.1002/med4.75.

Zeng, G., Guo, Y., Zhan, J., Wang, Z., Lai, Z., Du, X., Qu, X., Guo, D., 2021. A review on deep learning MRI reconstruction without fully sampled k-space. BMC Med. Imag. 21 (1). https://doi.org/10.1186/s12880-021-00727-9.

Zhu, X., Huang, Q., DiSpirito, A., Vu, T., Rong, Q., Peng, X., Sheng, H., Shen, X., Zhou, Q., Jiang, L., Hoffmann, U., Yao, J., 2022. Real-time whole-brain imaging of hemodynamics and oxygenation at micro-vessel resolution with ultrafast wide-field photoacoustic microscopy. Light Sci. Appl. 11 (1). https://doi.org/10.1038/s41377-022-00836-2.

3

Natural language technologies in biomedical domain

Mohammad "Sufian" Badar[1, 2, 3, 4], Uzma Khan[5], Mohd Shadab Alam[6], Zeenat Mirza[7], Afzal Ahmad[8], Khairol Amali bin Ahmad[9], Ehtesham Jameel[10], Syed Shad Mohsin[11], Imran Hussain[12], Syed Khurram Shahabuddin[13], Farheen Asaf[14] and Fakhr Ul-Islam[15]

[1]*DEPARTMENT OF BIOENGINEERING, UNIVERSITY OF CALIFORNIA, RIVERSIDE, CA, UNITED STATES;* [2]*UNIVERSAL SCIENTIFIC EDUCATION AND RESEARCH NETWORK (USERN), TEHRAN, IRAN;* [3]*SPI DARBHANGA, DARBHANGA, BIHAR, INDIA;* [4]*DEPARTMENT OF COMPUTER SCIENCE AND ENGINEERING (BIOINFORMATICS), SCHOOL OF ENGINEERING SCIENCES AND TECHNOLOGY (SEST), JAMIA HAMDARD UNIVERSITY, NEW DELHI, INDIA;* [5]*DEPARTMENT OF COMPUTER SCIENCE, JAMIA MILLIA ISLAMIA, NEW DELHI, INDIA;* [6]*DEPARTMENT OF COMPUTER SCIENCE AND ENGINEERING, SCHOOL OF ENGINEERING SCIENCES AND TECHNOLOGY, JAMIA HAMDARD, NEW DELHI, INDIA;* [7]*KING FAHD MEDICAL RESEARCH CENTRE, KING ABDULAZIZ UNIVERSITY, JEDDAH, SAUDI ARABIA;* [8]*DEPARTMENT OF ELECTRONICS, SCHOOL OF ENGINEERING SCIENCES AND TECHNOLOGY (SEST), JAMIA HAMDARD, NEW DELHI, INDIA;* [9]*NATIONAL DEFENCE UNIVERSITY OF MALAYSIA (NDUM), KUALA LUMPUR, MALAYSIA;* [10]*MEDICINAL & PROCESS CHEMISTRY DIVISION, CSIR-CENTRAL DRUG RESEARCH INSTITUTE, LUCKNOW, UTTAR PRADESH, INDIA;* [11]*APOLLO CRADLE & CHILDREN'S HOSPITAL, GHAZIABAD, UTTAR PRADESH, INDIA;* [12]*DEPARTMENT OF COMPUTER SCIENCE & ENGINEERING (CSE), SCHOOL OF ENGINEERING SCIENCES AND TECHNOLOGY (SEST), JAMIA HAMDARD UNIVERSITY, NEW DELHI, INDIA;* [13]*DEPARTMENT OF BIOSCIENCES, JAMIA MILLIA ISLAMIA (JMI), NEW DELHI, INDIA;* [14]*MAX HOSPITAL, NEW DELHI, INDIA;* [15]*FSGLOBAL, LLC, MARIETTA, GA, UNITED STATES*

What you will learn

In this chapter, you will learn how Biomedical Natural Language Processing (BioNLP) has evolved from early rule-based methods to modern transformer-based models. You will understand the structure and workflow of biomedical NLP pipelines, from data acquisition and preprocessing to information extraction. The chapter explains core techniques such as representation learning, traditional models, and transformer architectures. It also highlights real-world applications in drug discovery, clinical decision-making, clinical trials, and public health. Finally, you will explore the challenges of

Applying Artificial Intelligence to Computational Biology and Medical Informatics. https://doi.org/10.1016/B978-0-443-44915-4.00004-6

privacy, transparency, and biased data, along with future directions of BioNLP in advancing healthcare.

3.1 Introduction

Natural Language Processing is an interdisciplinary field of study that combines computer science, linguistics, artificial intelligence, and machine learning. It involves developing computational techniques that enable machines to interpret, analyze, and understand meaningful information from dense and complex text. NLP converts the unstructured texts into a structured, knowledgeable form that can be easily analyzed and used.

In the biomedical domain, NLP deals with a large amount of data such as research articles, clinical records and reports, medical guidelines, and patient records. It is very important for researchers to identify the relationships among diseases, drugs, genes and proteins, which can help to lead to various new scientific discoveries. BioNLP combines computer science techniques (NLP) with medical knowledge to unlock the information hidden in clinical reports, clinical notes, research articles, and more.

In simple terms, Biomedical Natural Language Processing (BioNLP) is like teaching computers to understand and use the complex language that doctors and researchers write in,like clinical notes, research papers, or, discharge summaries. It supports applications such as information retrieval, machine translation, automated report generation, and question answering. The scope of BioNLP extends across drug development, public health, biomedical research, and healthcare.

Biomedical NLP continues to evolve as new methods and larger datasets become available, making it an essential area of modern biomedical science. Its scope ensures long-term impact by bridging medical knowledge with intelligent computational solutions. It is the engine that transforms medical knowledge into computational language and bridges the gap between raw clinical narrative and actionable understanding Fig. 3.1.

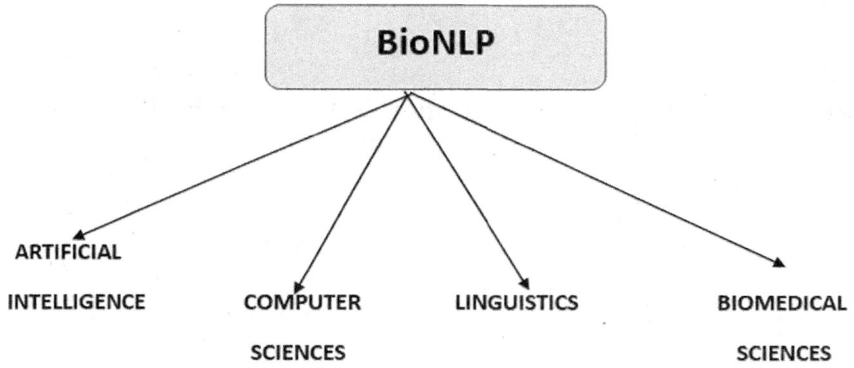

FIGURE 3.1 BioNLP is an interdisciplinary field of study that combines artificial intelligence, computer sciences, linguistics, and biomedical sciences.

3.1.1 Objectives

After studying this unit, you should be able to.

- explain the significance of biomedical NLP.
- describe the structure and workflow of biomedical NLP pipelines.
- differentiate between traditional models and transformer-based approaches.
- analyze real-world applications of biomedical NLP in research and healthcare.
- identify challenges and propose future directions for biomedical NLP.

3.2 Evolution of biomedical natural language processing

In the biomedical field, the journey of NLP has been long, starting from very simple methods and moving to todays powerful large language models (LLMs).

3.2.1 Early stage: rules and statistics

At first, computers worked with language using rules written by humans. For example, systems like ELIZA could give simple replies but had no real understanding. Later, statistical methods became popular. These used patterns, word counts, and probabilities to process text. Tools like Hidden Markov Models were used for tasks such as finding medical terms in text. But these methods needed a lot of human effort to design features, and they struggled with complex biomedical data.

3.2.2 Word embeddings

A big change came with word embeddings. Models like Word2Vec (2013) and GloVe (2014) turned words into numbers that captured their meaning. For example, the computer could now understand that "doctor" is closer in meaning to "nurse" than to "apple". In biomedicine, researchers trained special embeddings on PubMed articles and clinical records, creating tools that could recognize relations between diseases, drugs, and genes.

3.2.3 Transformers and contextual models

Static embeddings had one problem: they gave the same meaning to a word every time. But words change meaning depending on context. Thus, contextual models like ELMo and BERT (2018) solved all these problems. These models used the transformer architecture and could read sentences in both directions, giving a much better understanding. Biomedical versions such as BioBERT and ClinicalBERT improved tasks like finding medical concepts, reading research papers, and analyzing patient records.

3.2.4 Large language models

From 2020 onwards, Large Language Models (LLMs) like GPT-3, PaLM, and GPT-4 took NLP to a new level. They were trained on huge amounts of text and could translate, summarize, answer questions, and even pass medical exams with near-human accuracy. Unlike older models, they do not always need retraining—just giving instructions is enough. This made advanced NLP more accessible for doctors, researchers, and healthcare workers.

3.2.5 Future: multimodal and self-learning models

The future of biomedical NLP is moving toward multimodal systems, which can read not only text but also images, lab data, and even genetic sequences. New research, like AlphaEvolve, is exploring models that can improve themselves over time by learning from new data Fig. 3.2.

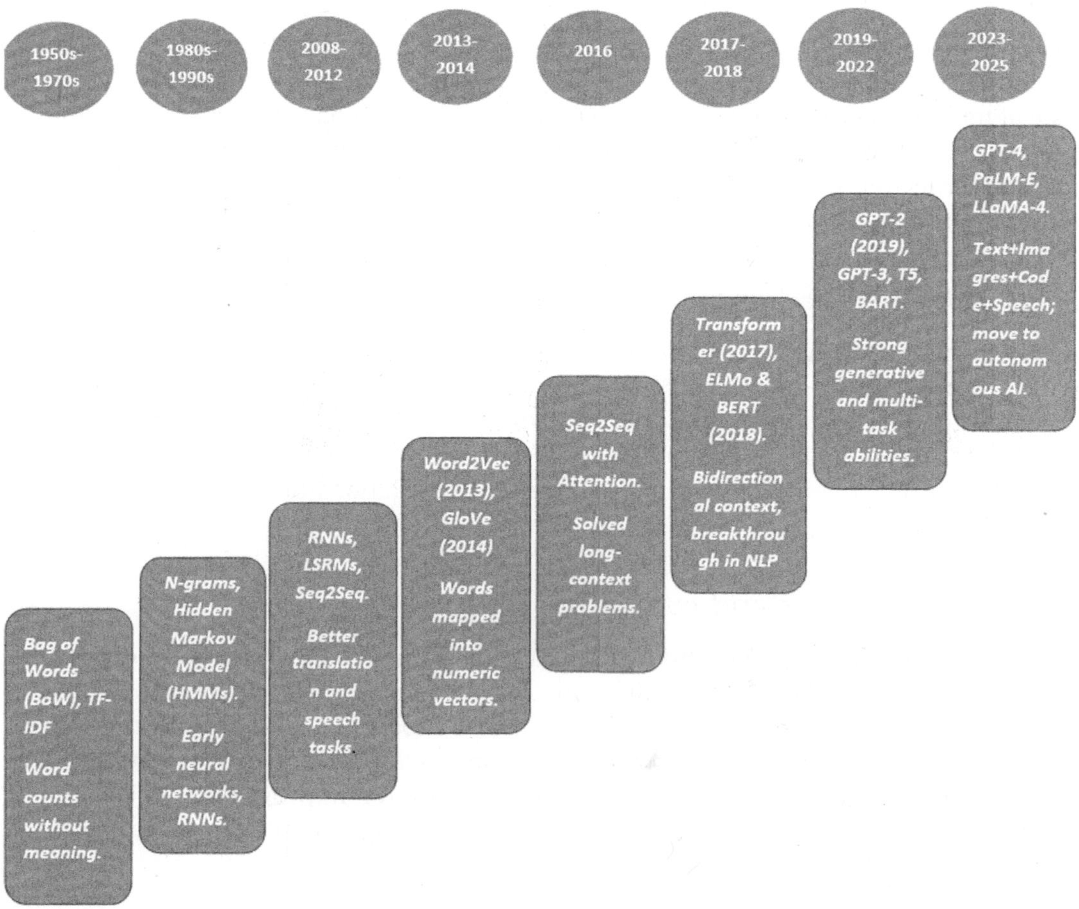

FIGURE 3.2 Timeline of biomedical natural language processing that shows key breakthroughs and model innovations shaping its evolution.

3.3 Importance of NLP in modern biomedical research and healthcare

BioNLP acts as a driving force in healthcare and biomedical research. There is a production of enormous volumes of unstructured text, such as from physician dictations and EHR entries to case reports, lab notes, and radiology findings. Natural Language Processing (NLP) addresses these challenges in clinical workflows and converts them into a simple, unstructured, knowledgeable form.

NLP tools convert raw text into actionable ones. They help doctors quickly find groups of patients who need special care, warn about possible bad drug combinations, and care for them better. For example, NLP tools can read doctors' notes to find symptoms that may mean a patient's health is getting worse, so help can be given on time.

BioNLP empowers researchers to navigate the biomedical collection of literature. It can quickly read and understand large amounts of medical text, spot important patterns and connections that would take humans way too long to find. It can discover new links between genes and diseases and do it much faster than traditional methods. It's very useful for public health and can help track disease outbreaks by looking through medical records, social media posts, and news reports. Some of the importance of NLP in the biomedical domain are.

- *Speeding Up Research*

  Medical research produces a huge amount of literature every single day, which is impossible for a human to review completely. NLP helps researchers by scanning, filtering, and summarizing large sets of studies quickly. It not only highlights key findings but also reveals fresh connections between diseases, treatments, and genes, making the research process faster and more effective.
- *Better Clinical Decisions*

  Doctors often need to make urgent choices based on patient records that are long and complex. NLP tools can extract important details such as lab reports, diagnostic findings, and treatment history, presenting them in a clear format. This quick access to accurate information supports more confident decision-making and helps tailor treatment plans to individual patients.
- *Saving Time in Hospitals*

  A lot of hospital time is lost in routine administrative tasks such as filling forms, writing notes, or coding bills. NLP systems can take over these repetitive jobs, which reduces paperwork for doctors and nurses. By cutting down clerical work, hospitals become more efficient, and staff can spend more time directly caring for patients.

- *Helping in Drug Discovery*

  Developing a new medicine is a slow and costly process, but NLP makes it easier by analyzing existing biomedical texts. It can detect new uses for already approved drugs, predict possible side effects, and even suggest new therapeutic targets. This ability to process huge databases of scientific knowledge accelerates the discovery of innovative drug candidates.
- *Managing Patient Risks*

  Every patient carries a different level of risk depending on their medical background and ongoing treatment. NLP can automatically scan records to detect warning signs of complications or chances of hospital readmission. This early identification allows doctors to act in time, preventing emergencies and improving overall patient safety.
- *Faster Clinical Trials*

  Recruiting patients for clinical trials is often one of the biggest delays in medical research. NLP can read through thousands of records to quickly identify people who match eligibility criteria for a study. By speeding up recruitment, it shortens the trial timeline and helps bring new treatments to patients sooner.
- *Personalized Care*

  Each patient describes their illness in their own words, and those narratives contain subtle clues about how they feel. NLP can interpret these personal descriptions, giving doctors a deeper understanding of both physical and emotional aspects of the illness. This leads to more personalized care plans that reflect the unique needs of every patient.

  Biomedical NLP is reshaping modern medicine by bridging the gap between unstructured text and actionable knowledge. From unlocking hidden information and accelerating research to reducing hospital workload and enabling personalized care, it touches almost every aspect of healthcare. Its role is becoming central in improving efficiency, safety, and innovation across the medical field.

3.4 Biomedical NLP pipelines

In biomedical NLP, a pipeline is more than a technical sequence of computational steps; it is a structured and knowledge-driven framework that systematically transforms raw, unstructured biomedical text into structured and actionable information. Biomedical language is notoriously complex, drawn from diverse sources such as electronic health records (EHRs), clinical trial reports, pathology notes, research articles, and even patient-generated content. Each of these sources introduces unique challenges—EHRs are filled with shorthand and inconsistencies, while research papers contain dense technical terminology. A pipeline integrates these heterogeneous inputs into a consistent workflow, resolving ambiguity and redundancy to produce outputs that are interpretable by both machines and clinicians.

Unlike generic NLP workflows, biomedical NLP pipelines must incorporate domain-specific modules to ensure semantic precision. For example, abbreviation expansion is essential since terms like "BP" may refer to blood pressure or binding protein, depending on context. Ontology-based normalization further ensures that extracted entities are aligned with standardized biomedical vocabularies such as UMLS, MeSH, or SNOMED CT. These specialized steps are vital, as errors in interpreting genes, drugs, or diseases can lead to critical mistakes in downstream tasks like drug–drug interaction prediction or clinical decision support.

Biomedical NLP pipelines also rely heavily on curated resources and domain-adapted models such as BioBERT, ClinicalBERT, and SciBERT, which provide contextual understanding beyond the reach of general NLP architectures. By combining algorithmic precision with biomedical knowledge, pipelines support large-scale literature mining, patient risk prediction, pharmacovigilance, and personalized medicine. Ultimately, they serve as the backbone of biomedical text analysis, ensuring reproducibility, scalability, and clinical utility in an ever-expanding domain of unstructured biomedical data.

3.4.1 Data acquisition and input sources

The strength of any biomedical NLP pipeline begins with the data it ingests. In medicine and life sciences, text does not arrive in a single neat format but rather in many streams, each shaped by its purpose and context. Capturing, harmonizing, and preparing these varied inputs is not a passive task but an active act of curation, where both quality and diversity directly determine the value of downstream insights.

Among the most central sources are electronic health records (EHRs). These documents—ranging from progress notes to discharge summaries and pathology reports—reflect the realities of day-to-day care. They are detailed, but often fragmented, filled with shorthand, and written under pressure. Despite their noise, they hold unmatched value because they capture the lived progression of disease, treatment decisions, and patient outcomes.

In contrast, biomedical literature offers structured and polished language. Journal articles, systematic reviews, and conference papers provide rigorously curated findings that anchor scientific evidence. While their precision makes them a reliable knowledge base, their sheer volume far exceeds human capacity to track manually, making automated mining essential.

Clinical trial documents add another dimension. These texts are formal and highly regulated, emphasizing compliance over narrative flow. Extracting eligibility criteria, dosage details, or outcome definitions demands algorithms tuned to their unique structure. Equally important are patient-generated sources, such as health forums, surveys, and social media posts. Though informal and linguistically diverse, they reveal lived experiences, side effects, and quality-of-life concerns often absent from formal records. Finally, regulatory and guideline documents—from agencies, associations, or patent repositories—bring authoritative standards and legal clarity.

Taken together, these sources create a complementary ecosystem. EHRs reveal practice, literature provides evidence, trials validate interventions, patient voices add perspective, and regulatory texts enforce standards. By integrating them, biomedical NLP pipelines construct a holistic and trustworthy knowledge fabric, transforming raw text into insights that advance science and improve patient care.

3.4.2 Text preprocessing

Text preprocessing is the foundation of every biomedical NLP pipeline. Before algorithms can recognize entities, extract relationships, or perform predictions, the raw text must be carefully cleaned, standardized, and structured. Unlike ordinary language data, biomedical text is unusually complex. Clinical notes are often written in haste, filled with shorthand, abbreviations, and fragmented grammar. Research articles, on the other hand, are packed with technical jargon, compound terms, and long, clause-heavy sentences. This variety creates inconsistencies that must be handled before analysis. Preprocessing ensures that what enters the pipeline is consistent, interpretable, and suitable for computational models.

The process can be understood as a sequence of steps, each addressing a different type of challenge.

- *Text Cleaning*

  Text cleaning is the first step of preprocessing, focused on removing irrelevant or misleading elements that interfere with analysis. Biomedical data often arrives with noise such as formatting errors, extraneous symbols, or repeated entries. Cleaning ensures text is free from inconsistencies while retaining information of clinical or scientific value. It may involve correcting irregular spellings, standardizing measurement units, and aligning document structures. Unlike superficial editing, this step carefully distinguishes between genuine biomedical information and artifacts. By creating a clean foundation, subsequent processes like tokenization or normalization can function effectively without distortion from unnecessary clutter.
- *Tokenization*

  Tokenization divides the text into smaller, analyzable units such as words, subwords, or complete sentences. This is a core requirement for most NLP tasks because models operate on structured units rather than continuous raw text. In biomedical NLP, tokenization must be precise, as incorrect segmentation can break meaningful terms or alter their interpretation. Tokenization is also the stage where context boundaries are set, determining how later tasks like part-of-speech tagging and entity recognition proceed. Proper tokenization enables both shallow analysis, such as word frequency, and deeper modeling tasks, such as embedding generation and contextual representation.

3.4.3 Sentence segmentation

Sentence segmentation identifies and separates sentences within biomedical documents, ensuring logical meaning is preserved. Biomedical texts often have irregular sentence structures, either very short and fragmented in clinical notes or long and complex in research publications. Accurate segmentation ensures algorithms interpret boundaries correctly, preventing loss of context. This step directly impacts higher-level tasks such as negation detection, relation extraction, and event recognition, all of which rely on correctly defined sentence units. Proper segmentation not only clarifies textual flow but also enhances computational efficiency by creating manageable units for analysis and processing.

- *Section Segmentation*
  Section segmentation organizes biomedical documents into structured divisions such as history, medications, outcomes, or experimental details. These divisions help downstream models focus only on relevant parts of the text, avoiding unnecessary processing of unrelated sections. By introducing document-level structure, section segmentation improves interpretability and context preservation, especially in lengthy clinical or research documents. It supports both targeted data extraction and domain-specific analysis, such as identifying drug mentions only in medication lists. This segmentation mirrors the way human readers interpret medical texts and ensures computational pipelines respect the natural logic of biomedical writing.
- *Abbreviation Expansion*
  Abbreviation expansion resolves the widespread use of acronyms and shortened terms in biomedical writing. Abbreviations, while efficient for human communication, create ambiguity for machines, as the same acronym can hold different meanings depending on context. Expansion clarifies these terms by replacing or mapping them to their full forms, ensuring consistency throughout the dataset. This process significantly improves the accuracy of entity recognition, concept mapping, and overall interpretability of biomedical text. Without abbreviation expansion, downstream tasks risk propagating errors caused by confusion between multiple possible meanings of a single shorthand.
- *Normalization and Concept Mapping*
  Normalization ensures that different textual expressions referring to the same biomedical concept are treated as a single entity. This step addresses linguistic variability by standardizing synonyms, alternative spellings, and abbreviations into a unified representation. Concept mapping further aligns these standardized terms with established biomedical vocabularies and ontologies, enabling interoperability across systems and datasets. By harmonizing terminology, normalization reduces ambiguity and ensures analytical consistency. This step is especially critical for cross-institutional research, clinical decision support, and integration of

heterogeneous biomedical data sources, where inconsistent terminology could otherwise fragment interpretation and weaken results.

- *Lemmatization and Stemming*

  Lemmatization and stemming reduce word variations to their root or base forms, simplifying analysis while preserving meaning. Lemmatization maps words to their dictionary form, while stemming applies algorithmic rules to strip suffixes or prefixes. These techniques prevent the fragmentation of conceptually related words, enabling models to group them under a single representation. In biomedical NLP, this step must be handled carefully to avoid distorting terms with precise meanings. When applied correctly, lemmatization and stemming enhance efficiency, improve retrieval, and strengthen the ability of models to detect semantic patterns across diverse biomedical texts.
- *Stop Word Handling*

  Stop word handling manages common words that may not add significant meaning to computational models. In biomedical NLP, however, these words must be processed with caution. While many function words can safely be filtered, some carry important contextual meaning, especially in clinical settings. Instead of indiscriminate removal, this step involves careful curation of lists to ensure essential terms are preserved. Proper handling reduces unnecessary computational load while safeguarding the accuracy of analysis. It exemplifies the balance required in preprocessing: removing redundancy without erasing subtle but crucial details embedded in the language.
- *Negation and Uncertainty Detection*

  Negation and uncertainty detection identify whether biomedical concepts are affirmed, denied, or expressed with doubt. This step is vital because misinterpreting negated or uncertain statements could lead to incorrect conclusions about a patient's condition or a scientific result. Specialized algorithms flag words and phrases that modify the status of entities, ensuring models capture not only what is mentioned but also how it is described. By preserving logical accuracy, negation and uncertainty detection prevent downstream errors and ensure extracted information reflects the intended meaning of the source text.
- *Deidentification*

  Deidentification safeguards patient privacy by removing or masking personally identifiable information from biomedical text. This includes names, dates, locations, and unique identifiers that could compromise confidentiality. The process ensures compliance with ethical standards and legal frameworks such as HIPAA, enabling safe use of clinical records for research or algorithm training. Importantly, deidentification preserves the clinical value of text by retaining relevant medical details while protecting sensitive information. As biomedical NLP increasingly relies on real-world data, deidentification stands as both a technical and ethical necessity in preprocessing.

- *Formatting and Feature Preparation*

  The final step of preprocessing involves transforming text into representations that models can directly process. This may include creating embeddings, generating parse trees, or encoding contextual features. By formatting the text into a structured computational form, this step ensures that downstream tasks such as classification, prediction, or entity extraction operate effectively. The quality of feature preparation determines how well models capture meaning and perform analysis. It serves as the bridge between linguistic structure and machine learning, finalizing the transition from raw biomedical text to reliable, analyzable data.

3.4.4 Core components of biomedical NLP pipelines

Biomedical NLP pipelines are like assembly lines that turn raw text into meaningful knowledge. Each stage of the pipeline does one specific job, and together they build a bridge from unstructured sentences to structured information that computers can work with. These steps are not just technical tricks; they are carefully designed to handle the unique way biomedical language is written—filled with abbreviations, compound terms, and long sentences. What follows is a detailed look at the most important stages, explained in a way that connects them together as a single process.

- *Tokenization*

  Every pipeline begins with tokenization. Imagine reading a paragraph without spaces between words; it would be almost impossible to make sense of. Tokenization is the process of breaking down the unstructured text into smaller, meaningful units called tokens. These tokens may be words, subwords, characters, numbers, or punctuation. Once the text is split into tokens, it becomes easier for machines to analyze, interpret, and perform higher-level tasks such as translation, sentiment analysis, or text classification. In biomedical NLP, this step is particularly important because biomedical texts contain highly specialized terminology, abbreviations, chemical compounds, genes, and protein names.
- *Part-of-Speech Tagging*

  After the text has been split into tokens, the next step is to understand the role each token plays in a sentence. Part-of-speech tagging does this by labeling words as nouns, verbs, adjectives, and so on. This might sound simple, but it is crucial in biomedical texts where the same word can play different roles depending on the context. Tagging brings order to sentences by showing which words are describing, which are acting, and which are being described. This structural clarity is essential before moving on to more advanced analysis.
- *Named Entity Recognition*

  Named Entity Recognition (NER) is where the pipeline starts to uncover real biomedical meaning. At this stage, the system identifies specific items in the text—things like diseases, medications, genes, or organs. NER is one of the fundamental tasks in NLP. It is the process of detecting and classifying important terms (called entities) in biomedical texts into predefined categories such as genes, proteins, drugs, diseases, and medications. In the biomedical field, NER has special importance because a huge amount of data is stored in the form of unstructured text, such as research articles, electronic health records, drug databases, and clinical trial reports. By applying NER, this unstructured data can be transformed into structured and machine-readable information, making it useful for medical research, clinical decision-making, and knowledge discovery. The challenge here is that biomedical language often refers to the same concept in different ways. For example, the same condition can be written in short form, long form, or even described in layman's terms. NER cuts through this variety to mark what counts as a biomedical concept. This step is vital because it creates the foundation for building structured knowledge from free-flowing sentences.

- *Normalization*

  Once concepts are recognized, they need to be made consistent. Normalization takes different mentions of the same thing and ties them to one common identity. Without this, a system might think three different terms describe three different conditions, when in fact they are all the same. Normalization removes this confusion and ensures that concepts are recorded in a uniform way. It is a step that brings harmony to diverse language and makes large-scale data analysis possible, especially when combining information from different sources.

- *Parsing*

  Parsing moves one step deeper into sentence structure. Instead of just listing words and their roles, parsing builds a kind of map that shows how words are connected. Sentences in biomedical writing are often long and layered, full of side notes and conditions. Parsing untangles these sentences and reveals the underlying logic—who is acting, what is being acted upon, and under what circumstances. This structural map allows later stages to extract accurate meaning rather than being misled by surface complexity.

- *Relation Extraction*

  Once entities and sentence structures are clear, the next step is to discover how these entities are connected. Relation extraction identifies links such as cause-and-effect, association, or interaction. For example, it can uncover that a certain drug reduces the severity of a particular condition. By capturing these relationships, the pipeline turns separate pieces of information into connected knowledge. This stage is what transforms biomedical text from isolated facts into networks of meaning that can support research and clinical decisions.

- *Coreference Resolution*

  Biomedical texts rarely describe things in just one way. A patient might be introduced by name in one sentence and later referred to simply as "the patient" or "he." A chemical substance might be given its full name once and then abbreviated afterward. Coreference resolution connects these references so that the text is understood as a continuous story, rather than as disconnected fragments. Without it, information would be scattered and incomplete. With it, the pipeline maintains coherence across paragraphs and entire documents.

- *Semantic Role Labeling*

  Semantic role labeling adds another layer of depth by asking: who did what, to whom, and how? Instead of just knowing that two things are related, this step clarifies the nature of their roles. It shows which entity the actor is, which is the receiver, and what kind of action is taking place. This richer understanding allows for more accurate modeling of biomedical events. For clinical or research purposes, it helps highlight not just associations but the full context of how those associations occur.

- *Embeddings and Representation*

  The final step prepares text for modern computational models. Words and phrases are converted into numerical forms called embeddings. These are like coordinates that place words into a mathematical space where similar meanings are close together. In the biomedical world, specialized embeddings are trained on large collections of scientific articles and clinical notes so that the system understands domain-specific language. With embeddings, computers can capture the subtle similarity between terms that humans naturally recognize. This step transforms language into a format ready for deep learning and predictive modeling.

  Together, these components form the backbone of biomedical NLP pipelines. Each stage performs a unique task—breaking text down, assigning structure, recognizing concepts, unifying meanings, uncovering relations, and finally converting everything into a machine-friendly form. Biomedical language is dense and complex, but with these steps, its richness is preserved while being reshaped into structured knowledge. This layered process allows computers not only to read but also to reason with biomedical text, creating opportunities for discovery, better clinical support, and stronger connections between science and practice.

3.4.5 Workflow of biomedical NLP pipelines

A biomedical NLP pipeline is more than a set of computational tools arranged in sequence—it is a carefully designed workflow that converts unstructured biomedical text into structured, actionable knowledge. Each stage refines the input, addressing specific challenges while preserving medical meaning. By the end of the process, raw language is transformed into insights that can guide research, clinical decision-making, and healthcare policy. The workflow can be seen as a journey, where each step builds upon the previous one, ensuring a smooth transition from text to knowledge Fig. 3.3.

- *Input Collection*

  The pipeline begins with input collection, the process of gathering biomedical text from multiple sources. These sources may include electronic health records (EHRs), scientific research articles, clinical trial reports, regulatory submissions, or even physician notes. Each of these inputs carries unique challenges—research

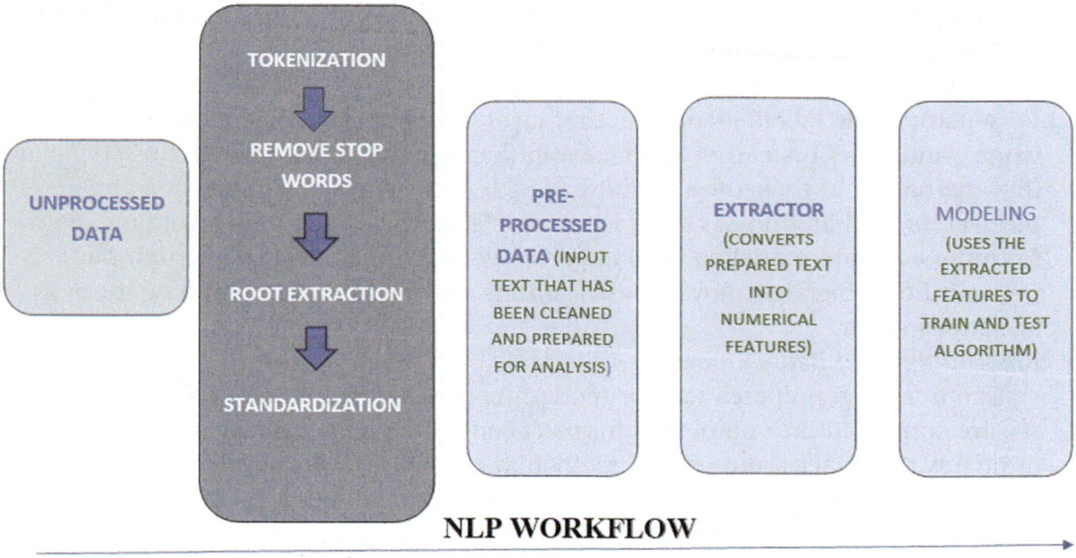

FIGURE 3.3 Step-by-step process of transforming unprocessed text into features for modeling and analysis.

papers are structured yet dense, while EHRs are rich but noisy, filled with abbreviations and shorthand.

The goal of this stage is not simply to collect data, but to collect it responsibly and selectively. Ethical and legal considerations, such as patient confidentiality, play a central role. Equally important is ensuring data relevance and diversity, so that the pipeline captures a representative view of biomedical information. In essence, the quality and reliability of the entire pipeline depend on the care taken during input collection, as poor input inevitably limits the accuracy of later stages.

- *Preprocessing*

  Once text is collected, it undergoes preprocessing, the critical stage where raw input is cleaned and standardized. Biomedical language is particularly challenging due to inconsistent styles, frequent use of acronyms, and technical vocabulary. Preprocessing addresses these problems systematically to make the text machine-readable without losing important meaning. This stage typically includes sentence segmentation, removal of noninformative symbols, and normalization of spelling variations. In biomedical settings, additional steps are often required, such as expanding acronyms (e.g., "BP" into "blood pressure"), aligning synonyms under controlled vocabularies, and ensuring chemical or gene names are handled correctly. The ultimate aim of preprocessing is to create a reliable foundation for further analysis, where text is consistent, clear, and semantically intact.

- *Linguistic Analysis*

  With the text cleaned, the pipeline advances to linguistic analysis, where the structural patterns of language are identified. At this point, text is no longer treated as a simple string of words but as a sequence with internal grammar and logical relationships.

Key processes include tokenization, which divides text into meaningful units; part-of-speech tagging, which assigns grammatical roles; and parsing, which reveals the connections between words. Biomedical sentences are often complex, filled with clauses and nested structures, and linguistic analysis untangles them into interpretable forms. By uncovering the skeleton of language, this stage provides the scaffolding necessary for recognizing biomedical concepts and their interactions.

- *Information Extraction*

  The most transformative part of the pipeline is information extraction, where text is converted into biomedical knowledge. This stage identifies relevant entities, ensures consistency, and maps relationships.

  Named Entity Recognition (NER) highlights biomedical concepts such as diseases, drugs, proteins, or genes. Normalization ensures that different mentions of the same concept (e.g., "myocardial infarction" and "heart attack") are unified under a single standard identity. Relation extraction then identifies connections among these entities, such as drug–disease associations or protein–protein interactions. At this point, unstructured language is converted into structured biomedical data i.e. ready for computational reasoning and downstream use.

- *Advanced Interpretation*

  Once entities and relations are identified, the pipeline moves to advanced interpretation, which enriches the extracted knowledge with deeper context. Biomedical text often spans multiple sentences, and meaning can only be preserved when these connections are maintained.

  Coreference resolution links different mentions of the same entity across a document, ensuring continuity (e.g., "TNF-alpha", "the cytokine", and "it"). Semantic role labeling adds further detail by identifying the roles of entities in events, such as who performed an action and what its outcome was. Finally, embedding methods convert words, entities, and sentences into numerical vectors that capture semantic meaning. Specialized models like BioBERT allow computers to process biomedical context more accurately. This stage ensures that knowledge is coherent, contextually rich, and ready for advanced tasks such as predictive modeling.

- *Output and Application*

  The final stage is output generation and application, where structured knowledge is transformed into usable results. The form of output depends on the specific goal of the pipeline.

  Outputs may include knowledge graphs that visualize networks of biomedical entities, summaries that condense complex research, clinical decision support systems that provide alerts or suggestions, or databases that store extracted information for large-scale analysis. What defines this stage is its focus on actionability—knowledge is not just extracted, but shaped into insights that can directly support biomedical research, healthcare delivery, or public health monitoring.

The workflow of biomedical NLP pipelines is a carefully structured journey, beginning with messy, unprocessed text and ending with refined, actionable insights. Each stage plays a distinct role: input collection provides the material, preprocessing cleans and standardizes it, linguistic analysis reveals structure, information extraction uncovers biomedical meaning, advanced interpretation deepens context, and output turns knowledge into practice.

Together, these steps transform language into knowledge, bridging the gap between human communication and computational intelligence. By systematically handling complexity, biomedical NLP pipelines ensure that critical information hidden within vast volumes of text can be uncovered, organized, and applied to improve science, medicine, and health outcomes.

Workflow of Biomedical NLP Pipeline:

Input Collection → Preprocessing → Linguistic Analysis → Information Extraction → Advanced Interpretation → Output and Application

3.5 Core NLP tasks in the biomedical domain

The biomedical field generates a massive and ever-growing amount of unstructured text, including research articles, clinical trial reports, hospital records, and electronic health records (EHRs). These texts contain critical information that can advance medical research, support clinical decision-making, and improve patient care. However, due to their unstructured nature, such documents are often difficult for humans to process, organize, and analyze efficiently. To overcome this challenge, Natural Language Processing (NLP) has become an essential tool in biomedicine. Core NLP tasks form the foundation of this transformation process. These tasks enable computers to read and interpret biomedical language, converting raw text into structured, machine-readable knowledge. Biomedical language presents unique challenges because of its complex terminology, use of abbreviations, and rapid growth of new knowledge. Therefore, specialized models and domain-specific tools are required. By addressing these challenges, core NLP tasks unlock knowledge hidden in biomedical texts, accelerate scientific discovery, streamline clinical workflows, and ultimately improve healthcare outcomes.

3.5.1 Representation learning

Representation learning is about teaching a computer to turn words, sentences, or even whole documents into numbers called vectors. These numbers are not random—they capture the meaning of the text so that the computer can understand and work with it. Once text is in this number form, it can be used for many tasks like finding disease names, linking drugs to treatments, or answering medical questions.

Word Embeddings are used as input representation for other downstream NLP models and often as the first data processing layer. The methods for learning word embeddings are:

- Word2vec
- GloVe
- fastText

There are two main types of word representations.

1. *Static embeddings* – The tools like Word2Vec, GloVe, and fastText give each word one fixed vector. This means the word has the same meaning in every sentence. In medicine, special versions like BioWordVec are trained on medical articles so they know rare terms and abbreviations.
2. *Contextual embeddings* – The tools like ELMo and BERT can change a word's meaning based on the sentence it appears in. For example, the word "cold" in "I caught a cold" means illness, but in "The water is cold" it means temperature. In healthcare, versions like BioBERT and ClinicalBERT are trained on medical texts to understand medical language better. Good representation learning makes biomedical NLP much more accurate. Choosing the right one depends on the task, the data, and the available resources. Imagine every word in a text as a tiny puzzle piece. To let the computer work with these pieces, we must turn words into numbers—this is what representation learning does.

At first, people just counted how many times a word appeared (bag-of-words). But this missed the meaning.

Word embeddings: Later, methods like Word2Vec and GloVe gave each word a place in a "map of meaning". Words with similar meanings stood close together on this map. For example, a doctor and nurse would be nearer than a doctor and a banana.

Biomedical embeddings: Since medical language is special, scientists trained embeddings on PubMed and other biomedical texts (BioWordVec, PubMed vectors) so the model really "understands" medical terms.

Contextual embeddings: Modern methods, like ELMo and BERT, can even adjust meaning based on the sentence. For example, "cold" in "common cold" versus "cold weather" are treated differently.

So, representation learning is like teaching the computer the vocabulary of medicine in the form of numbers. Before modern deep learning models became popular, two main types of models were widely used in NLP-Conditional Random Fields and Support Vector Machines. These models were very important for early medical text analysis Fig. 3.4.

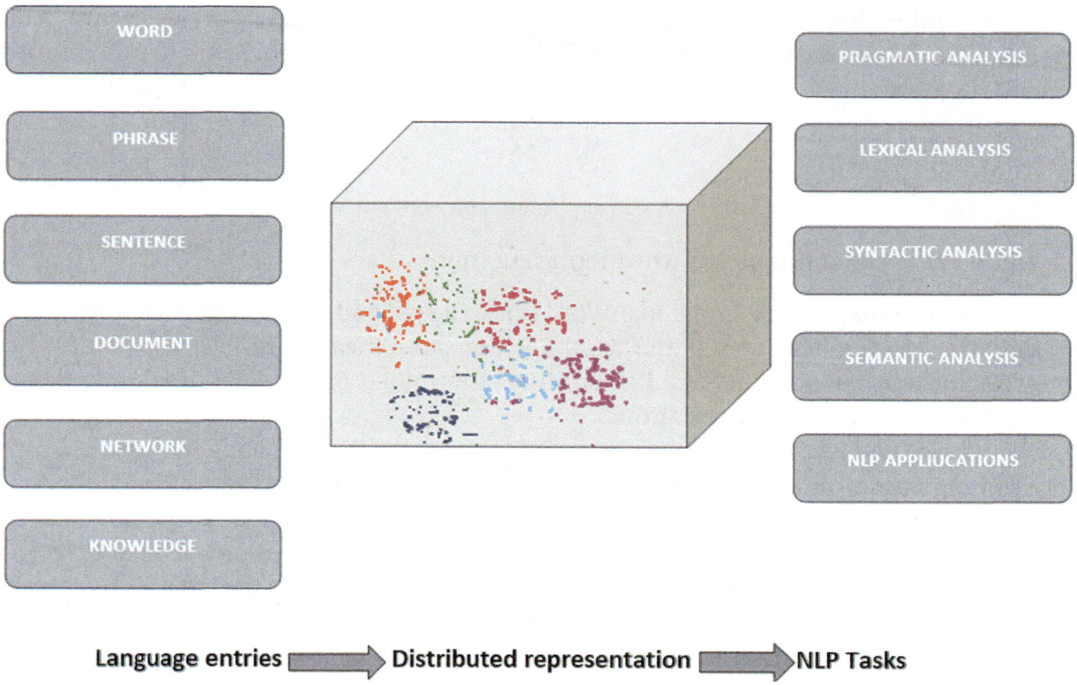

FIGURE 3.4 It shows how different language units can be converted into a single shared "semantic space" using distributed representation. In this space, similar items are placed close together, making it easier for NLP systems to perform tasks such as lexical, syntactic, and semantic analysis.

3.5.2 Traditional models

Before deep learning became popular, people used simpler but effective models.

SVMs (Support Vector Machines): These are like judges that draw a line between different categories. In biomedicine, they helped classify things—like which papers are about cancer and which are about diabetes. SVMs are models that classify data into different groups. They can decide whether a sentence is talking about a disease, a treatment, or something else. They do this by finding the best "boundary" between categories in the data.

CRFs (Conditional Random Fields): These are great at sequence tasks, like finding disease names in a sentence. For example, in "Aspirin reduces heart attack risk", CRFs can mark "Aspirin" as a drug and "heart attack" as a disease. CRFs are special models used for sequence labeling—that means they read sentences word by word and decide which "tag" each word should get. For example, in the phrase "severe chest pain", CRFs can mark it as a symptom. They are good at using both the word itself and the words around it to make accurate predictions.

In many cases, CRFs and SVMs worked as a team. CRFs would first find important parts of the text, such as medical terms. Then SVMs would classify these terms into the correct category. This two-step process gave good results, even with smaller datasets. CRFs and SVMs helped build early systems that could process doctors's notes, hospital records, and research papers. They were fast, reliable, and worked well when there wasn't a lot of labeled training data. Today, deep learning models often perform better, but CRFs and SVMs are still used in some projects because they are simple, effective, and need less computing power.

HMMs (Hidden Markov Models): HMMs are older tools but important models in NLP. They are mostly used for tagging words or detecting patterns. They try to guess hidden patterns behind a sequence of words. For example, while tagging sentences, the "hidden" states could be grammar tags like noun or verb, and the "observed" words were once used to find things like gene names or medical terms, but today they are mostly replaced by newer models like transformers.

Traditional models were useful, but they depended a lot on hand-made features—like noticing if a word ends in "-ase" (often an enzyme) or if it's capitalized like a gene name.

3.5.3 Transformer models (BERT and variants)

Transformer Models are deep learning model which are used to understand the language. The old models read sentences one word at a time, so they could miss the bigger picture. They look at words altogether and then understand how each word relates to every other word. They use self-attention—a method that decides which words are most important for understanding the meaning. Transformer-based language models are based on the transfer architecture (a network that uses self-attention mechanisms to process token sequences).

BERT stands for Bidirectional Encoder Representations from Transformers. It reads text in both the left-to-right and right-to-left directions. This helps it understand context better than older models. Example: In the sentence "He went to the bank to deposit money", BERT can tell that "bank" means a financial place (not a river bank) because of the words around it. Medical language is full of difficult words like "myocardial infarction" or "angioplasty." Normal BERT may not know these terms. That's why BioBERT was created — it's trained on millions of biomedical research papers and clinical notes, so it understands medical terms better. Other versions include ClinicalBERT (hospital data) and BlueBERT (scientific and clinical data combined) Fig. 3.5, Table 3.1.

3.5.3.1 Advantage

- Understands context deeply
- Works well for complex sentences
- Can be fine-tuned for specific medical tasks
- Often performs better than older models like SVMs and CRFs.

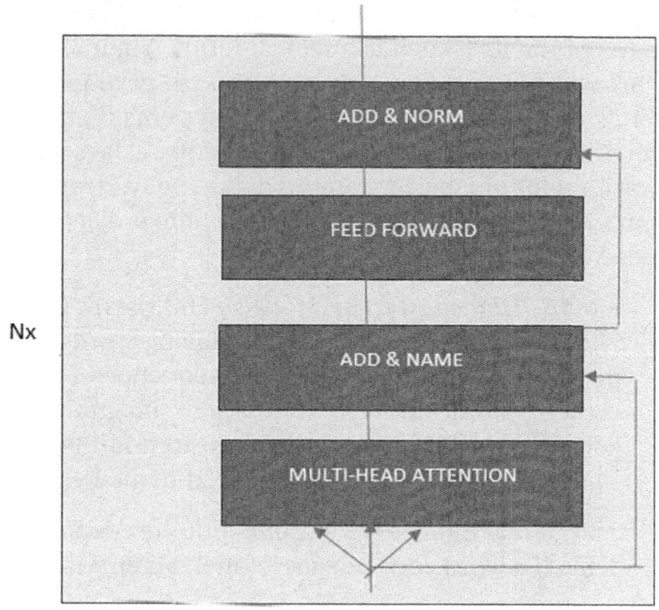

FIGURE 3.5 Architecture of the BERT model showing core structure: Embedding layers, multi-head self-attention, feed-forward networks, and normalization in stacked transformer encoders.

Table 3.1 Popular BERT in biomedicines.

Model name	Special feature	Main use
BioBERT	Trained on PubMed biomedical text	Disease/drug detection
ClinicalBERT	Trained on hospital records (MIMIC dataset)	Clinical decision support
BlueBERT	Trained on PubMed	Broad biomedical NLP
BioDistilBERT	Smaller and faster version of BioBERT	Mobile or low-compute applications

3.5.3.2 Challenges

- Needs a lot of data and computing power
- Requires careful training to avoid errors in sensitive medical tasks

Transformers like BERT are powerful tools in biomedical NLP. When adapted into BioBERT, ClinicalBERT, or BlueBERT, they can read and understand complex medical language, helping doctors, researchers, and healthcare systems turn huge amounts of text into useful, structured knowledge Fig. 3.6.

FIGURE 3.6 Applications of NLP in biomedical research and healthcare illustrating the key areas where it supports healthcare.

3.6 Applications in biomedical research and healthcare

Biomedical Natural Language Processing (BioNLP) is a branch of Artificial Intelligence that helps computers read, understand, and analyze medical text. Every day, hospitals, labs, and research centers produce a huge amount of information—patient notes, reports, and scientific articles. Much of this information is unstructured, meaning it is written in plain text, not in neat tables or databases.

3.6.1 Clinical decision support systems (CDSS)

Clinical Decision Support Systems are digital tools that help doctors make better medical decisions. Most patient information is stored in free-text notes, which are hard for computers to analyze. BioNLP solves this by reading those notes and turning them into structured insights.

- Extracts important details from patient files such as symptoms, allergies, and test results.
- Compares the patient's information with medical guidelines and research databases.
- Provides suggestions for possible diagnoses or treatment options.
- Issues safety alerts if a doctor is about to prescribe a drug that may cause harm due to allergies or drug interactions.

3.6.2 Drug discovery and repurposing

Developing a new drug is usually slow and very expensive, often taking more than 10 years. BioNLP helps speed this up by analyzing thousands of scientific articles, patents, and lab results to find useful patterns.

- Identifies connections between genes, proteins, and diseases to suggest new drug targets.
- Reads existing drug data to see if an approved drug can be reused for another disease (drug repurposing).
- Reduces the manual effort of reading thousands of pages, saving time and resources.

3.6.3 Clinical trial matching

Clinical trials are essential for testing new treatments, but one of the biggest challenges is finding the right patients. Many trials fail because they cannot recruit enough participants on time.

- Reads through electronic health records (EHRs), including doctor's notes and lab results.
- Compares patient details with trial eligibility requirements (age, medical condition, treatment history).
- Quickly identifies suitable candidates, reducing time and effort for doctors and researchers.

3.6.4 Public health surveillance

Public health depends on monitoring diseases and health trends in a population. Traditional methods, like official hospital reports, are often slow. BioNLP offers real-time surveillance by scanning large volumes of unstructured text online.

- Monitors social media posts, news articles, and online discussions for mentions of symptoms or outbreaks.
- Detects unusual spikes in health-related topics, serving as an early warning system.
- Tracks the spread of misinformation during health crises, helping authorities respond with accurate information Table 3.2.

Table 3.2 Applications of natural language processing in the biomedical domain.

Application	What NLP does	Main benefit
CDSS	Finds patient facts, sends alerts	Safer and faster
Drug discovery	Finds links in research papers	Quicker new treatment ideas
Trial matching	Matches patients to trial requirements	Easier recruitment
Public health surveillance	Tracks health trends	Early disease detection

3.7 Case studies

When we think of medicine, we usually imagine stethoscopes, X-rays, or blood tests. But behind every patient visit lies something less visible—words. Doctors write notes, researchers publish papers, hospitals store endless reports, and clinical trials fill thousands of pages with eligibility rules. Most of this information is unstructured text, difficult for computers to understand. This is where Biomedical NLP steps in. Think of it as a translator—taking messy, free-flowing words and turning them into structured insights that humans and machines can actually use. Let's explore some real stories where NLP has already made a difference.

3.7.1 Helping doctors focus on patients, not paperwork

Doctors often complain that they spend more time typing than actually talking to their patients. That's not only frustrating for them but also discouraging for patients. NLP-based speech recognition is changing that.

At several hospitals in the US and UK, doctors now use tools like Dragon Medical One. Instead of typing, they simply speak during consultations, and the system creates accurate clinical notes in real time.

Physicians at Nebraska Medicine said the tool saved them up to half an hour a day. A major hospital network in the US cut its transcription costs by millions of dollars every year. In the UK's National Health Service, follow-up letters that once took almost 2 weeks are now ready within 2–3 days.

Doctors can look patients in the eye rather than stare at a keyboard. Less stress for doctors, more attention for patients.

3.7.2 Reading between the lines of clinical notes

Not everything about a patient's condition fits neatly into a drop-down box in the hospital's computer system. A note like "patient struggles to walk one block without breathlessness" might be overlooked in structured data. But NLP can flag this as a potential red warning sign.

In fact, studies show that analyzing free-text notes with NLP can predict whether patients with conditions like heart failure are at risk of being readmitted—or worse. This means hospitals can step in earlier, possibly preventing an emergency.

3.7.3 Speeding up drug discovery

Every year, researchers publish millions of papers. Buried in those articles are hidden clues about how genes, diseases, and drugs are connected. Reading them all is humanly impossible, but NLP can scan through this ocean of text in hours. By mining literature, scientists can spot gene–disease links or identify biomarkers that show whether a treatment might work.

NLP is also helping with drug repurposing—finding new uses for old drugs. A medicine originally made for depression might, through text mining, be flagged as useful in certain cancers.

It also helps researchers scan thousands of clinical trials to spot promising candidates faster, saving both time and money.

3.7.4 Matching patients to clinical trials

Clinical trials are essential for developing new treatments, but enrolling patients is a nightmare. The rules are long, complicated, and written in medical jargon. Many patients miss out simply because no one realized they were eligible.

NLP and modern AI models are changing that. Tools now automatically read trial eligibility criteria and match them with patient records. Recent studies even showed that GPT-4-based systems can perform this task as well as trained clinicians.

One framework, TrialGPT, managed to find more than 90% of the relevant trials for patients while reducing screening time by nearly half. Earlier, platforms like Antidote made it easier for patients themselves to search for trials by answering simple questions. This means patients—especially those with rare diseases—can access cutting-edge treatments that might otherwise remain out of reach.

3.7.5 Cutting down the burden of medical billing

Medical billing may sound boring, but for hospitals, it's a massive workload. Doctors have to assign the right codes for every condition and treatment, and errors can cost both patients and providers. NLP-powered Computer-Assisted Coding (CAC) systems scan through medical notes and suggest accurate billing codes automatically. This reduces human error, speeds up claims, and saves money. More importantly, patients face fewer billing disputes and delays.

3.7.6 Building trust with explainable AI

AI in healthcare often faces a trust problem. Doctors hesitate to rely on a system if it simply says "yes" or "no" without explanation. That's where Explainable AI (XAI) comes in, supported by NLP. For example, researchers used a method called LIME to explain how a deep learning model classified cancer signals. Instead of just giving an answer, the system highlighted which words or patterns influenced its decision.

This transparency helps doctors feel confident that the system isn't a mysterious "black box" but a tool they can understand and rely on.

3.7.7 Looking beyond biology: social determinants of health

Health isn't just about genes or medicines. Social issues—like housing, income, or access to food—play a huge role. Doctors often note these things in passing, like "patient missed dialysis due to lack of transport".

NLP can pick up on these clues, allowing hospitals to step in. For example, a flagged note about transportation issues might trigger a social worker to arrange rides for the patient. This is a simple but powerful way to make care more human and holistic.

3.7.8 Digging into cancer vaccine trials

Another striking example: researchers used an NLP platform called Linguamatics I2E to go through more than 500 clinical trial records on mRNA cancer vaccines. They didn't just count trials—they extracted important details like dosage, delivery methods, and schedules. Doing this manually would have taken weeks; with NLP, it took hours.

This kind of efficiency speeds up the pace of cancer research and ensures that scientists don't miss valuable details hidden in trial descriptions.

Behind these applications are some powerful tools.

- Apache cTAKES, built by Mayo Clinic, which can pull out medical terms (like symptoms, drugs, or diagnoses) from unstructured text.
- Spark NLP for Healthcare, which comes with ready-to-use models like BioBERT and ClinicalBERT, designed for medical text Table 3.3, Fig. 3.7.

3.8 Challenges in biomedical NLP

Biomedical Natural Language Processing (BioNLP) is a powerful branch of artificial intelligence that allows computers to read, understand, and analyze medical and biological texts. It is an essential tool for healthcare and research, as it helps in organizing patient

Table 3.3 Case study of biomedical natural language processing.

Case study	Data used	Main benefit
Phenotyping	Hospital records	Fast patient search
ADR detection	Medical notes, social media	Safer medicines
PubMed mining	Research papers	Faster discoveries

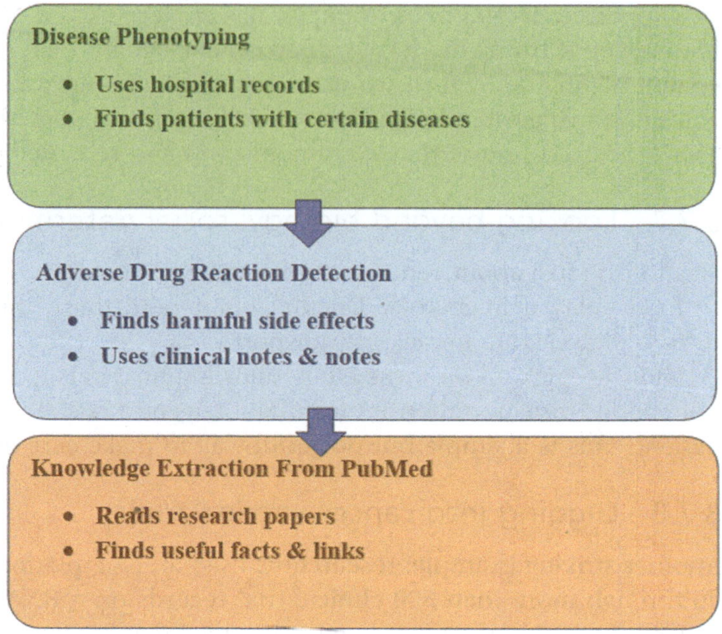

FIGURE 3.7 Flowchart of the case studies of biomedical natural language processing.

records, extracting information from scientific papers, and even assisting doctors in making clinical decisions.

However, despite its importance, BioNLP faces several serious challenges. These challenges come mainly from three areas: data privacy and security, lack of transparency in models, and issues with limited and biased data. If not addressed properly, these problems can slow down progress and reduce trust in BioNLP systems.

3.8.1 Privacy and security of data

One of the biggest challenges in BioNLP is ensuring the protection of sensitive medical information. Biomedical texts, especially clinical notes, often contain personal details about patients. Because this information is private, it must be handled with extreme care.

3.8.1.1 Patient information

Medical documents such as hospital notes, lab reports, and treatment histories contain what is called Protected Health Information (PHI). This includes names, dates of birth, addresses, and full medical histories. If this information is leaked, it can harm patients and violate strict privacy regulations.

3.8.1.2 Data access
To train effective BioNLP models, researchers need access to large amounts of medical data. However, laws like the Health Insurance Portability and Accountability Act (HIPAA) in the United States make data sharing difficult. While these laws are necessary for protecting patients, they limit the availability of data for AI research.

3.8.1.3 Data security
Even when data is collected, it must be protected from cyberattacks. Hackers are increasingly targeting healthcare data because it is extremely valuable. This means that strong encryption, firewalls, and strict access controls are necessary to prevent unauthorized use.

3.8.1.4 Anonymization
One common method to protect privacy is anonymization, where personal details are removed or replaced with fake information. For example, instead of using a real name, a dataset might use a code. But anonymization is not perfect—sometimes hidden clues in the data can still reveal a person's identity.

3.8.2 Model interpretability and black-box models

Another major challenge in BioNLP is the lack of clarity in how advanced models make their decisions. Many of the most powerful models, such as deep neural networks, are often described as "black boxes." This means that while they can give highly accurate predictions, it is very difficult to understand exactly how they arrived at those results.

3.8.2.1 The black box problem
In medicine, every decision must be justified, especially when it involves a patient's treatment or diagnosis. But if a model works like a black box, even its creators may not be able to explain how it reached a certain conclusion. This makes it difficult for doctors to rely on it fully.

3.8.2.2 Lack of trust
Doctors, patients, and healthcare organizations need to trust that the technology is safe and accurate. Without explanations, they may hesitate to use BioNLP tools, especially when the stakes are high.

3.8.2.3 Accountability
If a model makes a mistake—such as suggesting the wrong diagnosis—who is responsible? It is hard to answer this question when the model's inner logic is unclear. Accountability is critical in medicine, and black box models make this issue even more complicated.

3.8.2.4 Need for explanation

Unlike simple rule-based models, modern BioNLP systems require new techniques to explain their decisions. Researchers are working on "explainable AI" methods, which aim to make models more transparent and easier to interpret. However, this is still an ongoing challenge.

3.8.3 Limited and biased data

While medicine generates a large amount of information, not all of it is suitable for training BioNLP systems. Data can be limited in certain cases, and even when it is available, it may carry hidden biases.

3.8.3.1 Limited data

Some areas of medicine, such as rare diseases, do not have large datasets. Without enough examples, AI systems struggle to learn and may give poor results. Another limitation is language—most BioNLP resources are available in English, which creates difficulties for healthcare systems in nonEnglish-speaking regions.

3.8.3.2 Bias from data sources

If the training data mostly comes from a single group of patients—for example, one ethnic community—the model may not perform well for other populations. This can create unfairness and widen health inequalities.

3.8.3.3 Bias from human annotators

Medical datasets often need experts to label them, such as marking where a disease or drug name appears in text. But these experts may unconsciously bring their own biases into the dataset. The AI then learns those biases and may repeat them in its predictions.

3.8.3.4 Difficulty in fixing bias

Correcting bias is not straightforward. Researchers use methods like balancing datasets or adjusting algorithms, but these are often complex, time-consuming, and not always successful Table 3.4.

Table 3.4 Different challenges of natural language processing in the biomedical domain.

Challenge	Why It's a problem	Possible fix
Privacy	Risk of leaking patient info	Deidentification, synthetic data
Black box	Hard to trust the model	Show explanations
Bias/Data	Works only in some cases	Collect more data, test in new places

3.9 Future directions

Language is the most natural way humans communicate with each other, and teaching machines to understand it has always been a major goal of artificial intelligence. NLP is the field that makes this possible. Today, NLP powers many everyday applications—such as voice assistants, search engines, medical document analysis, chatbots, and translation tools. But what we see now is just the beginning.

The future of NLP is moving toward systems that do much more than simply "understand words". They will be able to sense context, pick up emotions, and interact in a way that feels closer to human conversation. With the growth of powerful computers, advanced algorithms, and large amounts of data, NLP is expected to transform how humans and machines connect in almost every area of life.

3.9.1 Key trends that will shape the future

3.9.1.1 Multimodal artificial intelligence

Language is not only spoken or written—it often comes with gestures, tone of voice, or even images and videos. Future NLP systems will combine different types of information (text, audio, video, images) to gain a deeper understanding of communication. For example, a smart assistant could listen to your words, analyze your facial expression, and then respond in a way that matches your mood. This will make interactions with technology more natural and intuitive.

3.9.1.2 Smaller and specialized models

While giant language models are popular today, they are expensive to run and sometimes too general for specific fields. The future will bring smaller models trained for focused areas like healthcare, law, or finance. These models will be faster, cheaper, and more accurate within their domain. They will also improve data privacy because organizations can train them on sensitive data without needing massive external resources.

3.9.1.3 Understanding context and intention

Humans rarely say things in a straightforward manner. We use sarcasm, humor, cultural references, and hidden meanings. Future NLP systems will be designed to pick up on these subtle cues. This means virtual assistants, customer service bots, and digital tutors will respond more intelligently and in a way that feels closer to talking with another human.

3.9.1.4 Emotion-aware language processing

Emotions are a big part of communication. A person who is sad, angry, or excited does not only use different words—they also express themselves in different tones. Emotion-aware NLP will allow machines to detect these emotional signals and adapt responses. For example, a healthcare chatbot could reply with empathy when talking to a patient, or a customer service bot could calm down an upset customer instead of giving robotic replies.

3.9.1.5 Explainable NLP

As NLP is used in sensitive fields like medicine, law, and governance, it becomes important for people to understand how and why a system makes a decision. Explainable NLP will focus on transparency, making sure that users can see the reasoning behind the output. This builds trust and ensures accountability in high-risk areas.

3.9.1.6 Quantum NLP

Although still experimental, quantum computing could 1 day change NLP completely. Some language problems are so complex that today's computers struggle with them. Quantum NLP aims to solve such problems faster and more efficiently, which could open the door to more advanced language understanding in the future.

The future of NLP is about creating systems that are smarter, more human-like, and more responsible. From multimodal understanding to emotion-aware responses and domain-specific applications, NLP is set to touch every part of our daily lives. At the same time, issues like bias, privacy, and energy use must be addressed with care.

If developed responsibly, NLP will not only improve human–machine interaction but also transform industries such as healthcare, education, finance, and beyond. In the coming years, NLP will act as a bridge between human thought and machine intelligence, opening a new era of communication where technology understands us better than ever before.

3.10 Summary of insights

This chapter provided an in-depth understanding of Biomedical NLP, emphasizing its growing role in research and healthcare. It traced the development of the field, explained its technical foundations, and showcased its practical applications, while also identifying the challenges and future opportunities.

Key Insights.

1. Evolution and Importance: Biomedical NLP has emerged as a vital tool for extracting knowledge from vast biomedical texts and clinical records.
2. Pipeline Structure: A biomedical NLP pipeline involves steps such as data collection, preprocessing, workflow integration, and analysis.
3. Core Techniques: Both traditional approaches and modern transformer-based architectures are central to understanding biomedical text.
4. Applications: Biomedical NLP is applied in research discovery, clinical decision-making, and patient care, with real-world case studies highlighting its impact.

5. Challenges: Privacy concerns, interpretability issues, and limited unbiased datasets remain critical barriers.
6. Future Directions: Emerging advancements aim at building ethical, transparent, and more effective NLP systems for healthcare and biomedical research.

3.10.1 Final remarks

- Biomedical NLP is a bridge between doctors' language and data analysis.
- It can make healthcare faster, safer, and smarter.
- In the future, it will mix text with images (X-rays, scans) and genetic data.
- Tools will also improve for languages with less data.
- Privacy and fairness must always be protected.

Final thought: If used carefully, Biomedical NLP can move from research labs into everyday hospital work, helping patients get better care.

Take-home messages

Biomedical NLP plays a vital role in transforming unstructured clinical and research text into actionable insights that support healthcare and scientific progress. It not only speeds up drug discovery and clinical trials but also improves decision-making, patient safety, and personalized care. While issues such as privacy concerns, black-box models, and biased datasets remain challenges, the field is rapidly progressing. Looking ahead, the integration of multimodal data and explainable AI will further enhance BioNLP's impact. If developed responsibly, BioNLP has the power to make healthcare smarter, safer, and more efficient for the future.

4

AI/ML in chemoinformatics

Mohammad "Sufian" Badar[1, 2, 3, 4], Taiba Shamim[5], Mohammed Abaan Kurikkal[5], Aamir Nehal[6], Naseem Rao[7], Fakroul Ridzuan Hashim[8], Md. Mustafa[9], Abdus Sattar[10], Kamran Khan[11] and Zubin Ingah[12]

[1]DEPARTMENT OF BIOENGINEERING, UNIVERSITY OF CALIFORNIA, RIVERSIDE, CA, UNITED STATES; [2]UNIVERSAL SCIENTIFIC EDUCATION AND RESEARCH NETWORK (USERN), TEHRAN, IRAN; [3]SPI DARBHANGA, DARBHANGA, BIHAR, INDIA; [4]DEPARTMENT OF COMPUTER SCIENCE AND ENGINEERING (BIOINFORMATICS), SCHOOL OF ENGINEERING SCIENCES AND TECHNOLOGY (SEST), JAMIA HAMDARD UNIVERSITY, NEW DELHI, INDIA; [5]DEPARTMENT OF COMPUTER SCIENCE, JAMIA MILLIA ISLAMIA, NEW DELHI, INDIA; [6]TOUCH N HEAL PHYSIOTHERAPY AND REHABILITATION CENTRE, MOTIHARI, BIHAR, INDIA; [7]DEPARTMENT OF COMPUTER SCIENCE AND ENGINEERING, SCHOOL OF ENGINEERING SCIENCES AND TECHNOLOGY (SEST), JAMIA HAMDARD UNIVERSITY, NEW DELHI, INDIA; [8]ARTIFICIAL INTELLIGENT, BIOMEDICAL ENGINEERING UNIVERSITI PERTAHANAN NASIONAL MALAYSIA, KUALA LUMPUR, MALAYSIA; [9]ALL INDIA INSTITUTE OF MEDICAL SCIENCES (AIIMS), NEW DELHI, INDIA; [10]AGROMETEOROLOGY, CENTRE FOR ADVANCED STUDIES ON CLIMATE CHANGE, DR. RAJENDRA PRASAD CENTRAL AGRICULTURAL UNIVERSITY, SAMASTIPUR, BIHAR, INDIA; [11]HIND INSTITUTE OF MEDICAL SCIENCES, SITAPUR, UTTAR PRADESH, INDIA; [12]GLOBAL NETWORK TECHNOLOGIES, SUNNYVALE, CA, UNITED STATES

What you will learn

- How to deploy AI and ML in chemoinformatics
- How to utilize and analyze the chemical data in the area of drug discovery
- It provides the fundamental and advanced techniques to understand the mechanism
- To represent the chemical data, including SMILES by ECFP, Mol2Vec, and MACCS
- It investigates the AI-driven applications in molecular design

4.1 Introduction

Due to exponential chemical and biological data growth, drug discovery and materials research face great promises and challenges. Computers encode, analyze, and interpret chemical structures, and chemoinformatics has long enabled virtual screening, QSAR

Applying Artificial Intelligence to Computational Biology and Medical Informatics. https://doi.org/10.1016/B978-0-443-44915-4.00005-8

modeling, and molecular property prediction. Chemical informatics has moved from rule-based techniques to data-driven predictive modeling that learns from molecular representations thanks to AI and ML (Vamathevan et al., 2019).

AI-driven chemoinformatics uses SMILES strings, molecular graphs, 3D conformers, and learned embeddings for structural and functional abstraction (David et al., 2020). Researchers can explore chemical space, identify molecules, and accurately predict molecular properties using supervised, unsupervised, and semisupervised algorithms. GNNs and transformer-based chemical language models are now popular molecular representation learning and generative modeling methods (Zhou et al., 2020).

The chapter addresses three interconnected dimensions:

- Exploring how molecules are encoded into computationally effective SMILES, SELFIES, 3D conformers, and embeddings.
- Algorithms and paradigms include classical ML, GNNs, transformers, model evaluation, and dimensionality reduction.
- Deep generative models (VAEs, GANs, and reinforcement learning), ADMET, bioactivity prediction, virtual screening, and hybrid physics–AI structure-based drug design.

In addition to these methodological foundations, the chapter examines chemoinformatics and AI integration into drug discovery pipelines, DMTA loop automation, robotics and LLM agents, and AlphaFold, Insilico Medicine, Recursion, and DeepMind cases. Data quality, interpretability, cross-disciplinary integration, and how AI will continue to improve chemoinformatics and expedite therapeutic and novel material discovery conclude the chapter.

4.1.1 What is chemoinformatics

Chemoinformatics, often known as cheminformatics, is a field that combines chemistry, computer science, mathematics, and information technology to analyze and understand chemical data. The basis of this is the digital representation of chemical structures and the use of computer tools to find patterns, correlations, and knowledge in large molecular datasets (O'Boyle et al., 2011). Some of the most important uses are virtual screening, quantitative structure–activity relationship (QSAR) modeling, molecular docking, drug discovery, and keeping track of chemical databases (Mitchell, 2014).

Chemoinformatics has become much more important since chemical data is growing rapidly in places like PubChem, ChEMBL, and ZINC. These resources provide a place to create computational workflows for drug development, toxicology, and materials research. Chemoinformatics is an important link between experimental chemistry and computational modeling, which makes it an essential tool for modern scientific study (Weininger, 1988).

4.1.2 Evolution of AI/ML in chemoinformatics

Chemoinformatics has changed the way chemical and pharmaceutical research is conducted by adding artificial intelligence (AI) and machine learning (ML). In the past, people used rule-based expert systems to predict how chemicals would react and how their properties would relate to each other. Over time, conventional ML methods like decision trees, support vector machines, and random forests were increasingly used. These methods worked better for QSAR modeling and predicting toxicity (Cortes and Vapnik, 1995).

Deep learning architectures, such as convolutional neural networks (CNNs) and graph neural networks (GNNs), are the most popular today. They can learn directly from chemical graphs or SMILES strings without needing to create descriptors by hand (Gilmer et al., 2017). Natural language processing (NLP) has been utilized to analyze chemical literature and patents, improving knowledge extraction (Krenn et al., 2020).

The evolution of AI and ML in chemoinformatics signifies a shift from hypothesis-driven to data-driven research, marked by substantial advancements in de novo molecular creation, retrosynthetic planning, and pharmacokinetic prediction (Vamathevan et al., 2019).

4.1.3 Significance, objective, and scope

Chemoinformatics, enhanced by AI and ML, plays a significant role in speeding up and improving the drug development process, which has a reputation for being slow and expensive. Researchers may use computer models to quickly go through large libraries of compounds, find possible lead molecules, and forecast biological activity, reducing the reliance on experimental trials (Lo et al., 2018).

The goal of combining AI/ML with chemoinformatics is to:

- Make models that can anticipate how drugs will interact with targets and how hazardous they will be (Mitchell, 2014).
- Automate molecular design and retrosynthetic strategies.
- Combine chemical and biological data for systems-level drug discovery (Vamathevan et al., 2019).
- Improve the understanding, reproducibility, and interpretability of computer predictions (David et al., 2020).

This field encompasses a wide range of disciplines: finding new drugs, repurposing old ones, virtual screening for new treatments, materials science for polymers and catalysts, environmental chemistry and toxicology for chemical safety, and personalized medicine for tailored therapeutic design (Zhou et al., 2020).

The combination of chemoinformatics with AI/ML is not only advancing computational chemistry but also driving innovations in medicine, biotechnology, and environmental sciences (Lo et al., 2018).

4.2 Representation of chemical data

Chemoinformatics relies on representing chemical structures in formats that machines can read. For molecular modeling, virtual screening, and ML, chemical information needs to be processed by computers, and encoding systems need to be both fast and accurate (O'Boyle et al., 2011). A strong representation enables the storage, access, and analysis of chemical knowledge on various computer systems. Common methods of representing chemical information include SMILES, SELFIES, and molecular graphs (Weininger, 1988).

4.2.1 SMILES, SELFIES, and molecular graphs

4.2.1.1 SMILES (simplified molecular input line entry system)

SMILES is a linear notation often used to represent molecular structures as text strings. A series of ASCII characters shows every atom, bond, and branching configuration of a molecule (Weininger, 1988). This simple but flexible method enables chemical databases and algorithms to store and retrieve substances quickly and easily.

4.2.1.2 SELFIES (self-referential embedded strings)

SELFIES is a recent concept that addresses the issues with SMILES. Each SELFIES string represents a valid chemical structure, making it very useful for creating new molecules and using AI for molecule generation (Krenn et al., 2020). This self-referencing property ensures that generated molecules are always chemically valid.

4.2.1.3 Molecular graphs

Molecular graphs use mathematical graphs to represent molecules, where atoms are the nodes and bonds are the edges (Gilmer et al., 2017). This representation works well with graph theory and is especially useful for GNNs.

4.2.2 3D representation and conformer generation

SMILES, SELFIES, and molecular graphs are all two-dimensional (2D) and string-based methods to represent molecules, which work well for storing and processing chemical information. However, these formats don't provide the three-dimensional (3D) shape of molecules, which is crucial for understanding their behavior in biological and chemical contexts (Friedrich et al., 2017). Molecular interactions, such as receptor binding, solubility, and permeability, are intrinsically 3D processes, so accurate 3D representation and conformer generation are critical components of cheminformatics and computational drug design (Hawkins and Nicholls, 2012).

4.2.2.1 Conformer generation

Molecules are not fixed; they exist as conformers, which are 3D shapes that can be reached by rotating around single bonds. Conformer generation aims to predict these shapes to explore the potential variations of a molecule in three-dimensional space (Rupp et al., 2015). Some important methods for conformer generation include:

- Systematic Search: thorough torsional sampling, accurate but computationally expensive for large molecules.
- Stochastic Methods: such as Monte Carlo, simulated annealing, and evolutionary algorithms, which explore the conformational space more efficiently.
- Knowledge-Based: Using torsional libraries from crystallographic databases to predict possible conformations (Ebejer et al., 2012).
- Force Field Optimization: Utilizing molecular mechanics force fields (MMFF94, UFF, OPLS) to find the best conformer (Friedrich et al., 2017).
- Quantum Chemical Refinement: Using ab initio or DFT methods to refine the conformers for highly accurate results (Hawkins and Nicholls, 2012).

4.2.3 Molecular embeddings: ECFP, Mol2Vec, MACCS

Molecular embeddings are vital for converting chemical structures into numerical attributes that are suitable for ML models. Extended Connectivity Fingerprints (ECFPs), Mol2Vec, and MACCS keys are some of the widely used methods for molecular embeddings.

- Extended Connectivity Fingerprints (ECFPs)

ECFPs provide robust molecular fingerprints that are used in similarity searching and quantitative structure-activity relationship (QSAR) tasks. ECFPs transform molecules into binary vectors based on their chemical substructure and are often used in ML tasks (Rogers and Hahn, 2010).

- Mol2Vec Mol2Vec is an unsupervised molecular embedding technique inspired by natural language processing (NLP) methods. It treats chemical substructures like "words" and molecules like "sentences," capturing contextual similarities between substructures to create informative low-dimensional embeddings (Jaeger et al., 2018). This method has been used in pharmacological research, including drug discovery and drug response prediction (Jaeger et al., 2018).
- MACCS Keys

The MACCS keys are one of the earliest and most commonly used types of molecular fingerprints. They represent molecular substructures as a binary vector, with each bit corresponding to the presence or absence of a specific structural feature, such as hydrogen bond donors or aromatic rings (Durant et al., 2025).

4.3 AI/ML algorithms and learning paradigms

Chemoinformatics leverages AI and ML to extract patterns and insights from large chemical databases. These methods enable the discovery of bioactive molecules, novel drugs, and an improved understanding of structure-property relationships. There are three main types of learning paradigms: supervised, unsupervised, and semisupervised

(Cortes and Vapnik, 1995). Recent advances in deep learning, particularly transformers and GNNs, have significantly improved predictive capabilities. Additionally, methods for model assessment and dimensionality reduction make these models more robust and interpretable, facilitating their application to a variety of datasets (Xia et al., 2013).

4.3.1 Supervised, unsupervised, and semisupervised learning

4.3.1.1 Supervised learning

Supervised learning is widely used to predict molecular attributes such as bioactivity, toxicity, and solubility. Algorithms such as support vector machines, random forests, and gradient boosting have shown excellent performance in tasks like virtual screening and QSAR modeling when labeled data is available (Cortes and Vapnik, 1995).

4.3.1.2 Unsupervised learning

Unsupervised learning techniques are employed when labels are not available. Clustering and dimensionality reduction methods uncover hidden relationships in the organization of chemical space, helping to categorize compounds and identify outliers (Sheridan, 2013).

4.3.1.3 Semisupervised learning

- Semisupervised learning techniques utilize a combination of labeled and unlabeled data. This approach improves model accuracy, especially when fully labeled datasets are not available, reducing the cost of experiments (Zhu, 2005).

4.3.2 Deep learning: graph neural network, transformer

4.3.2.1 Graph neural networks (GNNs)

GNNs represent molecules as graphs, with atoms as nodes and bonds as edges. These networks incorporate localized chemical information through message-passing techniques, facilitating the prediction of molecular properties and modeling drug-target interactions (Gilmer et al., 2017). GNNs have surpassed traditional descriptors in predicting bioactivity, quantum chemical properties, and toxicity (Zhang et al., 2025).

- Transformers Originally developed for natural language processing, transformers have been adapted to represent chemical sequences, such as SMILES strings. Pretrained chemical language models like ChemBERTa and MolBERT capture long-range dependencies and chemical context, making them effective in predicting chemical reactions, retrosynthesis, and molecular properties (Luong and Singh, 2024; Chithrananda et al., 2020).

4.3.3 Model evaluation and dimensionality reduction

Model evaluation is critical to ensure that AI and ML models generalize well and do not overfit. Techniques such as cross-validation and the use of external test sets are commonly employed. Performance metrics for classification models include precision, recall, AUC, F1-score, RMSE, and R^2 (Xia et al., 2013).

Molecular descriptors have numerous features, and dimensionality reduction techniques help simplify models and improve their performance. Methods like principal component analysis (PCA), t-distributed stochastic neighbor embedding (t-SNE), and uniform manifold approximation and projection (UMAP) help in understanding chemical space and making high-throughput virtual screening more efficient (Mitchell, 2014; McInnes et al., 2018).

4.3.4 AI in the design–make–test–analyze (DMTA) cycle

In the new age of drug discovery, there is a framework that helps in the primary operation, and that is the Design-Make-Test-Analyze (DMTA) cycle. This cycle works as an iterative process, and the molecules are grouped before designing, synthesizing, and even experimentally assaying to get data that understands the relationship between structure, activity, and structural properties of the molecule. The resulting information helps us to plan and design the compounds required for further activities (Ghiandoni et al., 2024). The most important goal when a program is designed to conduct a drug discovery is it's efficiency. When efficiency is foreseen, the program works to decrease the iteration value, which would be required to identify a particular drug. This step is very crucial as it could almost reach upto 30% of the whole program's expenses just for this iterative process(Ghiandoni et al., 2024). With the use of AI & ML, it could help us be more efficient, data-driven, and also partly automated to augment every part of the DMTA cycle, resulting in a fast and efficient drug discovery pipeline.

The current state-of-the-art DMTA cycle, which is supplemented by AI, works as a closed-loop workflow while even being known as "self-driven laboratories". When certain systems like this one turn more technologically superior, it helps in decision-making using AI with platforms that help in giving a chance for autonomous experimentation. There is a hub in this kind of algorithm that leads to a data lake, resulting in a smooth transition between the different stages in the cycle. This "hub" should be set on a standard and accessible data framework, which will help in the fast knowledge transfer (Seifrid et al., 2022).

4.3.4.1 Design: charting the chemical space

The 'Design' phase involves the conceptualization and tailoring of molecular structures to achieve a desired biological target activity while satisfying a complex matrix of property constraints, such as solubility, permeability, metabolic stability, and safety. Traditionally reliant on the intuition of medicinal chemists and knowledge gleaned from limited historical data, this stage is now heavily supported by a suite of computational tools. AI/ML predictive models, in particular, have become pivotal in guiding design algorithms to explore vast, uncharted regions of chemical space for de novo design, lead optimization, scaffold hopping, and library expansion (Ghiandoni et al., 2024).

This shift is exemplified by the "predict first" culture at AstraZeneca, which is supported by their internal Predictive Insight Platform (PIP). This cloud-native infrastructure provides scientists with on-demand access to nearly 250 predictive models covering a wide array of endpoints, including bioactivity, pharmacokinetics, and safety profiles. Before committing to costly and time-consuming synthesis, virtual compounds are thoroughly scrutinized by this battery of computational models, allowing teams to prioritize candidates with the highest probability of success. This predictive paradigm is not limited to traditional small molecules but has been extended to other therapeutic modalities, including peptides, PROteolysis TArgeting Chimeras (PROTACs), and antibody-drug conjugates (Ghiandoni et al., 2024).

A critical and nuanced challenge in the 'Design' phase is translating high-level project goals into effective, quantitative reward functions that can steer generative molecular design (GMD) algorithms (Urbonas et al., 2023). Manually configuring these functions—deciding how to normalize, weight, and aggregate multiple, often competing, objectives—can be an error-prone, trial-and-error process heavily biased by human intuition (Urbonas et al., 2023). To overcome this, novel automated, data-driven methods are being developed. One such approach leverages experimental data to construct preference rankings based on the principle of Pareto dominance. In this framework, one molecule is considered superior to another only if it is quantifiably better or equal across all measured objectives and strictly better in at least one. A neural network is then trained to learn a reward function whose rankings correlate with these objective, data-driven Pareto relationships. This removes subjective bias from the reward function and leads GMD tools more effectively and robustly toward optimal regions of the chemical space (Urbonas et al., 2023).

4.3.4.2 Make: from virtual concept to physical reality

The 'Make' phase, which encompasses the chemical synthesis and subsequent purification of designed compounds, has historically been a significant bottleneck, constraining the pace of discovery (Seifrid et al., 2022). Chemists traditionally depended on personal experience and published literature, which could bias compound selection toward molecules with known, easier synthetic routes rather than those with the most promising predicted profiles. AI is revolutionizing this stage through sophisticated, data-driven tools for both retrosynthesis and forward reaction prediction.

Platforms like AstraZeneca's PIP deploy models capable of predicting multi-step synthetic routes for novel compounds, accounting for complex factors like multiple reactive sites to improve the efficiency and feasibility of synthesis planning. Beyond planning, AI also provides crucial support for the purification process by predicting key physical properties like solubility, which helps prevent experimental failures such as column blockage during chromatography (Ghiandoni et al., 2024).

The ultimate vision is the fully autonomous synthesis of molecules within self-driving laboratories. These facilities are being developed to integrate robotic automated synthesis platforms directly into the DMTA loop. They often employ robust and versatile chemical strategies, such as iterative Suzuki-Miyaura cross-coupling using MIDA-protected boronate building blocks, to assemble complex molecules with minimal human intervention. This level of automation enables the rapid, parallelized synthesis of hundreds of unique compounds, dramatically increasing experimental throughput and the speed at which design hypotheses can be tested (Seifrid et al., 2022).

4.3.4.3 Test: generating high-quality data efficiently

The 'Test' phase, where compounds are assayed to measure their biological activity, properties, and potential liabilities, has evolved from a labor-intensive manual process to a highly automated one, largely due to robotics and high-throughput screening (HTS) (Ghiandoni et al., 2024). AI is now adding a layer of intelligence on top of this automation, primarily through smart experimental prioritization.

Predictive modeling, often coupled with active learning, can be used to strategically select which compounds from a larger set should be assayed to maximize the information gained from a limited experimental budget. For example, by identifying regions of chemical space where model uncertainty is high, AI can prioritize the testing of compounds that will provide the most valuable data for subsequent model improvement and training. This ensures that experimental resources are deployed with maximum efficiency to generate information-rich datasets. Furthermore, predictive models are being investigated to help mitigate and correct for confounding factors and systematic errors in HTS assays, such as plate edge effects, thereby improving the overall quality and reliability of the resulting data (Ghiandoni et al., 2024).

4.3.4.4 Analyze: closing the loop with intelligent insights

In the 'Analyze' stage, the experimental data from the 'Test' phase are interpreted to prioritize compounds for the next DMTA cycle and to formulate new design hypotheses. This process has advanced far beyond simple Structure-Activity Relationship (SAR) tables to integrated, federated data environments where experimental results are seamlessly combined with a rich layer of computational predictions (Ghiandoni et al., 2024).

AI platforms like PIP play a crucial role by complementing sparse experimental data with a comprehensive profile of predictions. Predicted properties can be correlated with experimental values to formulate and test new hypotheses; for instance, a trend in predicted cell permeability might be compared against bioactivity measurements to infer information about the target. As a project progresses and generates more data, the predictive models are continuously retrained on this new, project-specific data, which iteratively increases their accuracy and applicability within the relevant chemical space. This integration provides a more holistic view, allowing research teams to understand not just what worked but also to develop hypotheses about *why* certain compounds were successful. The ability to retrospectively analyze past predictions versus experimental outcomes is crucial for refining both the predictive models and the overarching design strategies, with the ultimate goal of intelligently guiding the discovery process and reducing the number of DMTA cycles needed to identify a clinical candidate (Ghiandoni et al., 2024).

4.3.5 Integration with robotics and LLM agents

The convergence of chemoinformatics with robotics and Large Language Models (LLMs) is catalyzing a paradigm shift from computer-aided design to autonomous, AI-driven discovery. While the previous section detailed how AI augments the individual steps of the DMTA cycle, this section explores how LLMs are being architected into intelligent "agents" that can orchestrate the entire research workflow. These agents act as a cognitive backbone, translating high-level scientific goals expressed in natural language into concrete, executable actions performed by robotic platforms, effectively bridging the gap between human intent and automated execution (Darvish et al., 2024; Ramos et al., 2024).

4.3.5.1 The LLM as an autonomous reasoning engine

At the heart of this new wave of automation is the LLM, which functions as the central "brain" or reasoning engine (Ramos et al., 2024). Unlike earlier automation systems that required rigidly programmed scripts, LLM-based agents can interpret ambiguous, high-level commands, decompose them into logical subtasks, and devise a coherent plan to achieve a final goal. For instance, the ORGANA robotic assistant uses an LLM to interact with chemists via text or speech, comprehending an experimental goal like "generate a Pourbaix plot for an unknown quinone solution" and subsequently formulating a multi-step experimental plan (Darvish et al., 2024).

A key limitation of LLMs, however, is that they lack access to real-time information and are prone to "hallucinating" factually incorrect details (M. Bran et al., 2024). To overcome this, the most effective agents do not rely on the LLM's internal knowledge alone. Instead, they employ a framework where the LLM's reasoning capabilities are augmented by a suite of external, expert-designed computational tools. Systems like ChemCrow operate on an iterative "reason-act" loop: the LLM first reasons about the problem (Thought), then selects an appropriate tool (Action), and provides the necessary input (Action Input). The program executes the tool, and the result is fed back to the LLM (Observation), which then informs the next cycle of reasoning and action (M. Bran et al., 2024). This architecture transforms the LLM from a fallible information source into a robust reasoning engine that intelligently queries reliable tools to solve complex scientific tasks.

This tool-augmented approach allows an agent to seamlessly integrate a wide range of chemoinformatic capabilities. For example, to fulfill a request to "plan and execute the synthesis of an insect repellent," the ChemCrow agent sequentially uses tools to search the web to identify DEET as a common insect repellent, convert its name to an SMILES string, devise a multi-step retrosynthetic plan, predict an executable experimental procedure, and finally, send the procedure to a robotic platform for physical execution (M. Bran et al., 2024).

4.3.5.2 From digital instructions to robotic execution

The crucial link between the agent's digital plan and the physical world is the interface with robotic hardware. This integration is being achieved through several sophisticated methods. One primary approach involves translating high-level experimental plans into a universal, hardware-agnostic programming language for chemistry. Languages like the Chemical Description Language (XDL) allow for the precise specification of unit operations (e.g., add, stir, heat, and filter) that can be executed on any compatible robotic platform (Leonov et al., 2024; Salley et al., 2023). The ORGANA system leverages this by using an LLM to translate natural language procedures into validated XDL scripts, which are then used as goals for its task and motion planner (Darvish et al., 2024).

Alternatively, agents can interface directly with the Application Programming Interfaces (APIs) of specific robotic systems. The Robot Operator agent within the ChemAgents framework, for instance, generates Python code that calls upon a library of high-level robot APIs for actions like move, locate, and put_rack (Song et al., 2024). Similarly, ChemCrow connects directly to the IBM RoboRXN platform to execute syntheses it has planned (M. Bran et al., 2024). A key capability demonstrated by these agents is self-correction; ChemCrow can autonomously query a synthesis platform's validation feedback, identify errors in a predicted procedure (e.g., "not enough solvent"), and iteratively modify the procedure until it is valid for execution, all without human intervention (M. Bran et al., 2024). This dynamic, feedback-driven approach represents a significant step toward true autonomy, where the agent can adapt to the specific constraints of the physical world.

4.3.5.3 Multi-agent systems for end-to-end research

For highly complex, multi-faceted research campaigns, a single agent may be insufficient. The emerging solution is the development of hierarchical, multi-agent systems where different LLM-powered agents with specialized roles collaborate to achieve a common goal. The "ChemAgents" system exemplifies this architecture, featuring a central "Task Manager" agent that coordinates the activities of four role-specific agents (Song et al., 2024):

1. Literature Reader: Mines a local database of scientific publications to extract relevant prior knowledge and inform experimental design.
2. Experiment Designer: Leverages a library of established protocols to generate detailed, step-by-step experimental procedures tailored to the research task.
3. Robot Operator: Translates the experimental procedures into robot-executable code and commands to operate the automated lab hardware.

4. Computation Performer: Uses a library of ML models to analyze data, make predictions, and guide optimization tasks, for instance, by running a Bayesian optimizer.

This distributed cognitive architecture allows the system to tackle an entire research workflow on demand. In a demonstration, ChemAgents was tasked with discovering a high-performance, five-metal high-entropy catalyst for the oxygen evolution reaction (OER). The system autonomously used its Literature Reader to identify the five most promising metals from the literature, designed and robotically executed 100 experiments on randomly sampled compositions, used its Computation Performer to train a predictive model on the results, and then ran a Bayesian optimization loop to identify a novel, high-performance composition, which it then synthesized and validated experimentally. This achievement showcases the power of multi-agent systems to conduct complex, on-demand research with minimal human input (Song et al., 2024).

4.3.5.4 Human-AI collaboration and the future of the lab

While the prospect of fully autonomous "lights-out" laboratories is compelling, the immediate future is centered on human-AI collaboration. Systems like ORGANA are designed not to replace chemists but to act as intelligent robotic assistants that reduce tedious, labor-intensive work and augment human capabilities (Darvish et al., 2024). The LLM interface allows for intuitive, natural language-based interaction, reducing the barrier for chemists to use complex robotic systems. ORGANA also enhances robustness through human-in-the-loop disambiguation; for example, it can present a visual of its workspace and ask the user to confirm the function of different beakers.

Furthermore, these systems close the feedback loop by automatically analyzing experimental data and generating comprehensive reports for the human researcher, including data visualizations and statistical analysis. A user study conducted with ORGANA found that it reduced user frustration and physical demand by over 50% and saved chemists over 80% of their time compared to manual execution of the same electrochemistry experiment (Darvish et al., 2024). This integration of robotics and LLM agents is not merely automating tasks but is creating a synergistic partnership that promises to accelerate the pace of discovery, improve reproducibility, and ultimately, democratize access to advanced experimental capabilities (Song et al., 2024).

4.3.6 Case studies in AI-driven structural biology and drug discovery

The theoretical advancements in chemoinformatics and AI are best understood through their real-world applications. Several organizations have pioneered the use of sophisticated AI platforms to tackle long-standing challenges in biology and medicine, moving from purely computational exercises to tangible outcomes, including the development of novel therapeutics now in human clinical trials. This section examines two landmark case studies: DeepMind's AlphaFold, which has revolutionized structural biology, and Insilico Medicine, which has demonstrated an end-to-end AI-driven drug discovery pipeline.

4.3.6.1 DeepMind and alphafold: deciphering the building blocks of life

The challenge of accurately predicting the three-dimensional structure of a protein from its amino acid sequence has been a grand challenge in biology for over 50 years (Abramson et al., 2024). A protein's function is intrinsically linked to its structure, and understanding this structure is critical for deciphering biological mechanisms and designing targeted drugs. While experimental methods like X-ray crystallography and cryo-electron microscopy are powerful, they are often slow, expensive, and not always feasible for every protein.

In 2021, DeepMind introduced AlphaFold 2, an AI system that achieved unprecedented accuracy in protein structure prediction, marking a revolutionary moment for structural biology (Abramson et al., 2024). The success of AlphaFold spurred further innovation, leading to the development of AlphaFold 3. This newer model represents a substantial leap forward, moving beyond single proteins to predict the joint structure of a vast range of biomolecular complexes. AlphaFold 3's diffusion-based architecture can accurately model interactions between proteins, nucleic acids (DNA and RNA), small molecules (ligands), ions, and even posttranslationally modified residues within a single, unified deep learning framework (Abramson et al., 2024). This was achieved by evolving the architecture to replace the original "evoformer" with a simpler "pairformer" module and, crucially, by predicting raw atom coordinates directly via a diffusion module. This diffusion approach allows the model to learn chemical structures at multiple scales, from local stereochemistry to the global arrangement of a complex, without needing complex, hand-tuned stereochemical violation penalties (Abramson et al., 2024).

The impact of this technology has been profound. AlphaFold 3 has demonstrated accuracy that far surpasses previous specialized tools. In the critical area of drug discovery, it achieves substantially higher accuracy for predicting protein-ligand binding poses compared to state-of-the-art docking tools like AutoDock Vina. It has also shown superior performance in modeling protein-nucleic acid interactions and antibody-antigen complexes (Abramson et al., 2024). This capability allows researchers to generate high-fidelity structural models for previously uncharacterized biological systems, providing crucial insights for hypothesis generation and rational drug design. By making high-accuracy structural modeling accessible across the entire biomolecular space, AlphaFold has democratized a field that was once the domain of specialized structural biologists and is now an indispensable tool in the modern chemoinformatics toolkit.

4.3.6.2 Insilico medicine: an end-to-end AI platform for drug discovery

While AlphaFold addresses the structural component, Insilico Medicine has pioneered an integrated, end-to-end AI-driven platform, Pharma.AI, designed to accelerate the entire drug discovery process, from identifying novel disease targets to designing and optimizing therapeutic candidates (Ivanenkov et al., 2023). This platform is composed of several key components that work in synergy:

- PandaOmics: A target discovery engine that analyzes vast amounts of multi-omics data to identify and prioritize novel disease-associated targets (Ivanenkov et al., 2023).
- Chemistry42: A generative chemistry platform for de novo small molecule design and optimization (Ivanenkov et al., 2023).
- inClinico: A tool for predicting the probability of success for clinical trials (Ivanenkov et al., 2023).

The power of this integrated approach was demonstrated with the discovery and development of rentosertib (formerly ISM001-055), a novel, first-in-class small-molecule inhibitor for idiopathic pulmonary fibrosis (IPF) (Xu et al., 2025). IPF is a progressive and fatal lung disease with limited treatment options. Using its AI platform, Insilico Medicine first identified a novel target, TNIK (Traf2- and Nck-interacting kinase), which had not previously been implicated in IPF (Xu et al., 2025). Following target identification, the Chemistry42 platform was used to generate novel small molecules specifically designed to inhibit TNIK (Ivanenkov et al., 2023).

Chemistry42 employs an ensemble of over 40 generative models, including generative autoencoders, generative adversarial networks (GANs), and language models, which work in parallel to explore chemical space (Ivanenkov et al., 2023). The generation process is guided by a sophisticated reward function that incorporates a multitude of medicinal chemistry filters (MCFs), physicochemical property predictions (e.g., Lipinski's Rule of Five), novelty scores, and synthetic accessibility scores like the Retrosynthesis Related Synthetic Accessibility (ReRSA) score. For structure-based projects, it further incorporates 3D modules that assess pharmacophore matching, shape similarity, and binding affinity to the target pocket (Ivanenkov et al., 2023). This multi-objective optimization process allows the system to efficiently design molecules that are not only potent against the target but also possess a desirable drug-like profile.

This AI-driven approach dramatically accelerated the preclinical drug discovery timeline. Insilico Medicine was able to move from novel target discovery to the nomination of a preclinical candidate for IPF in just 18 months, with phase 0 and 1 clinical testing completed in under 30 months from the project's start (Xu et al., 2025). Rentsertib subsequently became the first AI-discovered novel drug for an AI-discovered novel target to enter human clinical trials. The results from a Phase IIa clinical trial were promising, showing that rentosertib was safe, well-tolerated, and demonstrated a trend toward improving lung function in patients with IPF (Xu et al., 2025). The success of rentosertib provides a powerful proof-of-concept, demonstrating that generative AI platforms can not only accelerate drug discovery but can also successfully identify novel biology and generate clinically viable therapeutic candidates (Xu et al., 2025).

4.3.7 Key challenges and limitations

The integration of AI and ML into chemoinformatics has demonstrated transformative potential, accelerating timelines and expanding the horizons of what is computationally possible in drug discovery. However, the path from promising retrospective results to prospective real-world impact is fraught with significant challenges. The successful deployment of AI is not merely a matter of applying more complex algorithms or greater computational power; it is contingent on overcoming fundamental issues related to data, model interpretability, evaluation, and human factors. This section details the key challenges and limitations that currently constrain the broader application and adoption of AI in chemoinformatics.

4.3.7.1 *The data problem: foundation of AI*

The adage "garbage in, garbage out" is particularly resonant for AI in drug discovery. The performance, generalizability, and reliability of any model are fundamentally limited by the quality and characteristics of the training data. A recognized barrier to improving AI's application is the absence of standardized, large-scale, high-quality, and openly accessible datasets that are essential for training and validating robust models (Gold and Cook-Deegan, 2025). This overarching "data problem" manifests in several specific, interconnected challenges.

4.3.7.2 *Data quality, availability, and inconsistency*

A primary hurdle is the scarcity of high-quality, annotated data. Many crucial datasets are proprietary, locked behind paywalls, or exist in unstructured, nonmachine-readable formats within scientific literature, which limits the potential of AI applications (Kant et al., 2025). Even when data is available, it is often inconsistent. Data generated by different laboratories using varied protocols, equipment, or sample preparation methods can lead to poor model generalization. For example, studies have shown strong discrepancies in the measured activities of the same drug molecules on the same cancer cell lines when tested in different labs (Ghislat et al., 2024). This heterogeneity requires careful data curation and robust preprocessing techniques to harmonize disparate data sources into a unified format suitable for AI training (Kant et al., 2025).

4.3.7.3 *Data bias and skewness*

AI models learn from the data they are given, and if that data is biased, the model's predictions will be too. Sample bias occurs when training data unevenly samples the distribution of interest; for instance, models trained on known active compounds are more likely to be predictive for other drugs than for novel, untested molecules (Ghislat et al., 2024). This is a significant issue in pharmacokinetics, where clinical data is inevitably biased toward compounds with 'desirable' properties like lower clearance, as molecules with unfavorable profiles are weeded out early in development (Obrezanova, 2023).

A particularly challenging form of bias is data skewness, or class imbalance, which is the norm in early drug discovery. In virtual screening, molecules with affinity for a target (the minority class) are vastly outnumbered by inactive molecules (the majority class), with hit rates often below 1% (Ghislat et al., 2024). A model trained on such an imbalanced dataset may achieve high accuracy simply by predicting every compound to be inactive, rendering it useless for practical discovery campaigns.

4.3.7.4 Small and high-dimensional data

Many problems in chemoinformatics are characterized by a small number of data instances relative to the number of features describing them. In biomarker discovery, for example, a study might have molecular profiles for only a few hundred tumors, each described by hundreds of thousands of features (e.g., a DNA methylation profile) (Ghislat et al., 2024). This "curse of dimensionality" makes it difficult for supervised learning algorithms to infer meaningful patterns without overfitting. While techniques like self-supervised pretraining on large unlabeled datasets (e.g., millions of SMILES strings) can help, the challenge of learning from small, labeled datasets remains a significant obstacle (Ghislat et al., 2024).

4.3.7.5 The "Black Box" Problem: Model Interpretability and Transparency

Perhaps the most persistent challenge to the widespread adoption of advanced AI models in drug discovery is their inherent opacity. Deep learning models, with their millions of parameters and complex, nonlinear transformations, often function as "black boxes," making it difficult to understand the reasoning behind their predictions (Qadri et al., 2025; Wellawatte et al., 2023; Kant et al., 2025). This lack of interpretability creates a barrier to trust, especially in high-stakes fields like medicine and pharmaceutical development, where decisions must be justifiable and transparent (Arrieta et al., 2019).

The trade-off between a model's predictive accuracy and its interpretability is a central dilemma. While simpler, intrinsically interpretable models like linear regression or decision trees are easy to understand, they often lack the predictive power needed for complex biomedical applications. Conversely, highly accurate deep neural networks offer little insight into their decision-making processes (Qadri et al., 2025). This opacity can lead to several problems:

- Lack of Trust: Chemists and clinicians may be skeptical of using predictions from a model if they cannot understand *why* the prediction was made (Wellawatte et al., 2023).
- Hidden Biases: A black-box model might learn spurious correlations from the training data, such as basing predictions on experimental artifacts rather than true causal relationships, which is difficult to detect without interpretability (Wellawatte et al., 2023).
- Regulatory Hurdles: For AI systems to be used in clinical or regulatory settings, they must provide scientifically relevant explanations that align with expert knowledge and meet safety standards (Qadri et al., 2025).

To address this, the field of Explainable AI (XAI) has emerged, developing posthoc techniques like SHAP (SHapley Additive exPlanations) and LIME (Local Interpretable Model-agnostic Explanations) to provide insights into black-box models. These methods can, for example, highlight which molecular substructures or features contributed most to a prediction, bridging the gap between computational output and chemical intuition (Qadri et al., 2025; Wellawatte et al., 2023). However, ensuring that these explanations are faithful to the model, actionable for chemists, and truly reflect the underlying biology remains an active area of research (Wellawatte et al., 2023).

4.3.7.6 *Challenges in model evaluation and benchmarking*

The way AI models are evaluated can significantly misrepresent their true potential for prospective application. Several challenges in this area can confound the development and adoption of new methods.

4.3.7.7 *Performance misestimation and unrealistic benchmarks*

Model performance is often estimated retrospectively on benchmark datasets. However, the choice of metrics and the structure of the benchmark can lead to performance misestimation. For instance, using metrics like the Area Under the Receiver Operating Characteristic Curve (ROC-AUC) for highly imbalanced virtual screening tasks can be misleading, as a model can achieve a seemingly reasonable score while having poor early recognition of active compounds (Ghislat et al., 2024). Furthermore, many popular benchmarks are not representative of real-world discovery scenarios. They may have unrealistically low chemical diversity or may have been curated to remove "problematic" molecules (e.g., frequent hitters), which ironically also removes many known drugs (Ghislat et al., 2024). Models that excel on these sanitized benchmarks may struggle when applied to the vast and complex chemical space of a prospective screening campaign.

4.3.7.8 *The impact of overfitting*

A common misconception among domain experts is that any model that performs better on the training set than on the test set (i.e., overfits) will generalize poorly. While excessive overfitting is a problem, a certain degree of it is expected, and a model can still be highly predictive on a test set despite overfitting the training data. Overfitting is not a binary property of a model but rather a continuous property of the model in relation to a specific dataset (Ghislat et al., 2024). The more critical issue is the "distribution shift" between the training data and the data encountered in a prospective application. The larger this shift, the more likely a model's performance will degrade, regardless of its performance on a retrospective test set (Ghislat et al., 2024).

4.3.7.9 *Human and implementation challenges*

Finally, the successful integration of AI into drug discovery is not just a technical problem but also a human and organizational one.

4.3.7.10 Researcher bias

Both AI experts and drug discovery experts can bring biases that hinder progress. AI experts may focus on algorithmic novelty and benchmark performance without a deep understanding of the domain-specific nuances, leading to models that are technically sophisticated but practically irrelevant. Conversely, drug discovery experts may be overly critical or skeptical of AI, substituting deep collaboration with a superficial application of AI tools, leading to suboptimal results and frustration (Ghislat et al., 2024). Overcoming this requires genuine, multidisciplinary collaboration where both sides share leadership and learn the fundamentals of the other's domain (Ghislat et al., 2024).

4.3.7.11 Integration into existing workflows

Integrating AI tools into the rigid, established workflows of pharmaceutical R&D is a significant logistical and cultural challenge. It requires changes to infrastructure, processes, and expertise, and raises concerns about data privacy and intellectual property (Kant et al., 2025). As AI models become more integral to decision-making, clear ethical guidelines and regulatory frameworks will be essential to ensure their responsible and effective deployment (Qadri et al., 2025).

4.3.8 The need for multidisciplinary collaboration

The successful integration of artificial intelligence into the complex and multifaceted process of drug discovery is not solely a computational challenge; it is fundamentally a human and organizational one. The limitations discussed in the previous section—stemming from data quality, model interpretability, and evaluation—cannot be overcome by technological advances alone. Instead, they highlight an imperative need for deep, synergistic, and sustained collaboration between experts from disparate fields. The full potential of AI in chemoinformatics will only be realized when the distinct expertise of data scientists, computational chemists, medicinal chemists, biologists, clinicians, and software engineers is effectively integrated into a cohesive, goal-oriented framework (Paul et al., 2021; Vamathevan et al., 2019).

4.3.8.1 Bridging the expertise gap

One of the most significant barriers to the effective application of AI in drug discovery is the cultural and knowledge gap between different disciplines. AI and ML experts, while proficient in algorithms and data architecture, may lack a deep understanding of the nuanced complexities of chemical and biological systems. This can lead to the development of models that are statistically robust but biologically implausible or practically irrelevant (Bender and Cortes-Ciriano, 2021). For instance, a model might excel on a retrospective benchmark but fail in a prospective setting because the benchmark did not adequately capture the complexities of real-world biological variability or the stringent criteria for a viable drug candidate. Similarly, focusing on optimizing a proxy endpoint, like in vitro activity, without considering the multifaceted requirements for a successful drug (e.g., pharmacokinetics, safety, and synthesizability) is a common pitfall when domain expertise is lacking (Bender and Cortes-Ciriano, 2021).

Conversely, medicinal chemists and biologists, who possess deep domain knowledge, may lack the computational skills required to develop, validate, and deploy sophisticated AI models. This can lead to a skepticism of "black-box" models, a reluctance to adopt new data-driven methodologies, or a superficial application of AI tools without a rigorous understanding of their underlying assumptions and limitations (Vamathevan et al., 2019). The fear of being replaced by automation can also create resistance, hindering the adoption of technologies that are designed to augment, not supplant, human expertise (Paul et al., 2021).

To bridge this gap, a new paradigm of multidisciplinary teamwork is essential. This involves moving beyond a service-oriented model—where, for example, a biologist provides data to a data scientist for analysis—to a truly integrated approach where experts from different fields are involved throughout the entire project lifecycle. This includes codesigning experiments, codeveloping models, and collaboratively interpreting results. Such collaborations ensure that the AI models are not only computationally sound but are also grounded in scientific reality and aligned with the practical goals of a drug discovery project (Vamathevan et al., 2019). Furthermore, a mutual understanding and respect for the expertise of each discipline is crucial. AI experts must appreciate that biological systems are far more complex and less governed by explicit rules than domains like image recognition or games, where AI has seen dramatic success (Bender and Cortes-Ciriano, 2021). Likewise, chemists and biologists must recognize that modern ML methods can uncover nonobvious, high-dimensional patterns that are beyond the scope of traditional analysis or human intuition (Malheiro et al., 2025).

4.3.8.2 The role of collaboration in overcoming data challenges

As established in the previous section, data is the foundation of AI in drug discovery, and many of the key challenges are data-centric. Multidisciplinary collaboration is critical to addressing these challenges effectively.

- **Data Generation and Curation:** Biologists and chemists are essential for designing experiments that generate high-quality, relevant data. Their expertise is needed to select appropriate assays, control for experimental variability, and ensure that the data generated contains a strong signal relevant to the in vivo endpoint of interest (Bender and Cortes-Ciriano, 2021). Furthermore, the curation of large, complex datasets, such as annotating adverse events from clinical literature or standardizing histopathology data, requires a combination of domain knowledge and computational skill (Paul et al., 2021; Vamathevan et al., 2019).
- **Feature Engineering and Representation:** The way a molecule or a biological system is represented computationally has a profound impact on model performance. While AI experts can implement various representation schemes (e.g., molecular fingerprints, graph convolutions), medicinal chemists and biologists can provide

crucial insights into which features are most likely to be relevant for a given endpoint. For instance, a chemist's understanding of SAR can guide the selection of molecular descriptors, while a biologist can help identify the most relevant genes or pathways to include in a model of drug response (Bender and Cortes-Ciriano, 2021).

- **Labeling and Annotation:** As discussed previously, assigning meaningful labels to biological data is a nontrivial task. The "mode of action" of a drug, for example, can be defined at the target, pathway, or functional level, and these labels are often highly context-dependent (Bender and Cortes-Ciriano, 2021). Effective labeling requires a deep understanding of the underlying biology, highlighting the need for close collaboration between domain experts and data scientists to create annotations that are both accurate and computationally useful.

4.3.8.3 Building a collaborative ecosystem

To foster this level of collaboration, pharmaceutical organizations and academic institutions are increasingly creating integrated, multidisciplinary teams. These teams often include roles that bridge the gap between disciplines, such as computational biologists or chemoinformaticians, who are fluent in both the life sciences and data science. Moreover, the rise of precompetitive consortia and public-private partnerships is creating a collaborative ecosystem where data, tools, and expertise can be shared more freely (Vamathevan et al., 2019).

Ultimately, the successful application of AI in chemoinformatics is a team sport. It requires a shared vision, open communication, and a willingness to learn from and respect the expertise of others. By breaking down the traditional silos between disciplines, the field can move beyond simply developing novel algorithms to creating truly integrated, intelligent systems that can accelerate the discovery of new medicines for the benefit of patients worldwide (Paul et al., 2021).

Take-home message

- AI and ML are deployed in Chemoinformatics
- Chemical data has been used in the domain of drug discovery
- Advanced techniques have been used to understand the mechanism of chemoinformatics
- Chemical data like SMILES has been used by tools like ECFP, Mol2Vec, and MACCS
- Extensively used and analyzed the AI-driven applications in molecular design
- It is shown extensively by supervised, unsupervised, and semisupervised methods, and deep learning architectures like graph neural networks (GNNs)

References

Abramson, J., Adler, J., Dunger, J., Evans, R., Green, T., Pritzel, A., Ronneberger, O., Willmore, L., Ballard, A.J., Bambrick, J., Bodenstein, S.W., Evans, D.A., Hung, C.C., O'Neill, M., Reiman, D., Tunyasuvunakool, K., Wu, Z., Žemgulytė, A., Arvaniti, E., Beattie, C., Bertolli, O., Bridgland, A., Cherepanov, A., Congreve, M., Cowen-Rivers, A.I., Cowie, A., Figurnov, M., Fuchs, F.B., Gladman, H., Jain, R., Khan, Y.A., Low, C.M.R., Perlin, K., Potapenko, A., Savy, P., Singh, S., Stecula, A., Thillaisundaram, A., Tong, C., Yakneen, S., Zhong, E.D., Zielinski, M., Žídek, A., Bapst, V., Kohli, P., Jaderberg, M., Hassabis, D., Jumper, J.M., 2024. Accurate structure prediction of biomolecular interactions with AlphaFold 3. Nature 630 (8016), 493–500. https://doi.org/10.1038/s41586-024-07487-w.

Arrieta, A.B., Díaz-Rodríguez, N., Ser, J.D., Bennetot, A., Tabik, S., Barbado, A., Garcia, S., Gil-Lopez, S., Molina, D., Benjamins, R., Chatila, R., Herrera, F., 2019. Explainable artificial intelligence (XAI): concepts, taxonomies, opportunities and challenges toward responsible AI . arXiv, Spain arXiv. https://doi.org/10.48550/arxiv.1910.10045.

Bender, A., Cortes-Ciriano, I., 2021. Artificial intelligence in drug discovery: what is realistic, what are illusions? Part 2: a discussion of chemical and biological data. Drug Discov. Today 26 (4), 1040–1052. https://doi.org/10.1016/j.drudis.2020.11.037.

Bran, A.M., Cox, S., Schilter, O., Baldassari, C., White, A.D., Schwaller, P., 2024. Augmenting large language models with chemistry tools. Nat. Mach. Intell. 6 (5), 525–535. https://doi.org/10.1038/s42256-024-00832-8.

Chithrananda, S., Grand, G., Ramsundar, B., 2020. ChemBERTa: large-scale self-supervised pretraining for molecular property prediction.

Cortes, C., Vapnik, V., 1995. Support-vector networks. Mach. Learn. 20 (3), 273–297. https://doi.org/10.1007/bf00994018.

Darvish, K., Skreta, M., Zhao, Y., Yoshikawa, N., Som, S., Bogdanovic, M., Cao, Y., Hao, H., Xu, H., Aspuru-Guzik, A., Garg, A., Shkurti, F., 2024. Organa: a robotic assistant for automated chemistry experimentation and characterization. arXiv, Canada arXiv. https://doi.org/10.48550/arXiv.2401.06949.

David, L., Thakkar, A., Mercado, R., Engkvist, O., 2020. Molecular representations in AI-driven drug discovery: a review and practical guide. J. Cheminf. 12 (1). https://doi.org/10.1186/s13321-020-00460-5.

Durant, J.L., Leland, B.A., Henry, D.R., Nourse, J.G., Reoptimization of M Tetko, I.V., Clevert, D.-A., 2025. Advanced machine learning for innovative drug discovery. J. Cheminf. 17 (2002), 122–123.

Ebejer, J.P., Morris, G.M., Deane, C.M., 2012. Freely available conformer generation methods: how good are they? J. Chem. Inf. Model. 52 (5), 1146–1158. https://doi.org/10.1021/ci2004658.

Friedrich, N.O., De Bruyn Kops, C., Flachsenberg, F., Sommer, K., Rarey, M., Kirchmair, J., 2017. Benchmarking commercial conformer ensemble generators. J. Chem. Inf. Model. 57 (11), 2719–2728. https://doi.org/10.1021/acs.jcim.7b00505.

Ghiandoni, G.M., Evertsson, E., Riley, D.J., Tyrchan, C., Rathi, P.C., 2024. Augmenting DMTA using predictive AI modelling at AstraZeneca. Drug Discov. Today 29 (4). https://doi.org/10.1016/j.drudis.2024.103945.

Ghislat, G., Hernandez-Hernandez, S., Piyawajanusorn, C., Ballester, P.J., 2024. Data-centric challenges with the application and adoption of artificial intelligence for drug discovery . arXiv, United Kingdom arXiv. https://doi.org/10.48550/arXiv.2407.05150.

Gilmer, J., Schoenholz, S.S., Riley, P.F., Vinyals, O., Dahl, G.E., 2017. Neural message passing for quantum chemistry. In: 34th International Conference on Machine Learning. ICML, pp. 2053–2070.

Gold, E.R., Cook-Deegan, R., 2025. AI drug development's data problem. Science 388 (6743), 131. https://doi.org/10.1126/science.adx0339.

Hawkins, P.C.D., Nicholls, A., 2012. Conformer generation with OMEGA: learning from the data set and the analysis of failures. J. Chem. Inf. Model. 52 (11), 2919–2936. https://doi.org/10.1021/ci300314k.

Ivanenkov, Y.A., Polykovskiy, D., Bezrukov, D., Zagribelnyy, B., Aladinskiy, V., Kamya, P., Aliper, A., Ren, F., Zhavoronkov, A., 2023. Chemistry42: an AI-driven platform for molecular design and optimization. J. Chem. Inf. Model. 63 (3), 695–701. https://doi.org/10.1021/acs.jcim.2c01191.

Jaeger, S., Fulle, S., Turk, S., 2018. Mol2vec: unsupervised machine learning approach with chemical intuition. J. Chem. Inf. Model. 58 (1), 27–35. https://doi.org/10.1021/acs.jcim.7b00616.

Kant, S., Deepika, Roy, S., 2025. Artificial intelligence in drug discovery and development: transforming challenges into opportunities. Discover Pharmaceutical Sciences 1 (1). https://doi.org/10.1007/s44395-025-00007-3.

Krenn, M., Häse, F., Nigam, A.K., Friederich, P., Aspuru-Guzik, A., 2020. Self-referencing embedded strings (SELFIES): a 100% robust molecular string representation. Mach. Learn.: Sci. Technol. 1 (4), 045024. https://doi.org/10.1088/2632-2153/aba947.

Leonov, A.I., Hammer, A.J.S., Lach, S., Mehr, S.H.M., Caramelli, D., Angelone, D., Khan, A., O'Sullivan, S., Craven, M., Wilbraham, L., Cronin, L., 2024. An integrated self-optimizing programmable chemical synthesis and reaction engine. Nat. Commun. 15 (1). https://doi.org/10.1038/s41467-024-45444-3.

Lo, Y.C., Rensi, S.E., Torng, W., Altman, R.B., 2018. Machine learning in chemoinformatics and drug discovery. Drug Discov. Today 23 (8), 1538–1546. https://doi.org/10.1016/j.drudis.2018.05.010.

Luong, K.D., Singh, A., 2024. Application of transformers in cheminformatics. J. Chem. Inf. Model. 64 (11), 4392–4409. https://doi.org/10.1021/acs.jcim.3c02070.

Malheiro, V., Santos, B., Figueiras, A., Mascarenhas-Melo, F., 2025. The potential of artificial intelligence in pharmaceutical innovation: from drug discovery to clinical trials. Pharmaceuticals 18 (6), 788. https://doi.org/10.3390/ph18060788.

McInnes, L., Healy, J., Melville, J., 2018. UMAP: uniform manifold approximation and projection for dimension reduction . arXiv, undefined arXiv. https://doi.org/10.48550/arxiv.1802.03426.

Mitchell, J.B.O., 2014. Machine learning methods in chemoinformatics. WIREs Comput. Mol. Sci. 4 (5), 468–481. https://doi.org/10.1002/wcms.1183.

O'Boyle, N.M., Banck, M., James, C.A., Morley, C., Vandermeersch, T., Hutchison, G.R., 2011. Open Babel: an open chemical toolbox. J. Cheminf. 3 (10). https://doi.org/10.1186/1758-2946-3-33Ireland.

Obrezanova, O., 2023. Artificial intelligence for compound pharmacokinetics prediction. Curr. Opin. Struct. Biol. 79, 102546. https://doi.org/10.1016/j.sbi.2023.102546.

Paul, D., Sanap, G., Shenoy, S., Kalyane, D., Kalia, K., Tekade, R.K., 2021. Artificial intelligence in drug discovery and development. Drug Discov. Today 26 (1), 80–93. https://doi.org/10.1016/j.drudis.2020.10.010.

Qadri, Y.A., Shaikh, S., Ahmad, K., Choi, I., Kim, S.W., Vasilakos, A.V., 2025. Explainable artificial intelligence: a perspective on drug discovery. Pharmaceutics 17 (9). https://doi.org/10.3390/pharmaceutics17091119.

Ramos, M.C., Collison, C.J., White, A.D., 2024. A review of large language models and autonomous agents in chemistry. Chem. Sci. 16 (6), 2514–2572. https://doi.org/10.1039/d4sc03921a.

Rupp, M., Ramakrishnan, R., Von Lilienfeld, O.A., 2015. Machine learning for quantum mechanical properties of atoms in molecules. J. Phys. Chem. Lett. 6 (16), 3309–3313.

Salley, D., Manzano, J.S., Kitson, P.J., Cronin, L., 2023. Robotic modules for the programmable chemputation of molecules and materials. ACS Cent. Sci. 9 (8), 1525–1537. https://doi.org/10.1021/acscentsci.3c00304.

Seifrid, M., Pollice, R., Aguilar-Granda, A., Morgan Chan, Z., Hotta, K., Ser, C.T., Vestfrid, J., Wu, T.C., Aspuru-Guzik, A., 2022. Autonomous chemical experiments: challenges and perspectives on establishing a self-driving lab. Acc. Chem. Res. 55 (17), 2454–2466. https://doi.org/10.1021/acs.accounts.2c00220.

Sheridan, R.P., 2013. Time-split cross-validation as a method for estimating the goodness of prospective prediction. J. Chem. Inf. Model. 53 (4), 783–790. https://doi.org/10.1021/ci400084k.

Song, T., Luo, M., Chen, L., Huang, Y., Zhu, Q., Liu, D., Zhang, B., Zou, G., Zhang, F., Shang, W., Jiang, J., Luo, Y., 2024. A multi-agent-driven robotic AI chemist enabling autonomous chemical research on demand. ChemRxiv, China ChemRxiv. https://doi.org/10.26434/chemrxiv-2024-w953h-v2.

Urbonas, M., Ajileye, T., Gainer, P., Pires, D., 2023. Automating reward function configuration for drug design . arXiv, United Kingdom arXiv. https://doi.org/10.48550/arXiv.2312.09865.

Vamathevan, J., Clark, D., Paul, C., Dunham, I., Ferran, E., Lee, G., Li, B., Madabhushi, A., Shah, P., Spitzer, M., Zhao, S., 2019. Applications of machine learning in drug discovery and development. Nat. Rev. Drug Discov. 18 (6), 463–477. https://doi.org/10.1038/s41573-019-0024-5.

Weininger, D., 1988. SMILES, a chemical language and information system. 1. Introduction to methodology and encoding rules. J. Chem. Inf. Comput. Sci. 28 (1), 31–36. https://doi.org/10.1021/ci00057a005.

Wellawatte, G.P., Gandhi, H.A., Seshadri, A., White, A.D., 2023. A perspective on explanations of molecular prediction models. J. Chem. Theor. Comput. 19 (8), 2149–2160.

Xia, J., Broadhurst, D.I., Wilson, M., Wishart, D.S., 2013. Translational biomarker discovery in clinical metabolomics: an introductory tutorial. Metabolomics 9 (2), 280–299. https://doi.org/10.1007/s11306-012-0482-9.

Xu, Z., Ren, F., Wang, P., Cao, J., Tan, C., Ma, D., Zhao, L., Dai, J., Ding, Y., Fang, H., Li, H., Liu, H., Luo, F., Meng, Y., Pan, P., Xiang, P., Xiao, Z., Rao, S., Satler, C., Liu, S., Yuan, L., Zhao, H., Chen, S., Cui, H., Korzinkin, M., Gennert, D., Zhavoronkov, A., 2025. A generative AI-discovered TNIK inhibitor for idiopathic pulmonary fibrosis: a randomized phase 2a trial. Nat. Med. 31 (8), 2602–2610. https://doi.org/10.1038/s41591-025-03743-2.

Zhang, O., Lin, H., Zhang, X., Wang, X., Wu, Z., Ye, Q., Zhao, W., Wang, J., Ying, K., Kang, Y., Hsieh, C.Y., Hou, T., 2025. Graph neural networks in modern AI-aided drug discovery . arXiv, China arXiv. https://doi.org/10.48550/arXiv.2506.06915.

Zhou, J., Cui, G., Hu, S., Zhang, Z., Yang, C., Liu, Z., Wang, L., Li, C., Sun, M., 2020. Graph neural networks: a review of methods and applications. AI Open 1, 57–81. https://doi.org/10.1016/j.aiopen.2021.01.001.

Zhu, X., 2005. Semi-supervised learning literature survey . Computer Sciences Technical Report 1530.

Further reading

Altae-Tran, H., Ramsundar, B., Pappu, A.S., Pande, V., 2017. Low data drug discovery with one-shot learning. ACS Cent. Sci. 3 (4), 283–293. https://doi.org/10.1021/acscentsci.6b00367.

Artrith, N., Butler, K.T., Walsh, A., 2021. Best practices in machine learning for chemistry. Nat. Chem. 13 (6), 505–508. https://doi.org/10.1038/s41557-021-00716-z.

Askr, H., Elgeldawi, E., Aboul Ella, H., Elshaier, Y.A.M.M., Gomaa, M.M., Hassanien, A.E., 2023. Deep learning in drug discovery: an integrative review and future challenges. Artif. Intell. Rev. 56 (7), 5975–6037. https://doi.org/10.1007/s10462-022-10306-1.

Fabian, B., et al., 2020. Molecular representation learning with graph attention networks. J. Chem. Inf. Model. 60 (12), 5712–5724. https://doi.org/10.1021/acs.jcim.0c00659.

Jumper, J., Evans, R., Pritzel, A., et al., 2021. Highly accurate protein structure prediction with AlphaFold. Nature 596 (7873), 583–589. https://doi.org/10.1038/s41586-021-03819-2.

Kryshtafovych, A., Schwede, T., Topf, M., Fidelis, K., Moult, J., 2021. Critical assessment of methods of protein structure prediction (CASP)—round XIV. Proteins: Struct., Funct., Bioinf. 89 (12), 1607–1617. https://doi.org/10.1002/prot.26237.

Nag, S., Baidya, A.T., Mandal, A., Mathew, A.T., Das, B., Devi, B., Kumar, R., 2022. DL keys for use in drug discovery. J. Chem. Inf. Comput. Sci. 42 (6), 1273–1280. https://doi.org/10.1021/acs.jcim.7b00505.

Niazi, S.K., 2023. Recent Advances in Machine-Learning-Based Chemoinformatics: A Comprehensive Review. Explores a Wide Range of ML Techniques across Drug Discovery Stages. MDPI.

Schwaller, P., et al., 2019. Molecular transformer: a model for uncertainty-calibrated chemical reaction prediction. ACS Cent. Sci. 5 (9), 1572–1583. https://doi.org/10.1021/acscentsci.9b00576.

Senior, A.W., Evans, R., Jumper, J., et al., 2020. Improved protein structure prediction using potentials from deep learning. Nature 577 (7792), 706–710. https://doi.org/10.1038/s41586-019-1923-7.

Svetnik, V., Liaw, A., Tong, C., Culberson, J.C., Sheridan, R.P., Feuston, B.P., 2003. Random forest: a classification and regression tool for compound classification and QSAR modeling. J. Chem. Inf. Comput. Sci. 43 (6), 1947–1958. https://doi.org/10.1021/ci034160g.

Tu, X., Grisoni, F., et al. (2022). Structure-based drug discovery with deep learning. This article discusses deep learning's role in structure-based drug discovery pipelines. arXiv.

Varadi, M., Anyango, S., Deshpande, M., et al., 2022. AlphaFold protein structure database: massively expanding the structural coverage of protein-sequence space with high-accuracy models. Nucleic Acids Res. 50 (D1), D439–D444. https://doi.org/10.1093/nar/gkab1061.

Varnek, A., Tropsha, A., 2008. Chemoinformatics Approaches to Virtual Screening. RSC Publishing. https://doi.org/10.1039/9781847558650.

Wang, J., Zhou, Z., Li, T., 2009. Semi-supervised learning with multi-domain classification. In: Proceedings of the 2009 SIAM International Conference on Data Mining, pp. 665–676. https://doi.org/10.1137/1.9781611972795.57.

Xia, Z., Wu, H., Wei, L., Lei, B., 2021. Semi-supervised drug–drug interaction prediction with graph auto-encoders. Bioinformatics 37 (18), 2935–2942. https://doi.org/10.1093/bioinformatics/btab201.

Yang, K., et al., 2019. Analyzing learned molecular representations for property prediction. J. Chem. Inf. Model. 59 (8), 3370–3388. https://doi.org/10.1021/acs.jcim.9b00237.

5

Deep learning methods for network biology

Mohammad "Sufian" Badar[1, 2, 3, 4], Asma Maheen[5], Mohammad Rehan Badar[6], Noora[7], Aamir Nehal[8], Abdus Salam Sabir[9], Sarah Sufian Badar[10], Syed Zeeshan Hussain[11] and Hamid Omar[12]

[1]DEPARTMENT OF BIOENGINEERING, UNIVERSITY OF CALIFORNIA, RIVERSIDE, CA, UNITED STATES; [2]UNIVERSAL SCIENTIFIC EDUCATION AND RESEARCH NETWORK (USERN), TEHRAN, IRAN; [3]SPI DARBHANGA, DARBHANGA, BIHAR, INDIA; [4]DEPARTMENT OF COMPUTER SCIENCE AND ENGINEERING (BIOINFORMATICS), SCHOOL OF ENGINEERING SCIENCES AND TECHNOLOGY (SEST), JAMIA HAMDARD UNIVERSITY, NEW DELHI, INDIA; [5]NEW DELHI, INDIA; [6]APOLLO HOSPITAL, BADAR MEDICAL CENTRE (BMC), SALMON ARM, BC, CANADA; [7]DEPARTMENT OF PATHOLOGY, MADHUBANI MEDICAL COLLEGE, MADHUBANI, BIHAR, INDIA; [8]TOUCH N HEAL PHYSIOTHERAPY AND REHABILITATION CENTRE, MOTIHARI, BIHAR, INDIA; [9]CHARLES SCHWAB, LONETREE, CO, UNITED STATES; [10]SHARDA UNIVERSITY, GREATER NOIDA, UTTAR PRADESH, INDIA; [11]DEPARTMENT OF COMPUTER SCIENCE, JAMIA MILLIA ISLAMIA (JMI), NEW DELHI, INDIA; [12]UNIVERSITY OF WISCONSIN MADISON, MADISON, WI, UNITED STATES

What you will learn

- The fundamental concepts of biological networks and how they are represented graphically.
- The basic principles of graph neural networks (GNNs) and deep learning.
- The implementation of deep learning for data analysis and prediction in network biology.
- Applications in research on phytomedicine, drug-target interactions, and disease gene discovery.
- Significant challenges include validation, data quality, and model interpretability.
- Prospects for the future include precision medicine, multi-omics integration, and Indian biomedical contributions.

5.1 Introduction

To understand biological systems, it is important to take a broader view that shows the complex interactions among molecules, cells, and tissues, instead of looking at individual parts separately. Network biology has become an important field that models and

Applying Artificial Intelligence to Computational Biology and Medical Informatics. https://doi.org/10.1016/B978-0-443-44915-4.00006-X

studies these complex relationships. It uncovers hidden patterns and principles that influence health, disease, and evolution. The rise of high-throughput omics technologies has greatly increased the amount and complexity of biological data. This growth creates challenges for traditional computational methods, which often struggle to extract useful insights. In response, deep learning has emerged as a strong branch of machine learning. It can automatically spot hierarchical and nonlinear patterns, making it essential in modern network biology. This approach leads to better and more scalable interpretations of biological complexity (Zitnik et al., 2024).

5.1.1 What is network biology?

Through graph-based models, network biology studies biological systems. Nodes represent items such as genes, proteins, metabolites, or cells. Edges show how these entities interact or relate to each other. This approach provides a clear way to illustrate how biological processes are connected, encompassing: Gene regulatory networks (GRNs): charting the regulatory relationships between genes and transcription factors. Protein-protein interaction (PPI) networks: understanding functional protein relationships. Metabolic and signaling networks: tracking cellular communication and biochemical processes. Disease networks: linking phenotypes, molecular interactions, and genetic variations (Wu et al., 2022) (Fig. 5.1).

5.1.2 Why deep learning is useful in biology?

Because of their complexity, high dimensionality, and dynamic nature, biological systems are challenging to analyze by conventional computational algorithms. In this regard, deep learning is very advantageous. Firstly, by extracting automatically meaningful features from raw biological data, it obviates human prior analysis. Secondly, it performs remarkably well with large datasets generated by high-throughput omics technologies such as transcriptomics, proteomics, and genomics. Third, deep learning models are capable of modeling higher-order dependencies and non-linear interactions among genes, proteins, and pathways that are often skipped by traditional methods (Min et al., 2017). In addition, deep learning facilitates multi-omics data integration, which allows researchers to combine data across different layers of biology to gain a deeper insight into cellular systems. Finally, deep learning enhances predictive capability, which provides with opportunities such as network reconstruction, drug-target prediction, and disease-risk prediction, which redirect biological research from descriptive analysis to data-driven innovation (Baião et al., 2025).

5.2 Basics of biological networks

Biological networks provide a platform for visualizing the complex relationship between molecular and cellular components. Scientists can understand cellular processes, disease mechanisms, and regulatory processes at the systems level by modeling these

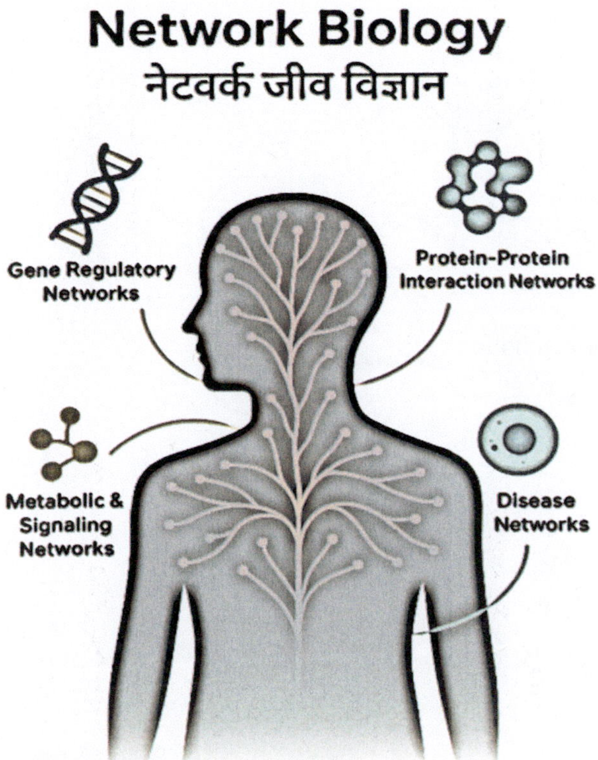

FIGURE 5.1 Network biology: Mapping interconnected biological systems.

interactions as networks. High-dimensional biological information can be systematically organized through networks, which also provide vital patterns, which remain hidden when individual components are analyzed (Mei et al., 2025).

5.2.1 Types of networks: gene, protein, and pathways

Biological networks can be categorized depending on the entity involved and the type of their interactions, such as genes, proteins, and pathways. Besides gene and protein interactions, other biological pathways such as metabolic and signaling networks, and disease networks give a more holistic systems-level view of cell processes.

1. **Gene Regulatory Networks (GRNs)**: GRNs give insight into the regulatory organization of gene expression by describing interactions between transcription factors and their target genes. Cellular differentiation, development, and responses to

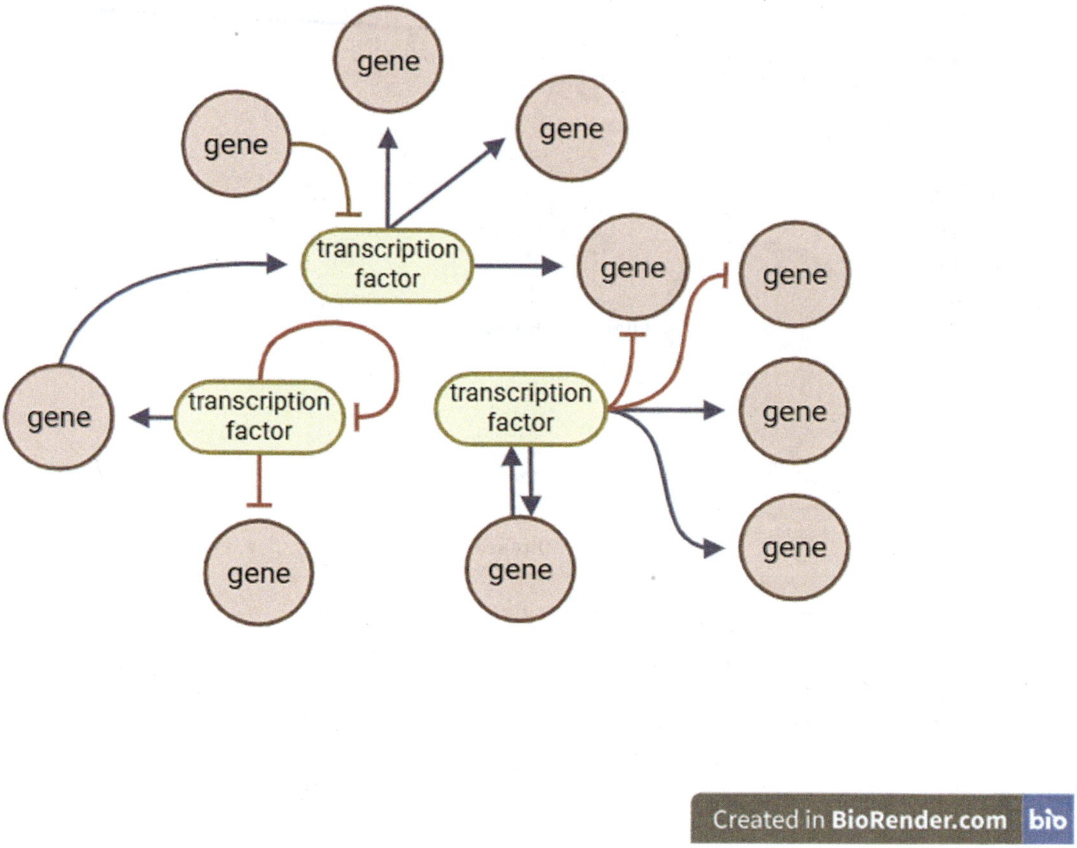

FIGURE 5.2 Gene regulatory network.

environmental stimuli are all accounted for by these networks. Scientists can identify significant regulatory genes that function as network hubs and control large numbers of downstream genes and pathways by analyzing GRNs (En Li et al., 2013) (Fig. 5.2).

2. **Protein-Protein Interaction (PPI) Networks**: Proteins seldom function independently; they often physically interact with other proteins to accomplish specific tasks. Protein complexes, signaling cascades, and molecular machines propelling essential cellular processes are unveiled by PPI networks, which catalog these interactions. When interactions between proteins are disrupted, PPI analysis can help to determine key proteins, potential therapeutic targets, and mechanisms of disease (Safari-Alighiarloo et al., 2014) (Fig. 5.3).
3. **Metabolic Networks**: Inside a cell, metabolic networks chart the biochemical reactions that convert substrates to products. They catalog processes such as lipid

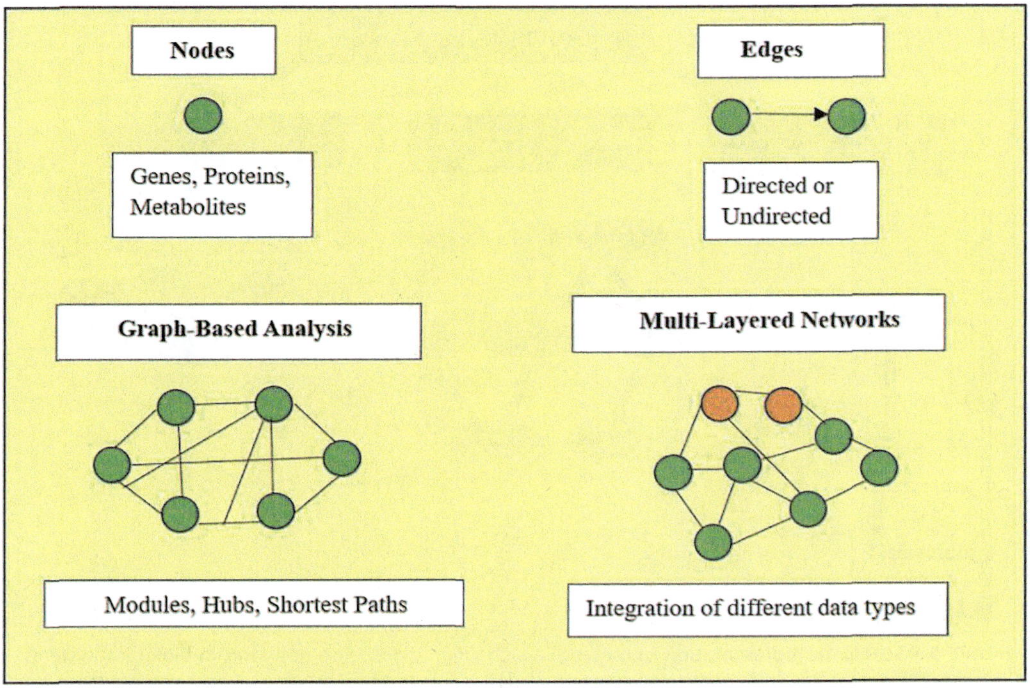

FIGURE 5.3 Protein-protein interaction network.

metabolism, the citric acid cycle, and glycolysis. By analyzing these networks, scientists can calculate metabolic bottlenecks, understand energy flow, and predict how genetic or environmental disruptions will influence cellular metabolism (Madrer et al., 2025).

4. **Signaling Networks**: Signaling networks show the flow of information through receptors, kinases, transcription factors, and secondary messengers inside cells. These networks are essential for comprehending how cells perceive and react to stimuli, control proliferation, differentiation, and apoptosis, and preserve homeostasis. Signaling networks are frequently disrupted in autoimmune disorders and cancer (Azeloglu and Iyengar, 2015).
5. **Disease Networks**: Disease networks integrate genetic, molecular, and phenotypic data to map the relationships that underlie diseased diseases. By linking genes, proteins, and cellular processes associated with specific diseases, these networks have the capacity to identify disease modules, predict comorbidities, and pinpoint potential treatment targets (Visonà et al., 2024) (Fig. 5.4).

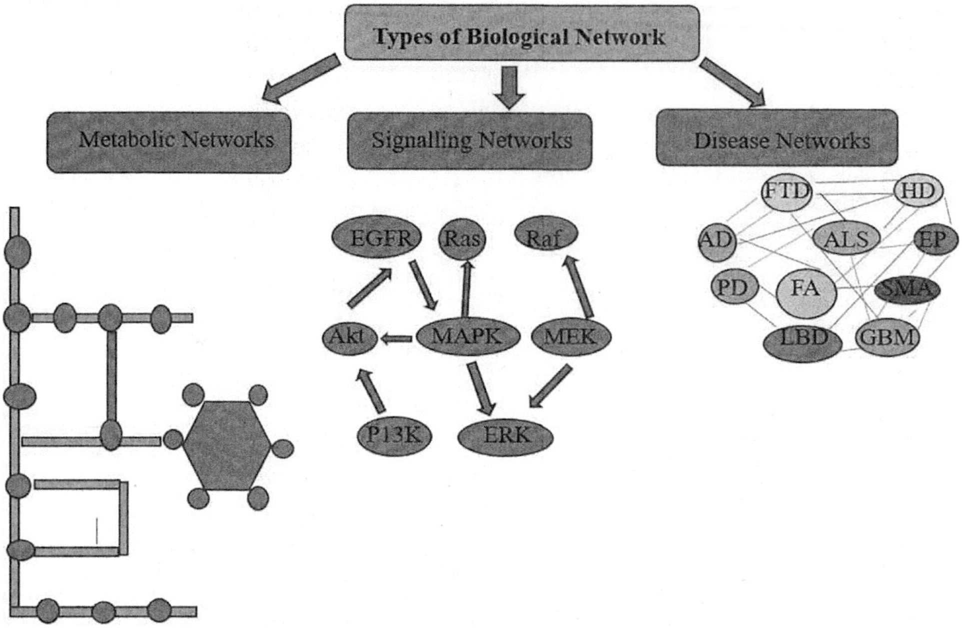

FIGURE 5.4 Schematic representation of metabolic, signaling, and disease networks in biological systems.

5.2.2 How to represent networks as graphs?

In network biology, graphs are a type of mathematical structure that describes relationships between objects are the best method to represent complicated biological interactions. Genes, proteins, and metabolites are some biological components depicted by nodes (vertices) of a graph, while interactions, associations, or regulation relationships between objects are depicted by edges (links). Edges may be undirected, meaning mutual association without direction, or directed, meaning causal or regulatory influence. In addition, edges may be weighted to denote biological relevance, confidence scores, or interaction strength (Koutrouli et al., 2020). Graph-based views offer a computational-analyzable framework that can be tuned. They allow graph-theoretic techniques to be applied to the discovery of network's important properties, such as modules, which are highly interconnected groups of nodes; hubs, which are highly connected nodes; and shortest paths, which are critical routes for transferring information or signals. understanding the structural and functional organization of biological structures, predicting novel relationships and finding key genes or proteins are all facilitated by these analyses (Jungck and Viswanathan, 2015). Additionally, graphs facilitate the integration of various data types, enabling multilayered networks that enable the analysis of various biological processes, including metabolic pathways, protein interactions, and transcriptional regulation. Advanced computational techniques

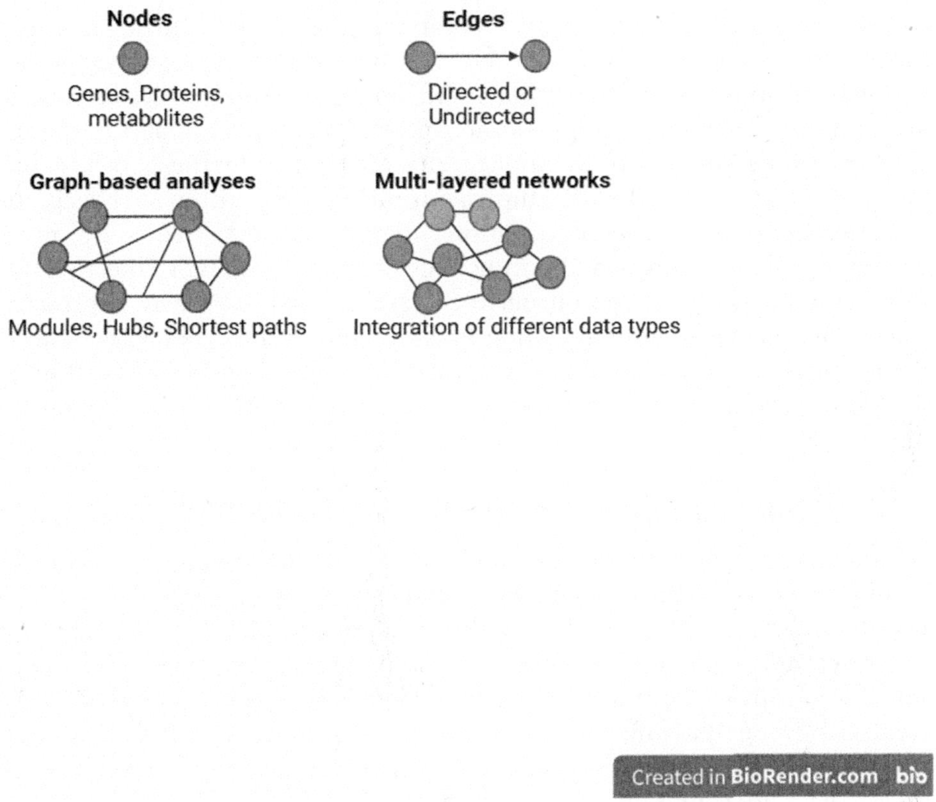

FIGURE 5.5 Graph-based framework for representing biological interactions.

that can uncover hidden patterns and draw predictions from complex biological data, for instance network-based machine learning and deep learning models, are built on top of this representation (Jiang et al., 2025) (Fig. 5.5).

5.3 Basics of deep learning

In biology, the analysis of high-dimensional and complex data has been completely transformed by deep learning, a branch of machine learning. Deep learning models are able to automatically learn hierarchical representations from raw data, which allows them to identify complex patterns and nonlinear relationships that are frequently overlooked by conventional techniques. Deep learning offers strong tools for modeling gene regulatory networks, protein interactions, and signaling pathways in the context of network biology. These tools make it easier to perform predictive analysis, reconstruct networks, and integrate multiple omics (Muzio, O'Bray and Borgwardt, 2021a).

5.3.1 Introduction to neural networks

Artificial neural networks (ANNs) are computational models whose algorithms are based on the structure and function of biological neurons. They are the foundation of deep learning. An input layer, an output layer, and one or more hidden layers make up a typical neural network. A nonlinear activation function is applied to the output after a weighted transformation of the input signals is applied by the nodes (neurons) in each layer (Eranti et al., 2025) By utilizing an optimization algorithm such as gradient descent in conjunction with backpropagation, neural networks learn by minimizing the discrepancy between expected and actual results by modifying the weights of connections between neurons. Neural network variations have shown impressive success in processing biological data, including imaging, time-series gene expression, and molecular sequences. These neural network variations include convolutional neural networks (CNNs) for spatial data and recurrent neural networks (RNNs) for sequential data (Selby et al., 2025).

5.3.2 Deep learning models used in network biology (e.g., GNNs)

Complex biological networks are now accessible for research with the help of deep learning models, especially Graph Neural Networks (GNNs). Since GNNs can work directly on graph-structured data, unlike traditional machine learning techniques, they are perfect for depicting disease networks, pathways, and molecular interactions. The model can capture both local node-level details and global network context in GNNs because message-passing mechanisms aggregate node features (e.g., gene expression levels and protein properties) and edge attributes (e.g., interaction types) (Zhang et al., 2021). GNNs are extensively utilized in network biology for the prediction of drug-target interactions, disease module detection, gene function annotation, and PPI prediction. By amalgamating multi-omics data, variants such as Graph Convolutional Networks (GCNs), Graph Attention Networks (GATs), and Graph SAGE enhance representation learning and allow scientists to produce predictions that have biological significance. These methods are especially useful for revealing hidden connections in complex biological systems, which aids in drug discovery and precision medicine (Zhou et al., 2020) (Fig. 5.6).

5.4 The role of deep learning in network biology

By unraveling hidden patterns, forecasting interactions, and combining various omics data, deep learning revolutionizes biological network analysis. It scales well to big, complicated biological datasets, increases prediction accuracy, and automates feature extraction. This makes it possible for scientists to more accurately model cellular processes, predict drug-target interactions, and find genes linked to disease (Muzio, O'Bray and Borgwardt, 2021b).

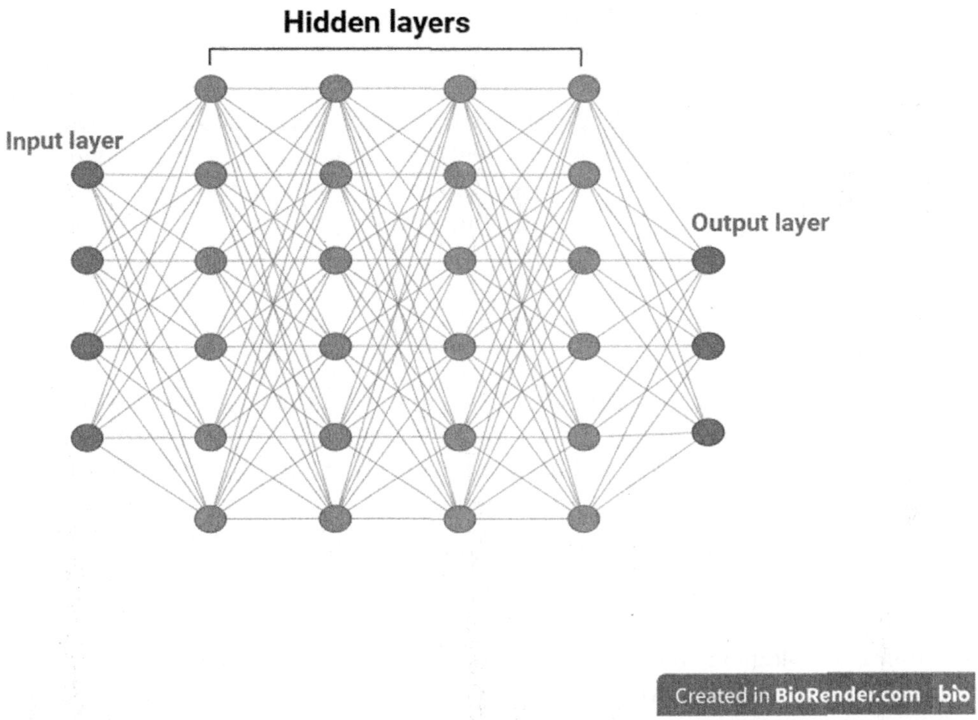

FIGURE 5.6 Deep learning model.

5.4.1 Data preparation and network construction

In network biology, deep learning significantly depends on high-quality, well-structured data. The first step in the process is collecting multi-omics datasets, which include transcriptomics, proteomics, metabolomics, and genomics. These datasets are frequently combined with interactions curated from the literature. To guarantee data consistency, preprocessing procedures include noise reduction, dimensionality reduction, and normalization. Then, biological entities (such as genes, proteins, and metabolites) are represented as nodes and their interactions as edges to create graphs, which are used to build networks. In order to prepare the data for deep learning models such as Graph Neural Networks (GNNs), sophisticated feature engineering techniques such as node embedding or graph representation learning are used to capture the intricate biological relationships within these networks (Yousef and Allmer, 2023).

5.4.2 Tools and platforms commonly used

Numerous platforms and technologies streamline deep learning applications in network biology. Prominent graph analysis tools include Cytoscape for network visualization and STRING for combining information on protein–protein interactions. Deep learning frameworks such as TensorFlow, PyTorch, and the Deep Graph Library (DGL) provide efficient model building and training. Two well-liked programs for putting GNNs into practice are Spektral and PyG (PyTorch Geometric). Large-scale biological data management and integration are made possible by platforms such as BioGRID, Reactome, and KEGG, which provide curated datasets. Deep learning is an efficient way to uncover biological insights since researchers can combine these technologies to develop repeatable pipelines that include everything from data pretreatment to network-based predictions (Kumar and Srivastava, 2024).

5.5 Applications

Deep learning has become an essential technique in network biology, assisting researchers in deriving novel biological insights from complex networks of molecular interactions. Scientists may achieve previously unheard-of precision in tasks such as detecting drug-target interactions, predicting disease-associated genes, and examining biochemical pathways by using models such as deep autoencoders and graph neural networks (GNNs).

5.5.1 Disease gene prediction

By identifying topological and functional patterns, deep learning models trained on multi-omics data and PPI networks can identify genes associated with diseases. On the disease-gene association prediction front, GNN-based methods surpass traditional methods by effectively capturing the context within the network. For instance, Prioritization of rare disease candidate genes has been achieved successfully through methods such as Deep Walk and GCN (Zidane et al., 2023).

5.5.2 Drug–target interaction and drug repositioning

By fusing chemical structures, omics data, and biological networks, deep learning facilitates more efficient prediction of drug–target interactions (DTIs). This promotes drug repurposing of existing compounds and accelerates drug discovery. NeoDTI and other network-based deep learning methods encode molecular and network features jointly to attain high accuracy (Talukder et al., 2025).

5.5.3 Studying pathways and molecular interactions

Deep learning facilitates the understanding of signaling dynamics, the identification of key regulators, and the reconstruction of molecular pathways. These models provide insight into pathway crosstalk and cryptic relationships between biomolecules for disease mechanism study. Techniques such as variational graph autoencoders make it easier to trace chemical interactions in a methodical manner (Xie et al., 2024).

5.6 Case studies

Case studies in network pharmacology provide tangible examples of how experimental and computational techniques can be integrated to find potential therapeutic mechanisms and drug candidates. They stress how crucial it is to integrate biological data such as proteomics, metabolomics, and genomics into meticulously planned interaction networks in order to derive useful insights. For example, research has used disease–gene and drug–target networks to predict off-target effects, identify biomarkers for complex diseases, and guide repurposing strategies. These techniques provide evidence for the development of hypotheses before expensive and time-consuming laboratory validations (Noor et al., 2022). One well-known example is the development of disease-specific protein-protein interaction (PPI) networks to find new therapeutic targets in neurodegenerative and cancerous diseases. The multi-target nature of traditional medicines has also been understood through the use of network-based strategies, which show how herbal formulations cooperate through a number of pathways. These case studies illustrate how network pharmacology can help close the gap between systems biology and translational medicine and make it easier to develop customized treatment plans (Zaidh et al., 2025).

5.6.1 Real-world research examples

- **Disease–Gene Prediction with Enhanced Accuracy**: A new model called ModulePred speculates which genes are associated with which diseases by using graph augmentation and graph embeddings on protein interaction networks. Tested on the DisGeNet dataset, ModulePred demonstrated significant performance gains. This demonstrates the value of deep learning in comprehending intricate network architectures. (Jia et al., 2024)
- **Transparent Biomarker Identification** via **PathFormer:** A new GNN-based architecture called PathFormer is ideal for identifying biomarkers and diagnosing illnesses because it is designed for signaling pathway networks. It consistently assigned biomarkers across datasets and increased predictive accuracy by up to 30% when applied to transcriptomic datasets for cancer and Alzheimer's disease (Dong et al., 2023).
- **Multilayer Graph Neural Networks for Cancer Gene Discovery:** The Explainable Multilayer Graph Neural Network (EMGNN) approach uses pan-cancer multi-omics data and multiple gene-gene interaction networks to address the intricacy of carcinogenesis. The area under the precision-recall curve was improved by an average of 7.15% by EMGNN in comparison to single-network models. The explanations it provided for its forecasts also made biological sense (Chatzianastasis et al., 2023).

5.6.2 Example from herbal/phytomedicine-based network analysis

5.6.2.1 Predicting medicinal uses of natural compounds through deep learning

Researchers developed a deep learning approach that takes into account heterogeneous variables, such as molecular interactions and chemical properties, for over 4500 natural substances and almost 3000 recognized pharmaceuticals. By autonomously predicting therapeutic uses of chemicals produced from plants with high accuracy, sensitivity, and specificity, the model creates new opportunities for drug discovery from natural products (Cao, 2024).

5.6.2.2 Predicting herbal medicine efficacy with deep neural networks

- In experiments aimed at understanding the therapeutic mechanisms of Unani herbal compositions, a deep neural network outperformed traditional machine learning methods (random forest, SVM). By effectively identifying 118 metabolites connected to nine therapeutic uses, it highlighted the effectiveness of neural networks in metabolite therapy connection prediction (Nasution et al., 2024).

5.6.2.3 Network pharmacology based on licorice and deep learning

Using a deep learning-based network pharmacology model called DeepDGC, which integrated fingerprint data and molecular maps, a study investigated the potential of licorice against COVID-19. This method demonstrated the translational potential of fusing deep learning with traditional herbal knowledge by identifying two novel compounds and two new targets with promising therapeutic effects (Fu et al., 2023).

5.7 Challenges

Notwithstanding its revolutionary potential, network pharmacology has a number of obstacles that prevent it from being widely used and having a significant translational impact.

5.7.1 Data limitations and quality issues

Publicly accessible datasets, including omics datasets, drug-target repositories, and protein–protein interaction (PPI) databases, are essential to network pharmacology. Nevertheless, these datasets frequently have inconsistent curation standards, experimental bias, and insufficient coverage. For instance, because natural products are chemically complex, herbal compound annotations are frequently lacking, and high-throughput screening data may contain false positives or negatives. Furthermore, differences in experimental platforms and metadata quality make it difficult to integrate data from heterogeneous sources (transcriptomics, proteomics, and metabolomics). These elements restrict reproducibility and introduce uncertainty into network construction (Zhang et al., 2024).

5.7.2 Model interpretability and black box problems

Many advanced computational models, especially deep learning tools used for network prediction, are often described as "black boxes" because their decision-making process is difficult to interpret. While these models can achieve very high accuracy, their lack of transparency makes it challenging for researchers to extract clear mechanistic insights, which in turn affects trust in their outcomes. This limitation becomes particularly critical in areas such as drug repurposing and herbal pharmacology, where understanding molecular pathways and providing mechanistic evidence are essential for regulatory validation. To address this challenge, current research is increasingly focused on explainable AI (XAI) approaches, which aim to make network-based drug discovery pipelines more transparent, interpretable, and reliable (Li et al., 2024).

5.7.3 Biological validation of computational results

In network pharmacology, computational predictions ought to be experimentally verified to ensure their biological significance. However, this validation frequently takes a lot of time and resources, especially when it comes to multi-component herbal formulations. Although promising, high-throughput screening techniques, molecular docking, and CRISPR-based validation are still not scalable for complex network predictions. One of the biggest obstacles to converting in silico results into clinical applications is still bridging the gap between computational modeling and wet-lab validation (Panossian, 2025).

5.8 Future scope

The implementation of network pharmacology in the biomedical sciences is developing quickly, and it is anticipated that this method will become more complex, multidisciplinary, and translational in the future. Network pharmacology has the potential to revolutionize precision medicine, disease modeling, and drug discovery by utilizing developments in computational biology, artificial intelligence, and high-throughput experimental platforms.

5.8.1 Integrating deep learning with multi-omics data

Deep learning combined with multi-omics data has enormous potential for comprehending intricate biological systems. Deep learning models can better predict diseases by capturing complex biological interactions by integrating data from genomics, transcriptomics, proteomics, metabolomics, and epigenomics. By identifying connections that might not be noticeable using single-omics techniques, these models can help identify new biomarkers, enhance diagnostics, and direct therapeutic interventions (Sartori et al., 2025). Future developments in algorithm design and processing power are going to render it possible to handle ever larger and more complicated datasets.

Multi-omics integration will become more feasible and available to biomedical researchers globally with the development of standardized data formats and enhanced computational pipelines.

5.8.2 Personalized and precision medicine applications

By providing accurate disease diagnosis and treatment plans based on a patient's genetic and clinical characteristics, deep learning is transforming personalized medicine. Researchers can minimize trial-and-error in treatment plans by designing targeted therapies, optimizing dosages, and predicting drug responses using patient-specific models (Ullah et al., 2025). As wearable technology, genetic sequencing data, and patient health records become more widely available, deep learning will become increasingly important for early disease detection and real-time patient monitoring. This can greatly enhance the quality of healthcare, increasing the affordability and applicability of precision medicine (Shajari et al., 2023).

5.8.3 Growing role in Indian biomedical research

Deep learning is becoming increasingly prevalent in India's biomedical research industry for drug development, disease prediction, and agricultural biotechnology. Infrastructure is expanding thanks to government programs such as Digital India and the National Supercomputing Mission, which enable Indian researchers to apply cutting-edge AI-based techniques (Doshi et al., 2025). The growth of joint ventures between the pharmaceutical industry and prestigious universities such as AIIMS, IISc, and IITs underscores India's potential as a center for biomedical AI research. With continued investments in data-driven healthcare and bioinformatics, deep learning applications should boost innovation and solidify India's place in the world's biomedical breakthroughs (Li et al., 2025).

5.9 Conclusion

By facilitating quicker data analysis, precise forecasting, and creative methods for disease diagnosis and treatment development, artificial intelligence and machine learning, or AI/ML, are revolutionizing the field of biomedical research. AI/ML tools have proven to be extremely helpful in managing the complexity of biological systems, from integrating large-scale multi-omics datasets to discovering new drug targets. Notwithstanding obstacles such as the scarcity of high-quality datasets, problems with model interpretability, and the requirement for experimental validation, there is a lot of promise in the expanding collaboration between laboratory research and computational tools (Medicines Agency, 2024). The impact of AI/ML is anticipated to be further amplified in the future by developments in deep learning, precision medicine, and India's growing biomedical research ecosystem. AI-driven solutions can speed up medical advancements and improve accessibility, efficiency, and personalization of healthcare by filling existing gaps and encouraging interdisciplinary collaborations (Abbas et al., 2025). The integration of deep learning and network biology promises a more comprehensive understanding of cellular systems, disease mechanisms, and therapeutic interventions as these disciplines continue to advance. The capacity to accurately predict outcomes and model intricate molecular interactions not only speeds up research but also closes the gap between conventional biological knowledge and contemporary computational techniques. These technologies have the potential to play an integral part in regional and international biomedical research, which will ultimately spur innovation and enhance patient care, provided that data availability, algorithm design, and experimental validation continue to improve (Cui et al., 2025).

Take-home messages

Biology is no longer about studying one gene, one protein, or one pathway in isolation. Network biology offers the framework for examining the web of interactions that makes up life. Now that deep learning has become popular, we have strong tools that can find hidden patterns in this web that are frequently missed by conventional approaches. We are changing how we predict disease genes, create medications, and even comprehend the therapeutic potential of natural remedies thanks to the combination of network biology and deep learning.

However, there are still challenges to overcome, including those related to data quality, model transparency, and experimental validation. But there is optimism regarding the future. Deep learning holds promise for bringing science closer to practical healthcare solutions as it drives personalized medicine and integrates with multi-omics data. This area presents an opportunity for nations such as India to assert the lead in biomedical innovation.

Deep learning and network biology are essentially about advancing patient-centered, accurate, and comprehensive healthcare, not just about devising better algorithms. It is a clear message: connections are the way of the future in biology, and the more we can learn from them, the healthier our world will be.

References

Abbas, S.R., Abbas, Z., Zahir, A., Lee, S.W., 2025. Advancing genome-based precision medicine: a review on machine learning applications for rare genetic disorders. Briefings Bioinf. 26 (4). https://doi.org/10.1093/bib/bbaf329.

Azeloglu, E.U., Iyengar, R., 2015. Signaling networks: information flow, computation, and decision making. Cold Spring Harbor Perspect. Biol. 7 (4), a005934. https://doi.org/10.1101/cshperspect.a005934.

Baião, A.R., Cai, Z., Poulos, R.C., Robinson, P.J., Reddel, R.R., Zhong, Q., Vinga, S., Gonçalves, E., 2025. A technical review of multi-omics data integration methods: from classical statistical to deep generative approaches. Briefings Bioinf. 26 (4). https://doi.org/10.1093/bib/bbaf355.

Cao, E.L., 2024. Natural product based anticancer drug combination discovery assisted by deep learning and network analysis. Front. Nat. Prod. 2. https://doi.org/10.3389/fntpr.2023.1309994.

Chatzianastasis, M., Vazirgiannis, M., Zhang, Z., Luigi Martelli, P., 2023. Explainable multilayer graph neural network for cancer gene prediction. Bioinformatics 39 (11). https://doi.org/10.1093/bioinformatics/btad643.

Cui, J., Yang, S., Yi, L., Xi, Q., Yang, D., Zuo, Y., 2025. Recent advances in deep learning for protein-protein interaction: a review. BioData Min. 18 (1). https://doi.org/10.1186/s13040-025-00457-6.

Dong, Z., Zhao, Q., Payne, P.R.O., Province, M.A., Cruchaga, C., Zhang, M., Zhao, T., Chen, Y., 2023. Highly accurate disease diagnosis and highly reproducible biomarker identification with PathFormer. Research Square 16. https://doi.org/10.21203/rs.3.rs-3576068/v1.

Doshi, R.V., Badhiye, S.S., Pinjarkar, L., 2025. Deep learning approach for biomedical image classification. J. Imaging Inform. Med. 8. https://doi.org/10.1007/s10278-025-01590-8.

En Li, B.W., Wang, C., Wang, Y., Li, P., Liu, Z.-P., 2013. RegNetwork 2025: an integrative data repository for gene regulatory networks in human and mouse. Nucleic Acids Res. 1, 13–14. https://doi.org/10.1093/NAR/GKAF779.

Eranti, B., Premika, S., Geetha, M.R.M.S., Krishna, A.V., Rakshith, N., Turaga, K., 2025. Artificial neural networks in pharma: revolutionizing drug development, optimization and smart manufacturing-a comprehensive review. Orient. J. Chem. 41 (3), 873–879. https://doi.org/10.13005/ojc/410320.

Fu, Y., Fang, Y., Gong, S., Xue, T., Wang, P., She, L., Huang, J., 2023. Deep learning-based network pharmacology for exploring the mechanism of licorice for the treatment of COVID-19. Sci. Rep. 13 (1). https://doi.org/10.1038/s41598-023-31380-7.

Jia, X., Luo, W., Li, J., Xing, J., Sun, H., Wu, S., Su, X., 2024. A deep learning framework for predicting disease-gene associations with functional modules and graph augmentation. BMC Bioinf. 25 (1). https://doi.org/10.1186/s12859-024-05841-3.

Jiang, W., Ye, W., Tan, X., Bao, Y.J., 2025. Network-based multi-omics integrative analysis methods in drug discovery: a systematic review. BioData Min. 18 (1). https://doi.org/10.1186/s13040-025-00442-z.

Jungck, J.R., Viswanathan, R., 2015. Graph Theory for Systems Biology: Interval Graphs, Motifs, and Pattern Recognition Algebraic and Discrete Mathematical Methods for Modern Biology. Elsevier, United States, pp. 1–27. https://doi.org/10.1016/B978-0-12-801213-0.00001-0.

Koutrouli, M., Karatzas, E., Paez-Espino, D., Pavlopoulos, G.A., 2020. A guide to conquer the biological network era using graph theory. Front. Bioeng. Biotechnol. 8. https://doi.org/10.3389/fbioe.2020.00034.

Kumar, N., Srivastava, R., 2024. Deep learning in structural bioinformatics: current applications and future perspectives. Briefings Bioinf. 25 (3). https://doi.org/10.1093/bib/bbae042.

Li, B., Tan, K., Lao, A.R., Wang, H., Zheng, H., Zhang, L., 2024. A comprehensive review of artificial intelligence for pharmacology research. Front. Genet. 15. https://doi.org/10.3389/fgene.2024.1450529.

Li, R., Rscpharma, Preet, J., Saini, S., Thakur, A., Yadav, D., 2025. AI-driven innovations in pharmaceuticals: optimizing drug discovery and industry operations. RSC Pharmaceutics 2 (3), 437–454. https://doi.org/10.1039/d4pm00323c.

Madrer, N., Perera, N.D., Uccelli, N.A., Abbondanza, A., Andersen, J.V., Carsana, E.V., Demmings, M.D., Fernandez, R.F., de Fragas, M.G., Gbadamosi, I., Kulshrestha, D., Lima-Filho, R.A.S., Marian, O.C., Markussen, K.H., McGovern, A.J., Neal, E.S., Sarkar, S., Šimončičová, E., Soto-Verdugo, J., Yandiev, S., Fernández-Moncada, I., 2025. Neural metabolic networks: key elements of healthy brain function. J. Neurochem. 169 (6). https://doi.org/10.1111/jnc.70084.

E. Medicines Agency, Review of AI/ML applications in medicines lifecycle (2024) www.ema.europa.eu/contact.

Mei, H., Wang, Z., Yang, H., Li, X., Xu, Y., 2025. Network analysis of multivariate time series data in biological systems: methods and applications. Briefings Bioinf. 26 (3). https://doi.org/10.1093/bib/bbaf223.

Min, S., Lee, B., Yoon, S., 2017. Deep learning in bioinformatics. Briefings Bioinf. 18 (5), 851–869. https://doi.org/10.1093/bib/bbw068.

Muzio, G., O'Bray, L., Borgwardt, K., 2021. Biological network analysis with deep learning. Briefings Bioinf. 22 (2), 1515–1530. https://doi.org/10.1093/bib/bbaa257.

Nasution, A.K., Alqaaf, M., Islam, R.M., Wijaya, S.H., Ono, N., Kanaya, S., Altaf-Ul-Amin, M., 2024. Identifying potential natural antibiotics from unani formulas through machine learning approaches. Antibiotics 13 (10). https://doi.org/10.3390/antibiotics13100971.

Noor, F., Qamar, M.T.U., Ashfaq, U.A., Albutti, A., Alwashmi, A.S.S., Aljasir, M.A., 2022. Network pharmacology approach for medicinal plants: review and assessment. Pharmaceuticals 15 (5). https://doi.org/10.3390/ph15050572.

Panossian, A., 2025. Trends and pitfalls in the progress of network pharmacology research on natural products. Pharmaceuticals 18 (4), 538. https://doi.org/10.3390/ph18040538.

Safari-Alighiarloo, N., Taghizadeh, M., Rezaei-Tavirani, M., Goliaei, B., Peyvandi, A.A., 2014. Protein-protein interaction networks (PPI) and complex diseases. Gastroenterol. Hepatol. Bed Bench 7 (1), 17–31.

Sartori, F., Codicè, F., Caranzano, I., Rollo, C., Birolo, G., Fariselli, P., Pancotti, C., 2025. A comprehensive review of deep learning applications with multi-omics data in cancer research. Genes 16 (6), 648. https://doi.org/10.3390/genes16060648.

Selby, D.A., Jakhmola, R., Sprang, M., Großmann, G., Raki, H., Maani, N., Pavliuk, D., Ewald, J., Vollmer, S., 2025. Visible neural networks for multi-omics integration: a critical review. Front. Artif. Intell. 8. https://doi.org/10.3389/frai.2025.1595291.

Shajari, S., Kuruvinashetti, K., Amin, K., Sundararaj, U., 2023. The emergence of AI-based wearable sensors for digital health technology: a review. Sensors 23 (23), 9498. https://doi.org/10.3390/s23239498.

Talukder, M.A., Kazi, M., Alazab, A., 2025. Predicting drug-target interactions using machine learning with improved data balancing and feature engineering. Sci. Rep. 15 (1). https://doi.org/10.1038/s41598-025-03932-6.

Ullah, R., Sarwar, N., Alatawi, M.N., Alsadhan, A.A., Salamah Alwageed, H., Khan, M., Ali, A., 2025. Advancing personalized diagnosis and treatment using deep learning architecture. Front. Med. 12. https://doi.org/10.3389/fmed.2025.1545528.

Visonà, G., Bouzigon, E., Demenais, F., Schweikert, G., 2024. Network propagation for GWAS analysis: a practical guide to leveraging molecular networks for disease gene discovery. Briefings Bioinf. 25 (2). https://doi.org/10.1093/bib/bbae014.

Wu, S., Chen, D., Snyder, M.P., 2022. Network biology bridges the gaps between quantitative genetics and multi-omics to map complex diseases. Curr. Opin. Chem. Biol. 66, 102101. https://doi.org/10.1016/j.cbpa.2021.102101.

Xie, X., Lin, G., Qiao, B., Wang, G., Huang, S., Zhao, Y., Sun, S., 2024. Deep learning in template-free de novo biosynthetic pathway design of natural products. Briefings Bioinf. 25 (6). https://doi.org/10.1093/bib/bbae495.

Yousef, M., Allmer, J., 2023. Deep learning in bioinformatics. Turk. J. Biol. 47 (6), 366–382. https://doi.org/10.55730/1300-0152.2671.

Zaidh, S.M., Vengateswaran, H.T., Habeeb, M., Aher, K.B., Bhavar, G.B., Irfan, N., Lakshmi, K.N.V.C., 2025. Network pharmacology and AI in cancer research uncovering biomarkers and therapeutic targets for RALGDS mutations. Sci. Rep. 15 (1). https://doi.org/10.1038/s41598-025-91568-x.

Zhang, P., Zhang, D., Zhou, W., Wang, L., Wang, B., Zhang, T., Li, S., 2024. Network pharmacology: towards the artificial intelligence-based precision traditional Chinese medicine. Briefings Bioinf. 25 (1). https://doi.org/10.1093/bib/bbad518.

Zhang, X.M., Liang, L., Liu, L., Tang, M.J., 2021. Graph neural networks and their current applications in bioinformatics. Front. Genet. 12. https://doi.org/10.3389/fgene.2021.690049.

Zhou, J., Cui, G., Hu, S., Zhang, Z., Yang, C., Liu, Z., Wang, L., Li, C., Sun, M., 2020. Graph neural networks: a review of methods and applications. AI Open 1, 57–81. https://doi.org/10.1016/j.aiopen.2021.01.001.

Zidane, M., Makky, A., Bruhns, M., Rochwarger, A., Babaei, S., Claassen, M., Schürch, C.M., 2023. A review on deep learning applications in highly multiplexed tissue imaging data analysis. Front. Bioinform. 3. https://doi.org/10.3389/fbinf.2023.1159381.

Zitnik, M., Li, M.M., Wells, A., Glass, K., Gysi, D.M., Krishnan, A., Murali, T.M., Radivojac, P., Roy, S., Baudot, A., Bozdag, S., Chen, D.Z., Cowen, L., Devkota, K., Gitter, A., Gosline, S.J.C., Gu, P., Guzzi, P. H., Huang, H., Jiang, M., Kesimoglu, Z.N., Koyuturk, M., Ma, J., Pico, A.R., Przulj, N., Przytycka, T.M., Raphael, B.J., Ritz, A., Sharan, R., Shen, Y., Singh, M., Slonim, D.K., Tong, H., Yang, X.H., Yoon, B.J., Yu, H., Milenkovic, T., 2024. Current and future directions in network biology. Bioinform. Adv. 4 (1). https://doi.org/10.1093/bioadv/vbae099.

6

Probabilistic optimization of ML for heart disease prediction

Aabida Khan[1], Syed Nazia Shah[1], Md E.A. Raghib Khan[1], Yousuf Siddiqui[2], Belal Bin Asaf[3], Md. Tabrez Nafis[4], Asghar Hussain[5], Ashif Raza[1] and Mohammad "Sufian" Badar[6, 7, 8, 9]

[1]DEPARTMENT OF COMPUTER SCIENCE, JAMIA MILLIA ISLAMIA, NEW DELHI, INDIA; [2]JNMC, AMU, ALIGARH, UTTAR PRADESH, INDIA; [3]DEPARTMENT OF CHEST SURGERY AND CHEST ONCOSURGERY AND LUNG TRANSPLANTATION, MEDANTA HOSPITAL NOIDA, NOIDA, UTTAR PRADESH, INDIA; [4]DEPARTMENT OF COMPUTER SCIENCE & ENGINEERING, JAMIA HAMDARD (DEEMED UNIVERSITY), NEW DELHI, INDIA; [5]CAPITALONE BANK, MCLEAN, VA, UNITED STATES; [6]DEPARTMENT OF BIOENGINEERING, UNIVERSITY OF CALIFORNIA, RIVERSIDE, CA, UNITED STATES; [7]UNIVERSAL SCIENTIFIC EDUCATION AND RESEARCH NETWORK (USERN), TEHRAN, IRAN; [8]SPI DARBHANGA, DARBHANGA, BIHAR, INDIA; [9]DEPARTMENT OF COMPUTER SCIENCE AND ENGINEERING (BIOINFORMATICS), SCHOOL OF ENGINEERING SCIENCES AND TECHNOLOGY (SEST), JAMIA HAMDARD UNIVERSITY, NEW DELHI, INDIA

What you will learn

This chapter will provide you with a thorough grasp of how probabilistic optimization improves machine learning techniques for heart disease prediction. You will first examine the increasing use of machine learning (ML) as a medical decision-support tool and the worldwide health burden of cardiovascular disease. In addition to introducing popular algorithms like decision trees, random forests, Naive Bayes, and clustering, the chapter emphasizes the value of ensemble learning in managing complicated, unbalanced medical data.

You will discover the fundamentals of probability-driven optimization, including how addressing misclassification and feature weighting can strengthen the robustness of the model. Best practices in data preparation are demonstrated through the discussion of practical topics such as dataset sources, preprocessing procedures, normalization, and correlation analysis. The integration of several classifiers with probabilistic weighting and adaptive reclassification is demonstrated step-by-step in an explanation of a suggested optimization framework.

Applying Artificial Intelligence to Computational Biology and Medical Informatics. https://doi.org/10.1016/B978-0-443-44915-4.00007-1

Furthermore, you will observe how various metrics—accuracy, recall, precision, F1, ROC-AUC, PR-AUC, and MCC—are used to evaluate performance and how they stack up against current techniques. Your understanding of the future role of explainable AI (XAI) in cardiovascular prediction will also be enhanced by case studies, literature reviews, and a discussion of issues such as bias, overfitting, and poor data quality.

6.1 Introduction

6.1.1 Heart disease as a global health challenge

Cardiovascular diseases (CVDs) are the leading cause of death worldwide, with about 18 million deaths per year (Cardiovascular diseases (CVDs), 2023). In addition to the devastating toll on human suffering, CVDs are associated with a heavy economic and clinical burden on health care systems. Studies from around the world highlight the fact that it is possible to prevent many of these deaths by tackling modifiable risk factors such as hypertension, obesity, and unhealthy lifestyles. However, conventional risk prediction models, such as the Framingham risk score, tend to underestimate the multiple biological and lifestyle determinants of cardiovascular health with their complex and nonlinear interactions (Kumar et al., 2025). This discrepancy highlights a need for more advanced and data-driven solutions to support early detection and effective intervention.

6.1.2 Role of machine learning in cardiovascular risk prediction

In the last decade, machine learning (ML) has become a compelling tool to complement and extend traditional statistical models for heart disease prediction. By using large, high-dimensional datasets—encompassing clinical records, imaging, wearable device outputs, and even genomic data - ML models can uncover subtle, hidden patterns that may not be detectable by other methods. The superiority of common machine learning algorithms, such as decision trees, random forests, and support vector machines, has always proven to be superior in terms of discrimination and accuracy to traditional regression models (Ahmad et al., 2023). In addition, combining longitudinal data from EHRs and digital health devices can facilitate real-time monitoring and early warning systems by ML systems, which can result in proactive disease management (Li et al., 2024).

6.1.3 Why probabilistic optimization?

Although ML models exhibit extraordinary potential, they still face a number of challenges that prevent their clinical reliability. These are class imbalances (healthy cases far outnumbering the diseased ones), population shift of data, and overconfident predictions. The design of these models (the way in which the uncertainty is represented, the way in which the hyperparameters are tuned, and the way in which the decision thresholds are set) is essentially probabilistic in nature. Probabilistic optimization algorithms, in particular Bayesian Optimization (BO), offer a principled way to deal with these challenges. Unlike a grid-based or exhaustive search, Bayesian methods explore the parameter space efficiently, simultaneously balancing exploration and exploitation, while also estimating uncertainty in predictions (Awal et al., 2021).

This has obvious healthcare advantages: models are better calibrated, more transparent, and safer to deploy in the real world, where false negatives can have devastating consequences. Probabilistic models also facilitate ensemble methods, in which multiple models are used and weighted in accordance with probabilistic evidence. These hybrid methods have undergone a significant improvement in robustness, calibration, and explainability without impacting the accuracy (Chen et al., 2020).

6.1.4 Toward clinically relevant interpretations of predictions

As ML for heart disease prediction gets closer to the clinic, it's no longer enough to just measure accuracy. Metrics like recall, precision, F1 score, and AUC are still necessary, but calibration and interpretability are highly important as well. Clinicians require models that not only provide accurate risk prediction, but one that is accompanied by the rationale behind the prediction to promote trust and allow for actionable clinical decision-making. Recent progress in explainable AI (XAI) for cardiovascular prediction highlights the need for standardized clinically interpretable metrics in addition to predictive performance.

Taken together, these insights are the basis for this chapter, which discusses probabilistic optimization of ML pipelines for heart disease prediction. We will cover all aspects of data curation and preprocessing, model and ensemble design, Bayesian search strategies, and thorough performance evaluation with the objective of constructing clinically trustworthy and interpretable decision-support systems.

6.2 Machine learning for the prediction of heart diseases

ML has gradually evolved from being a research novelty to an integral part of contemporary cardiovascular risk assessment. While classic clinical score systems such as Framingham or SCORE are based on a few predefined risk factors, ML is able to take in dozens—even hundreds—of heterogeneous variables at once. These include standard clinical data such as blood pressure, cholesterol, and ECG, but also imaging biomarkers, outputs from wearable devices, and genomic or metabolomics profiles. The power of ML is in its capacity to extract complex nonlinear patterns in these data sources that get oversimplified in traditional models (Ahmad et al., 2023).

6.2.1 Common algorithms

In addition, these ML algorithms have been widely used in the prediction of heart diseases:

Decision Trees are intuitive, rule-based structures that can reflect clinical reasoning. The networks are highly interpretable but are prone to overfitting when the data is noisy.

To counter this risk, Resampling methods such as Random Forest, which are collections of decision trees, have continuously proven reliable in terms of accuracy on benchmark datasets such as the UCI Cleveland dataset (Ahmad et al., 2023).

Naive Bayes classifiers are surprisingly effective for categorical variables with limited training data, and, as a consequence, they can be applied to small clinical datasets.

Support Vector Machine (SVM) is efficient at dealing with high-dimensional data and discriminating complex decision boundaries, which is a common challenge in discriminating between healthy and at-risk patients.

K-Means clustering and other unsupervised methods have been applied to cluster patients into new subgroups, identifying latent clusters of risk profiles that do not seem to be so easily reducible to traditional diagnostic classifications.

Each algorithm has trade-offs: while decision trees and Naive Bayes classifiers are interpretative, ensemble and kernel-based methods are likely to give the best predictions.

6.2.2 Ensemble learning and its relevance to medical data

Ensemble learning is becoming the method of choice for cardiovascular ML. By combining the predictions of multiple base learners, ensemble techniques such as bagging, boosting, and stacking reduce variance and bias, leading to more stable and accurate predictions. For instance, bagging-based (random forest) and gradient boosting machines have been shown to outperform individual classifiers in several comparative studies of heart disease prediction (Ashika and Hannah Grace, 2025).

In particular, stacked ensembles, where parallel training is done with several models, and their outputs are used as input to a meta-learner, constitute a very powerful method. These ensembles not only increase accuracy but also offer robustness in the face of dataset variability, a common problem in medical research. When paired with explainable AI methods like SHAP values, ensemble predictions can be broken down to feature-level contributions where the clinicians can interpret why the model classified a patient as high risk (Ganie et al., 2025).

6.2.3 Clinical relevance

The utility of such ML models goes beyond metrics of performance. For clinicians, ML's ability to combine heterogeneous data streams ranging from EHR entries to wearable-derived lifestyle data provides a more comprehensive view of cardiovascular well-being. Imagine something that learns from a patient's blood pressure readings, daily activity levels, and lipid profiles: such a tool could offer dynamic, personalized predictions, warning physicians that risk is increasing before symptoms show.

However, the clinical utility of ML requires careful development pipelines, proper class imbalance management, and assessment against clinically relevant endpoints. This is where probabilistic optimization, to be discussed later in this chapter, comes in: it makes sure the selected ML models are accurate, but also calibrated, reliable, and safe to use in real-world decision-making.

6.3 Concept of probabilistic optimization

Many applications of machine learning in healthcare must deal with uncertainty. Patient data are noisy, incomplete, and unbalanced, in contrast to physics or chemistry experiments where variables could be tightly controlled. Probabilistic optimization builds on the idea of accepting rather than neglecting uncertainty and provides a reasoned approach to model building that can result in models that are not only more accurate but are safer and more reliable for clinical decision making.

6.3.1 Fundamentals of probability-driven model improvement

"Probabilistic optimization is a deliberate step away from deterministic 'trial and error' methods, which are replaced by probability distributions that guide the search for the optimal models." Rather than trying to exhaustively search the space of all possible hyperparameter combinations (grid search) or randomly sample hyperparameters (random search), Bayesian Optimization (BO) treats the performance of an ML algorithm as an uncertain function that can be characterized by a model.

For instance, BO employs a Gaussian Process or other surrogate model to approximate the value of which hyperparameter values are most promising and then updates its belief after each trial (Awal et al., 2021). This introduces a feedback loop in which the search for new solutions is balanced by the exploitation of solutions that are known to be good. In a heart disease prediction, this is especially useful since datasets like the UCI Cleveland are relatively small. BO makes sure that the available computational resources are well utilized without overfitting, and yet obtains good models.

In addition, probabilistic optimization naturally allows calibration (fitting predicted probabilities to observed results). A model calibrated for 70% risk of heart disease should, in a population, be correct about 70% of the time. Such a property is critical for clinical trust (Chen et al., 2020).

6.3.2 Feature weighting and misclassification handling

Healthcare data are seldom in balance. In the vast majority of heart disease datasets, the number of patients without disease is greater than the number of positive patients. A naive model predicting "no disease" for all patients would get superficially high accuracy, but this would be very clinically dangerous as it would miss important positive cases.

Probabilistic optimization reduces ignorance with feature weighting and cost-sensitive learning. By coding the uncertainty about features, the system can weigh more heavily those features that are most predictive of the minority class (patients with disease). For example, morphologies such as the type of chest pain, exercise-induced angina, and maximum heart rate have disproportionate prediction power for identifying genuine disease individuals (Noroozi et al., 2023).

Furthermore, it is possible to weigh the loss functions of probabilistic models so that false negatives are penalized more than false positives. This moves the decision boundary to favor detection of diseased patients - at the expense of increased false alarms. In healthcare, this trade-off is usually permissible, since the cost of missing a diagnosis is much greater than having to run a few extra, precautionary tests.

6.3.3 Probabilistic ensembles

Ensemble methods aggregate several models to achieve robustness, and not all models should be equally weighted. Probabilistic optimization allows weighted ensembling, in which each base learner has a probability weight that scales with its reliability.

For instance, in a stacked ensemble for heart disease prediction, a logistic regression model might be allotted 20% of the final prediction, a random forest 50%, and a gradient boosting model 30%, according to their posterior probabilities of being the correct model (Ganie et al., 2025). This makes sure those models that have a proven history of better performance on similar patients are trusted more in the final decision.

Such ensembles are also useful for quantifying uncertainty. If the different models fail to agree about a patient's risk, the system can report the case as uncertain rather than confidently outputting a binary prediction. This uncertainty-aware ensembling reflects the way clinicians seek second or third opinions when making difficult diagnoses.

6.3.4 Probabilistic thinking for clinical use

The final test of probabilistic optimization is not technical performance but clinical translation. Physicians seldom think in absolutes; they think in terms of probabilities. A cardiologist may say: "Given the symptoms and tests I did, there is a 70% chance that this patient may have coronary artery disease." Probabilistic ML systems are consistent with this logic by outputting calibrated probabilities instead of black box labels.

For example, a system may estimate that a patient has a 72% probability of heart disease plus or minus 5%. Such output gives the clinician the ability to quantify the risk, consider further testing, and better explain this to patients. Studies have highlighted that probabilistic outputs increase trust in and facilitate shared decision-making between doctors and patients (Chen et al., 2020).

In the context of limited resources, calibrated probabilities provide a means to rank care. This means that patients who are most likely to have the highest risk can be prioritized for urgent intervention, while those with moderately high risk can be monitored closely to ensure that we are maximizing the efficiency of healthcare.

6.4 Datasets & preprocessing techniques

Any ML system is as good as the data it's trained on. In the field of cardiovascular prediction, where datasets are usually small and heterogeneous, not only are data quality and preprocessing not ancillary tasks,they are key to model performance and clinical credibility. This section discusses the most popular datasets, the data cleaning and normalization procedures needed to preprocess them, feature selection, and the importance of statistical analysis in building robust predictive pipelines.

6.4.1 Publicly available datasets (kaggle, UCI, etc.)

The UCI Heart Disease dataset continues to be the standard dataset for heart disease prediction research. Compiled by(Detrano et al., 1989), it combines clinical records of the Cleveland, Statlog, Hungarian, and Switzerland cohorts, of which the Cleveland cohort is the most commonly studied. It has 303 patients and 76 attributes, although researchers usually use fewer than 13–14 clinically relevant ones such as age, sex, cholesterol level, blood pressure, maximum heart rate, and exercise-induced angina. This dataset has been the standard testbed for comparing algorithms and establishing conventions for preprocessing and evaluation.

Platforms like Kaggle, for example, have made the UCI dataset more easily accessible by hosting cleaned and reformatted versions of it, enabling easier experimentation by students and researchers. Kaggle competitions also have given rise to synthetic or derived datasets to experiment with ensemble methods and probabilistic optimization (Noroozi et al., 2023).

However, the main issue with these datasets is that they are small, and the minority is not large enough. While these are useful places to play around with method development, their representativeness for global populations is limited. This can introduce risks of overfitting and a fallback when models are tested on real-world patients.

6.4.2 Data cleaning and normalization

Medical data are seldom clean. Missing entries may be caused by a patient skipping tests, variations in diagnostic protocols, or mere human error. Data cleaning is therefore the first step of preprocessing data. Missing values are most often imputed with median or mean substitution or regression-based methods or more sophisticated methods such as k-nearest neighbor imputation (Noroozi et al., 2023). In other cases, records with an excess of missing data are excluded to avoid bias in the results.

Normalization brings attributes that have different units of measurement to similar scales; for instance, the cholesterol (measured in mg/dL) and age (measured in years) have wildly different ranges, and this can skew algorithms like support vector machines or clustering techniques. Min-max normalization (rescaling to the [0, 1] range) and standardization with the z-score (rescaling from mean and standard deviation) are the most widely used techniques in cardiovascular ML pipelines (Pudjihartono et al., 2022).

By using these steps, researchers decrease the noise and are able to avoid certain features from dominating the learning process because of their magnitude.

6.4.3 Feature selection and dimensionality reduction

Not all features in a data set are the same in terms of contributing to a prediction. In particular, irrelevant or redundant features can add noise, lead to higher computational complexity, and decrease generalization. Feature selection methods seek to find the subset of attributes that is most informative.

The most important risk factors reported in the studies were chest pain type, resting blood pressure, serum cholesterol, maximum heart rate, and exercise-induced angina (Noroozi et al., 2023). Techniques such as Recursive Feature Elimination (RFE) iteratively eliminate less important features, while filters (e.g., chi-square tests) quantify the statistical association between features and the outcome. Principal Component Analysis (PCA), a-dimensionality reduction technique, is also applied in order to convert correlated features to a smaller number of uncorrelated components while preserving as much of the variance as possible.

The value of effective feature selection is not only to improve model accuracy but also to improve model interpretability to allow clinicians to visualize which variables contribute to the prediction.

6.4.4 Correlation and statistical analysis

Correlation analysis is used to identify the relationship between variables before models are constructed. For example, age is typically positively associated with resting blood pressure, and exercise capacity is negatively associated with the incidence of heart disease. Knowing these relations directs feature engineering and avoids redundancy.

Statistical methods such as chi-square, ANOVA, and Pearson correlation coefficients are usually used during preprocessing for evaluating features' significance (Pudjihartono et al., 2022). Beyond technical advantages, these analyses offer clinical insight by confirming which risk factors are biologically meaningful and creating a bridge between the findings of data-based analyses and existing medical knowledge.

6.4.5 Clinical significance of preprocessing

The reason why preprocessing is not only about optimizing accuracy metrics, but patients safety. A low dataset can cause models to be biased, and a feature that is poorly scaled can skew risk calculations. In clinical applications, such errors can correspond to missed diagnoses or false alarms. Thus, strict preprocessing not only guarantees technical robustness but also yields trustworthy models, forming the foundation for models that can be confidently embraced by clinicians.

6.5 Proposed optimization framework

Following the principles of machine learning, probabilistic optimization, and careful preprocessing, this section describes a proposed system of heart disease prediction. This is aimed at combining probabilistic optimization tools with ensemble learning to ensure that models are accurate, calibrated, interpretable and actionable in clinical settings. In Fig. 6.1, a general probabilistic optimization framework for heart disease has been developed.

6.5.1 Model architecture and workflow

Section 4 introduces the framework of dataset curation and preprocessing. The cleaned and normalized data are input into a series of processing stages consisting of feature selection, correlation analysis, and dimensionality reduction as needed (Noroozi et al., 2023).

Secondly, several classifiers are trained in parallel, including decision trees, random forests, logistic regression, and gradient boosting machines. Rather than a single model that is better than the others, this architecture accepts variety. The models represent various facets of the data, and these outputs combined give the basis of a stronger ensemble (Ahmad et al., 2023).

An optimization strategy with a focus on Bayesian Optimization lies at the heart of the architecture and is applied to hyperparameters of all candidate models. Probabilistically modeling the objective function, Bayesian Optimization finds the optimal parameter sets, taking into consideration the uncertainty so that it will not overfit small datasets such as UCI Cleveland (Awal et al., 2021).

6.5.2 Integration of multiple classifiers with probabilistic weighting

Ensuring that more powerful models not only help performance, but also naive averaging can water down performance in case weaker models perform equally. This

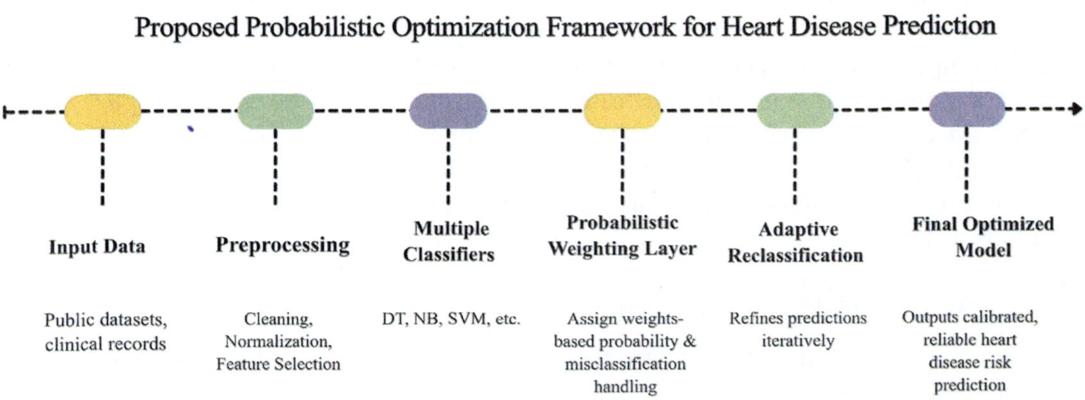

FIGURE 6.1 Proposed probabilistic optimization framework for heart disease prediction.

structure uses probabilistic weighting, with each base learner given a weight based on its posterior probability of being correct.

As an example, a random forest that is always good during cross-validation folds will be assigned more weight than a weaker logistic regression model. By doing so, the ensemble output is a manifestation of the combined intelligence of the many classifiers, shaped by probabilistic evidence (Ganie et al., 2025).

Dynamic weighting can also be facilitated by this form of integration: the weights of the model can be updated to reflect changing levels of reliability as new information is made available, so that the system can be adaptive through time.

6.5.3 Adaptive reclassification strategy

Sometimes even the best-optimized models are uncertain, particularly in cases where patients have borderline values or conflicting risk factors. The framework uses the adaptive reclassification strategy to deal with this. The system can:

- When confidence in prediction is less than some predefined value, use a human expert.
- Order extra items (e.g., order a stress test, order lab work, where the uncertainty remains).
- Dynamic classification based on the constantly gathered data.

This corresponds to the idea of a rejection in machine learning, where cases that are unsure are identified, instead of misclassified (Cesa-Bianchi et al., 2021). These adaptive strategies are of particular value in areas of healthcare, where the safe bet is always to defer rather than predict incorrectly.

6.5.4 Clinical integration and workflow

The framework is intended to fit into the established clinical processes. Projections will be offered not only in yes/no terms but on a scale of probabilities with uncertainty ranges, as well as in feature-wise explanations such as SHAP or LIME (Ganie et al., 2025). As an example, a patient may be diagnosed with a 65% risk of heart disease due to high cholesterol and low maximum heart rate.

This openness develops confidence in clinicians who are able to reason with the model and apply it to their knowledge. Additionally, the system is flexible enough to scale to varying populations and clinical settings through probabilistic weighting and adaptive reclassification.

6.5.5 Advantages of the framework

- **Efficiency**: Bayesian Optimization allows one to avoid brute-force searches on the hyperparameters.

- **Robustness**: Compared to individual models, weighted ensembles are more accurate and still calibrated.
- **Adaptability**: Adaptive reclassification and dynamic weighting safely cope with uncertainty.
- **Interpretability**: Clinician trust is boosted by probabilistic results and explainable artificial intelligence methods.

Overall, this framework will be a heart disease prediction pipeline that is holistic: it takes clean data and makes probabilistically optimized, ensemble-based, interpretable predictions that can be translated into clinical use.

6.6 Performance evaluation

6.6.1 Metrics for medical prediction (accuracy, recall, precision, F1, and AUC)

In medical prediction tasks, including disease diagnosis, prognosis, or therapy response categorization, the selection of evaluation metrics is crucial, as various measures highlight distinct facets of predictive performance. A solitary metric can be deceptive, particularly in significantly skewed clinical datasets, where instances of disease are substantially outnumbered by healthy controls. This document addresses various commonly utilized metrics, emphasizing their advantages and drawbacks for biomedical applications. An overview of advantage and disadvantages of the performance matrices for medical prediction has been discussed in Table 6.1.

6.6.1.1 Accuracy

Accuracy is the most instinctive and commonly employed metric. It denotes the ratio of accurately identified samples (both positive and negative) compared to the total number of samples:

$$Accuracy = \frac{TP + FN}{TP + TN + FP + FN}$$

Limitation: In imbalanced datasets (e.g., 95% healthy vs. 5% diseased), a basic classifier that consistently predicts "healthy" will attain 95% accuracy, despite its complete inability to identify cancer patients. Consequently, accuracy often yields overly hopeful and deceptive outcomes in clinical research (Akosa, 2017).

6.6.1.2 Recall (sensitivity/True Positive Rate)

Recall measures the proportion of actual positive cases that are correctly identified:

$$Recall = \frac{TP}{TP + FN}$$

Elevated recall is crucial in life-threatening conditions, since the omission of a positive example (false negative) may result in dire repercussions.

Table 6.1 Summary of metric characteristics in medical prediction tasks.

Metric	Advantages	Inconsistencies in medical data
Accuracy	Uncomplicated, instinctive	Deceptive due to asymmetry
Recall	Identifies overlooked patients	Disregards false negatives
Precision	Minimizes erroneous alerts	Disregards false negatives
F1 score	Achieves equilibrium between precision and recall	Disregards true negatives
ROC AUC	Beneficial for comprehensive threshold analysis	Deceptive in disparity
PR AUC	More advantageous for the imbalance	Infrequently documented
MCC	Equitable and dependable across scenarios	Less commonly recognized

Limitation: Recall alone disregards false positives. A test that categorizes all individuals as "diseased" attains flawless recall yet is clinically ineffectual (Powers DMW, 2011).

6.6.1.3 Precision (positive predictive value)

Precision measures the ratio of accurately anticipated positives to the total predicted positives:

$$Precision = \frac{TP}{TP + FP}$$

The importance of high accuracy is its capacity to reduce the occurrence of false alarms (false positives), which is vital in situations where unnecessary treatment entails considerable expenses or hazards.

Limitation: Precision alone overlooks missed detections (false negatives), which may present dangers in screening programs (Powers DMW, 2011).

6.6.1.4 F1 score

The F1 score is the **harmonic mean** of precision and recall:

$$F1Score = \frac{2 \cdot Precision \cdot Recall}{Precision \times Recall}$$

Extensively utilized in machine learning for medical predictions, particularly when balancing the risk of missing patients (FN) against the occurrence of false positives (FP).

Limitation: The F1 score has a limitation as it disregards true negatives (TN), hence neglecting the model's efficacy in detecting healthy people. This may result in deceptively elevated ratings in skewed datasets (Hripcsak and Rothschild, 2005).

6.6.1.5 Area under the curve

Two major variants are used:

ROC AUC (Receiver Operating Characteristic): Illustrates the True Positive Rate (recall) against the False Positive Rate. Assesses the model's capacity to differentiate between classes at various levels.

PR AUC (Precision–Recall Curve): Illustrates precision against recall across varying thresholds. Especially beneficial in significantly skewed medical datasets, where ROC AUC may present an excessively positive assessment.

Limitations: Both AUC measurements remain susceptible to imbalance sensitivity, with PR AUC frequently providing more pertinent information for medical diagnoses (Hand, 2010).

6.6.1.6 Matthews Correlation Coefficient

The MCC is a statistic based on correlation that considers all four terms of the confusion matrix (TP, TN, FP, and FN):

$$MCC = \frac{TP \cdot TN - FP \cdot FN}{\sqrt{(TP + FP)(TP + FN)(TN + FP)(TN + FN)}}$$

Range: 1 (complete disagreement) to +1 (ideal classification), with 0 signifying random chance.

In predicting gene expression for colon cancer, the Matthews correlation coefficient (MCC) revealed deficiencies in classifiers that were obscured by other metrics such as accuracy and F1 score.

Matthews Correlation Coefficient (MCC) offers a comprehensive assessment, even with imbalanced datasets, rendering it especially appropriate for medical and clinical research.

Limitation: Undefined in rare instances when a row or column of the confusion matrix is zero, notwithstanding the availability of workarounds (Jurman et al., 2012).

6.6.2 Comparative analysis with existing approaches

Recent studies on heart disease prediction assessed models utilizing several performance metrics have been discussed in Table 6.2.

Conventional measures (Accuracy, Precision, Recall, F1 Score) are frequently cited but may be deceptive when applied to imbalanced clinical data.

ROC AUC is frequently utilized to evaluate classifier discrimination capability, but PR AUC, albeit less prevalent, is particularly beneficial in scenarios where positive cases (patients with heart disease) are infrequent.

The Matthews Correlation Coefficient (MCC) is frequently cited in recent literature and is increasingly advocated for its ability to deliver a fair assessment across all areas of the confusion matrix(Chi et al., 2025). is significant as it presents all seven measures, including PR AUC and MCC, rendering it the most thorough comparison assessment.

Table 6.2 Comparative analysis of heart disease prediction studies based on standard evaluation metrics.

Task/data	Acc	Recall	Precision	F1	ROC AUC	PR AUC	MCC	Paper (year)
UCI cardiovascular disease; compares GBDT + LR versus RF/SVM/LR; explicitly reports AUPR (PR-AUC) and MCC among others	✓	✓ (sensitivity)	✓	✓	✓	✓	✓	Chi et al. (2025)
Heart-disease classification with feature selection	✓	✓ (sensitivity)	—	✓	✓	—	✓	Muhammad et al. (2020)
Heart failure dataset (Kaggle); compares LR, RF, DT, KNN, MLP; reports MCC and AUC; RF best (AUC = 0.97, MCC = 0.83)	✓	—	—	—	✓	—	✓	Chulde-Fernández et al. (2025)
CVD risk-factor prediction: evaluates accuracy, AUC, sensitivity, specificity, F1, MCC	✓	✓	—	✓	✓	—	✓	Mauya et al. (2025)
CVD diagnosis explicitly defines/uses MCC alongside standard metrics	✓	✓	✓	✓	✓	—	✓	Adanur Dedeturk, Dedeturk and Bakir-Gungor (2024)

6.7 Literature insights and case examples

67.1 Summary of notable studies on ML for heart disease

6.7.1.1 AutoML on UK biobank (AutoPrognosis)

This extensive study utilized an automated machine learning (AutoML) pipeline named AutoPrognosis on the UK Biobank dataset to predict cardiovascular events. AutoPrognosis demonstrated markedly superior predictive accuracy relative to traditional clinical scores like Framingham and QRISK, while also incorporating a variety of clinical, demographic, and lifestyle risk factors. An important finding was that the enhancement was not exclusively attributed to complex models, but rather to the capacity to capture and utilize more comprehensive sets of variables and their interactions. This highlights that richness and preprocessing are as crucial as the classifier selection, particularly in population-scale cohorts where nuanced patterns can enhance long-term disease risk stratification (Alaa et al., 2019).

6.7.1.2 ML versus classical Framingham risk score

A direct comparison employed machine learning algorithms, including XGBoost, in conjunction with the commonly utilized Framingham Risk Score (FRS). The results indicated that the machine learning-based approach attained an AUC of approximately 0.83, significantly surpassing the Framingham Risk Score, which had an AUC of around 0.78. The research indicated that machine learning models exhibit superior efficacy in managing nonlinear feature interactions and high-dimensional data, aspects that classical regression-based scores fail to address. This suggests that machine learning methods may enhance or supplant traditional clinical risk scores, contingent upon external validation (Yuda et al., 2025).

6.7.1.3 *New UK biobank risk model*

Researchers developed the UK Cardiovascular Risk Prediction model (UKCRP) by applying machine learning techniques to data from the UK Biobank. The model employed a mere 10 meticulously chosen predictors, yet it surpassed conventional multivariable models and several commonly utilized risk calculators. The research highlights that efficient machine learning models can maintain clinical relevance when feature selection is meticulously optimized. This study demonstrates the ability of machine learning to produce models that are clinically interpretable while maintaining computational efficiency, achieved through a balance of parsimony and predictive power (You et al., 2023).

6.7.1.4 *EHR-based CVD risk from irregular longitudinal data*

Machine learning has been utilized in electronic health records (EHRs) that include irregularly sampled, repeated patient measurements. Standard models typically consider health data as static snapshots; however, this study demonstrated that cardiovascular disease risk prediction is enhanced when the temporal and irregular characteristics of clinical records are acknowledged. Methods that effectively address time-varying signals and missing data demonstrate enhanced predictive performance, highlighting the importance of adapting machine learning techniques to the complexities of real-world healthcare data instead of solely depending on curated, balanced datasets.

6.7.1.5 *CHD incidence models with calibration emphasis*

In coronary heart disease (CHD) prediction research, certain studies have extended their analysis beyond discrimination metrics, such as accuracy or ROC AUC, to include calibration measures, including the Brier score and calibration plots. In clinical practice, well-calibrated probabilistic predictions are essential for guiding treatment decisions, including the initiation of statins. The incorporation of calibration underscores a transition in contemporary research from merely assessing whether a model can distinguish between cases and controls to evaluating whether the model's probability output accurately represents reality. This method signifies an elevated benchmark for practical translation (Vu et al., 2025).

6.7.1.6 *Reviews and horizon papers*

Multiple systematic reviews, such as those published in EHJ Digital Health and Frontiers in AI, have synthesized the applications of machine learning in cardiovascular disease. The reviews evaluated the performance of models applied to structured electronic health record data, imaging modalities such as ECG, PCG, and echocardiography, as well as wearable sensor data. A prevalent issue is that although machine learning consistently enhances discrimination performance, challenges related to external validation, interpretability, and calibration continue to hinder clinical adoption. Therefore, machine learning is regarded as promising; however, its success hinges on addressing issues of generalizability and transparency, rather than solely prioritizing raw accuracy (Li et al., 2024).

6.7.1.7 Deep learning on ECG for mortality risk

Deep neural networks utilizing standard 12-lead ECGs demonstrated a high accuracy in predicting 1-year mortality risk in patients. This study is notable for its model's ability to detect subtle ECG signal patterns that are not discernible to human clinicians, thus offering prognostic information that extends beyond conventional ECG interpretation. The output constituted a probability distribution regarding mortality risk, facilitating its incorporation into clinical decision-making frameworks. This example illustrates the enhancement of prognostic capabilities in existing diagnostic tools through the integration of deep learning probabilistic predictions (Raghunath et al., 2020).

6.7.2 Key observations on probabilistic approaches

6.7.2.1 Bayesian networks for cardiovascular risk

Bayesian networks (BNs) are utilized to represent conditional dependencies among various risk factors, including age, smoking, cholesterol, and hypertension. In contrast to black-box models, Bayesian Networks offer a transparent graphical framework that allows clinicians to observe the interactions among risk factors. Bayesian Networks (BNs) provide risk probabilities and facilitate "what-if reasoning," enabling the estimation of how smoking cessation could decrease an individual's future cardiovascular disease (CVD) probability. The interpretability of Bayesian Networks renders them particularly appealing for medical applications, where explainability and clinical trust are critical factors (Suo et al., 2024).

6.7.2.2 Calibration as a first-class metric

Numerous contemporary studies on coronary heart disease (CHD) prediction have included calibration plots and Brier scores in addition to the area under the curve (AUC). Calibration evaluates the alignment between predicted risks, such as a 20% risk of CHD over a decade, and actual observed outcomes. Models that are poorly calibrated can exhibit high AUC values yet remain significantly misleading in the context of clinical decision-making. Overestimated probabilities may result in overtreatment, whereas underestimated probabilities could postpone necessary interventions. Probabilistic calibration is increasingly recognized as a standard requirement for machine learning models in the field of medicine (Vu et al., 2025).

6.7.2.3 Gaussian processes and probabilistic surrogates
Gaussian processes (GPs) and analogous probabilistic surrogates are increasingly utilized for time-series health data, including ECG and wearable signals. They offer a risk score along with a distribution of potential outcomes that include quantified uncertainty. This is particularly beneficial in clinical triage, where a physician may opt for a conservative approach in the presence of significant uncertainty. General practitioners have been utilized as emulators in contexts where patient data is limited and costly to obtain, rendering them valuable in small-cohort cardiology research (Nnyaba et al., 2024).

6.7.2.4 Probabilistic versus deterministic risk scores
Comparative analyses of probabilistic machine learning methods and traditional deterministic risk calculators (such as Framingham and QRISK) indicate that although machine learning frequently enhances discrimination, the most significant advantages arise when models exhibit both accuracy and probabilistic calibration. This guarantees the reliability of their predictions for individual-level clinical decisions. Researchers are increasingly utilizing probability calibration methods, such as Platt scaling, isotonic regression, and Bayesian calibration, prior to clinical deployment (Suliman et al., 2024).

6.7.2.5 Longitudinal EHR data and probabilistic modeling
Clinical records in real-world settings frequently exhibit irregularities, incompleteness, and a longitudinal nature. Probabilistic models, such as time-aware Bayesian frameworks and recurrent models with uncertainty estimates, effectively address these challenges. They facilitate the integration of noisy, heterogeneous signals while generating reliable probabilistic predictions. This field underscores that probabilistic thinking is both an academic enhancement and a practical requirement for real-world healthcare data.

6.8 Challenges and future directions

6.8.1 Data quality, model bias, and overfitting risks

- **Data quality and shift:** Cardiology data from real-world sources, including electronic health records, electrocardiograms, imaging, and wearable devices, exhibit characteristics of noise, incompleteness, and temporal irregularity. Reviews highlight that models trained on curated datasets may overfit and subsequently underperform when applied across different sites, time periods, or devices. Addressing missingness, irregular sampling, and domain shift is crucial. (Liu et al., 2025)
- **Sampling bias and representativeness:** Numerous surveys indicate performance disparities in cardiovascular AI across sex, race/ethnicity, age, and socioeconomic subgroups. A synthesis from 2024 indicates that the majority of published cardiovascular disease AI models demonstrate accuracy disparities among racial and

ethnic groups, highlighting the necessity for subgroup reporting, stratified validation, and bias audits. (Cau et al., 2025)

- **Equity and remediation:** Effective methodologies encompass balanced and augmented training cohorts, fairness-aware learning, reweighting, subgroup threshold optimization, and ongoing monitoring postdeployment. Health-AI reviews, such as those published in The Lancet Digital Health and PLOS Digital Health, delineate strategies for mitigation throughout the phases of data, training, validation, and implementation. (Mihan et al., 2024)
- **Overfitting and data leakage:** Reviews on ECG and cardiovascular disease caution against overly optimistic outcomes resulting from patient overlap in data splits, inadequate preprocessing leading to leakage, and extensive hyperparameter tuning conducted on small cohorts. It is advisable to implement stronger temporal and external validation, utilize nested cross-validation, and report calibration metrics, such as the Brier score, to mitigate overfitting.(Somani et al., 2021)
- **Reporting that extends beyond AUC:** In the context of imbalanced cardiac cohorts, it is recommended that authors report PR-AUC, MCC, calibration plots, and decision-curve analysis in addition to ROC-AUC and accuracy to better represent clinical utility and mitigate bias associated with metrics. (Liu et al., 2025)

6.8.1.1 Prospects for explainable AI in cardiovascular prediction

- **Transitioning from saliency to actionable explanations:** Traditional posthoc tools such as saliency, Grad-CAM, and SHAP are now enhanced by counterfactual and concept-based methods that more effectively address the question, "What change would alter risk?" Recent ECG research illustrates the use of generative counterfactuals to alter waveforms, thereby examining model reasoning in a manner that aligns more closely with clinical "what-if" scenarios. (Wagner et al., 2024)
- **Playbooks for cardiac imaging and ECG XAI:** Cardiac imaging reviews examine modality-specific XAI, such as attention maps for echocardiography, computed tomography, and magnetic resonance imaging. They advocate for standardized evaluation of XAI based on fidelity, stability, and clinical usefulness, rather than relying solely on screenshots of heatmaps. Systematic assessments indicate that numerous cardiac studies continue to lack quantitative evaluations of their explanations, underscoring an important avenue for future research.(Salih et al., 2023)
- **Interpretability that holds clinical significance:** Novel tools are designed to enhance clinician-facing, resource-level interpretability of 12-lead ECGs by correlating features and waveform segments with associated risks, facilitating bedside review. Preliminary research indicates the practicality of incorporating model rationales into workflows and documentation. (Gliner et al., 2025)
- **Hybrid and explainable AI pipelines:** Emerging hybrid explainable systems integrate calibrated machine learning with rule-based or physiology-aware components to enhance trust, personalization, and triage decisions in cardiovascular disease risk prediction. Initial findings indicate improvements in transparency while maintaining accuracy levels.(Talaat et al., 2024)

- **Standards and deployment:** Future directions encompass: (i) prospective, multi-site trials incorporating subgroup XAI analyses; (ii) the establishment of documentation standards, including data sheets, model cards, and fairness/XAI reports; and (iii) human-factors studies aimed at verifying that explanations effectively enhance clinician decision-making rather than merely improving model aesthetics. (Chinta et al., 2025)

6.9 Conclusion

Cardiovascular disease continues to pose a worldwide health challenge, and machine learning has demonstrated considerable potential in enhancing risk assessment. This chapter highlighted that although prevalent algorithms like decision trees, random forests, and Naive Bayes serve as valuable baselines, their efficacy is significantly improved when integrated with probabilistic optimization. By assigning weights to features, addressing misclassification, and amalgamating several classifiers within a probabilistic framework, models can get more equitable and dependable results, especially crucial in imbalanced medical datasets.

Robust prediction is fundamentally reliant on effective data preprocessing techniques and the application of varied evaluation metrics, including accuracy, recall, precision, F1 score, ROC-AUC, PR-AUC, and MCC. Literature indicates that probabilistic models routinely enhance calibration and reliability, providing trustworthy risk estimations for doctors.

Finally, despite ongoing obstacles like data quality, bias, and overfitting, the future resides in the integration of probabilistic optimization with explainable AI, facilitating transparent and clinically interpretable predictions. Consequently, probabilistic optimization is an essential advancement in establishing machine learning as a reliable ally in cardiovascular care.

Take-home messages

Probabilistic optimization, which combines many classifiers, intelligently weights them, and repeatedly refines findings, offers a potent method for enhancing machine learning models for heart disease prediction. Strict data preprocessing, fair performance evaluation, and cautious uncertainty management are all necessary for success in this field, in addition to algorithm selection. More calibrated, transparent, and clinically trustworthy predictions can be achieved by combining probabilistic techniques with explainable AI, even while issues like bias, overfitting, and poor data quality still exist. In the end, this method helps improve cardiovascular care decision-making by bridging the gap between mechanical effectiveness and medical application.

References

Adanur Dedeturk, B., Kagan Dedeturk, B., Bakir-Gungor, B., 2024. CSA-DE-LR: enhancing cardiovascular disease diagnosis with a novel hybrid machine learning approach. PeerJ Comput. Sci. 10, e2197. https://doi.org/10.7717/peerj-cs.2197.

Ahmad, A.A., Rahman, R., Ismail, S., Noor, N.M., 2023. Prediction of heart disease based on machine learning algorithms. Comput. Intell. Neurosci. 2023, 1–12. https://doi.org/10.1155/2023/9256245.

J.S. Akosa, Predictive accuracy : a misleading performance measure for highly imbalanced data . (2017).

Alaa, A.M., Bolton, T., Angelantonio, E.D., Rudd, J.H.F., van der Schaar, M., 2019. Cardiovascular disease risk prediction using automated machine learning: a prospective study of 423,604 UK Biobank participants. PLoS One 14 (5). https://doi.org/10.1371/journal.pone.0213653.

Ashika, T., Hannah Grace, G., 2025. Enhancing heart disease prediction with stacked ensemble and MCDM-based ranking: an optimized RST-ML approach. Front. Digit. Health 7. https://doi.org/10.3389/fdgth.2025.1609308.

Awal, M.A., Rahman, M.A., Sohel, F., Bennamoun, M., 2021. A novel Bayesian optimization-based machine learning framework for medical classification. Diagnostics 11 (12), 2183. https://doi.org/10.3390/diagnostics11122183.

Cardiovascular diseases (CVDs). (2023), Available from: https://www.who.int/news-room/fact-sheets/detail/cardiovascular-diseases-(cvds).

Cau, R., Pisu, F., Suri, J.S., Saba, L., 2025. Addressing hidden risks: systematic review of artificial intelligence biases across racial and ethnic groups in cardiovascular diseases. Eur. J. Radiol. 183, 111867. https://doi.org/10.1016/j.ejrad.2024.111867.

Cesa-Bianchi, N., Gentile, C., Vitale, F., 2021. Learning with a reject option. Mach. Learn. 110 (2), 229–257. https://doi.org/10.1007/s10994-020-05929-3.

Chen, I.Y., Szolovits, P., Ghassemi, M., 2020. Can AI help reduce disparities in general medical and mental health care? AMA Journal of Ethics 22 (9), E803. https://doi.org/10.1001/amajethics.2020.803.

Chi, Z., Liu, L., Yi, L., Shi, L., 2025. Prediction of cardiovascular diseases based on GBDT+LR. Sci. Rep. 15 (1). https://doi.org/10.1038/s41598-025-04921-5.

Chinta, S.V., Wang, Z., Palikhe, A., Zhang, X., Kashif, A., Smith, M.A., Liu, J., Zhang, W., 2025. AI-driven healthcare: fairness in AI healthcare: a survey. PLOS Digit. Health 4 (5). https://doi.org/10.1371/journal.pdig.0000864.

Chulde-Fernández, B., Enríquez-Ortega, D., Guevara, C., Navas, P., Tirado-Espín, A., Vizcaíno-Imacaña, P., Villalba-Meneses, F., Cadena-Morejon, C., Almeida-Galarraga, D., Acosta-Vargas, P., 2025. Classification of heart failure using machine learning: a comparative study. Life 15 (3), 496. https://doi.org/10.3390/life15030496.

Detrano, R., Janosi, A., Steinbrunn, W., Pfisterer, M., Schmid, J.J., Sandhu, S., Guppy, K.H., Lee, S., Froelicher, V., 1989. International application of a new probability algorithm for the diagnosis of coronary artery disease. Am. J. Cardiol. 64 (5), 304–310. https://doi.org/10.1016/0002-9149(89)90524-9.

Ganie, S.M., Manhas, J., Rather, G.M., 2025. Ensemble learning with explainable AI for improved heart disease prediction. PLoS One 20 (1), e0300123. https://doi.org/10.1371/journal.pone.0300123.

Gliner, V., Levy, I., Tsutsui, K., Acha, M.R., Schliamser, J., Schuster, A., Yaniv, Y., 2025. Clinically meaningful interpretability of an AI model for ECG classification. npj Digit. Med. 8 (1). https://doi.org/10.1038/s41746-025-01467-8.

Hand, D.J., 2010. Evaluating diagnostic tests: the area under the ROC curve and the balance of errors. Stat. Med. 29 (14), 1502–1510. https://doi.org/10.1002/sim.3859UnitedKingdom.

Hripcsak, G., Rothschild, A.S., 2005. Agreement, the F-measure, and reliability in information retrieval. J. Am. Med. Inform. Assoc. 12 (3), 296–298. https://doi.org/10.1197/jamia.M1733.

Jurman, G., Riccadonna, S., Furlanello, C., Biondi-Zoccai, G., 2012. A comparison of MCC and CEN error measures in multi-class prediction. PLoS One 7 (8), e41882. https://doi.org/10.1371/journal.pone.0041882.

Kumar, R., Garg, S., Kaur, R., Johar, M.G.M., Singh, S., Menon, S.V., Kumar, P., Hadi, A.M., Hasson, S.A., Lozanović, J., 2025. A comprehensive review of machine learning for heart disease prediction: challenges, trends, ethical considerations, and future directions. Front. Artif. Intell. 8. https://doi.org/10.3389/frai.2025.1583459.

Li, C., Liu, X., Shen, P., Sun, Y., Zhou, T., Chen, W., Chen, Q., Lin, H., Tang, X., Gao, P., 2024. Improving cardiovascular risk prediction through machine learning modelling of irregularly repeated electronic health records. European Heart Journal - Digital Health 5 (1), 30–40. https://doi.org/10.1093/ehjdh/ztad058.

Liu, T., Krentz, A.J., Huo, Z., Ćurčin, V., 2025. Opportunities and challenges of cardiovascular disease risk prediction for primary prevention using machine learning and electronic health records: a systematic review. Kingdom Reviews in Cardiovascular Medicine 26 (4). https://doi.org/10.31083/RCM37443.

Mauya, J., Sahriar, S., Akther, S., Amin, R., Ruhi, S., Reza, M.S., 2025. Missing risk factor prediction in cardiovascular disease using a blended dataset and optimizing classification with a stacking algorithm. Eng. Rep. 7 (1). https://doi.org/10.1002/eng2.13034.

Mihan, A., Pandey, A., Van Spall, H.G., 2024. Mitigating the risk of artificial intelligence bias in cardiovascular care. Lancet Digit. Health 6 (10), e749. https://doi.org/10.1016/S2589-7500(24)00155-9.

Muhammad, Y., Tahir, M., Hayat, M., Chong, K.T., 2020. Early and accurate detection and diagnosis of heart disease using intelligent computational model. Sci. Rep. 10 (1). https://doi.org/10.1038/s41598-020-76635-9.

Nnyaba, U.V., Shemtaga, H.M., Collins, D.W., Muyskens, A.L., Priest, B.W., Billor, N., 2024. Enhancing Electrocardiography Data Classification Confidence: A Robust Gaussian Process Approach (MuyGPs). arXiv, United States. https://doi.org/10.48550/arXiv.2409.04642.

Noroozi, Z., Jalali, S., Ghasemi, M., 2023. Impact of feature selection on heart disease prediction using machine learning. Sci. Rep. 13 (1), 1256. https://doi.org/10.1038/s41598-023-12345-7.

Pudjihartono, N., Fadason, T., Kempa-Liehr, A.W., O'Sullivan, J.M., 2022. A review of feature selection methods for machine learning-based disease risk prediction. Frontiers in Bioinformatics 2 (8). https://doi.org/10.3389/fbinf.2022.927312.

Raghunath, S., Ulloa Cerna, A.E., Jing, L., vanMaanen, D.P., Stough, J., Hartzel, D.N., Leader, J.B., Kirchner, H.L., Stumpe, M.C., Hafez, A., Nemani, A., Carbonati, T., Johnson, K.W., Young, K., Good, C. W., Pfeifer, J.M., Patel, A.A., Delisle, B.P., Alsaid, A., Beer, D., Haggerty, C.M., Fornwalt, B.K., 2020. Prediction of mortality from 12-lead electrocardiogram voltage data using a deep neural network. Nat. Med. 26 (6), 886–891. https://doi.org/10.1038/s41591-020-0870-z.

Salih, A., Boscolo Galazzo, I., Gkontra, P., Lee, A.M., Lekadir, K., Raisi-Estabragh, Z., Petersen, S.E., 2023. Explainable artificial intelligence and cardiac imaging: toward more interpretable models. Circ., Cardiovasc. Imaging 16 (4), E014519. https://doi.org/10.1161/CIRCIMAGING.122.014519.

Somani, S., Russak, A.J., Richter, F., Zhao, S., Vaid, A., Chaudhry, F., De Freitas, J.K., Naik, N., Miotto, R., Nadkarni, G.N., Narula, J., Argulian, E., Glicksberg, B.S., 2021. Deep learning and the electrocardiogram: review of the current state-of-the-art. Europace 23 (8), 1179–1191. https://doi.org/10.1093/europace/euaa377.

Suliman, A., Masud, M., Serhani, M.A., Abdullahi, A.S., Oulhaj, A., 2024. Predictive performance of machine learning compared to statistical methods in time-to-event analysis of cardiovascular disease: a systematic review protocol. BMJ Open 14 (4). https://doi.org/10.1136/bmjopen-2023-082654.

Suo, X., Huang, X., Zhong, L., Luo, Q., Ding, L., Xue, F., 2024. Development and validation of a bayesian network-based model for predicting coronary heart disease risk from electronic health records. J. Am. Heart Assoc. 13 (1). https://doi.org/10.1161/jaha.123.029400.

Talaat, F.M., Elnaggar, A.R., Shaban, W.M., Shehata, M., Elhosseini, M., 2024. CardioRiskNet: a hybrid AI-based model for explainable risk prediction and prognosis in cardiovascular disease. Bioengineering 11 (8). https://doi.org/10.3390/bioengineering11080822.

Vu, T., Kokubo, Y., Inoue, M., Yamamoto, M., Mohsen, A., Martin-Morales, A., Dawadi, R., Inoue, T., Tay, J.T., Yoshizaki, M., Watanabe, N., Kuriya, Y., Matsumoto, C., Arafa, A., Nakao, Y.M., Kato, Y., Teramoto, M., Araki, M., 2025. Machine learning model for predicting coronary heart disease risk: development and validation using insights from a Japanese population–based study. JMIR Cardio 9. https://doi.org/10.2196/68066.

Wagner, P., Mehari, T., Haverkamp, W., Strodthoff, N., 2024. Explaining deep learning for ECG analysis: building blocks for auditing and knowledge discovery. Comput. Biol. Med. 176, 108525. https://doi.org/10.1016/j.compbiomed.2024.108525.

You, J., Guo, Y., Kang, J.J., Wang, H.F., Yang, M., Feng, J.F., Yu, J.T., Cheng, W., 2023. Development of machine learning-based models to predict 10-year risk of cardiovascular disease: a prospective cohort study. Stroke and Vascular Neurology 8 (6), 475–485. https://doi.org/10.1136/svn-2023-002332.

Yuda, E., Kaneko, I., Hirahara, D., 2025. Machine-learning insights from the framingham heart study: enhancing cardiovascular risk prediction and monitoring. Appl. Sci. 15 (15), 8671. https://doi.org/10.3390/app15158671.

7

The need for interpretable and explainable deep learning data in health care

Mohammad "Sufian" Badar[1, 2, 3, 4], Md Arshad[5], Labeebah Rizwan Badar[6], Jitendra Kumar[7], Jishnu Anilkumar[8], Shireen Siddiqui[9], Omar A. Alamoudi[10] and Hussam Bin Mehare[11]

[1]DEPARTMENT OF BIOENGINEERING, UNIVERSITY OF CALIFORNIA, RIVERSIDE, CA, UNITED STATES; [2]UNIVERSAL SCIENTIFIC EDUCATION AND RESEARCH NETWORK (USERN), TEHRAN, IRAN; [3]SPI DARBHANGA, DARBHANGA, BIHAR, INDIA; [4]DEPARTMENT OF COMPUTER SCIENCE AND ENGINEERING (BIOINFORMATICS), SCHOOL OF ENGINEERING SCIENCES AND TECHNOLOGY (SEST), JAMIA HAMDARD UNIVERSITY, NEW DELHI, INDIA; [5]DEPARTMENT OF COMPUTER SCIENCE, JMI, NEW DELHI, INDIA; [6]BADAR MEDICAL CENTRE (BMC), NEW DELHI, INDIA; [7]DEPARTMENT OF FORENSIC MEDICINE, MAULANA AZAD MEDICAL COLLEGE, NEW DELHI, INDIA; [8]NAAS (NETOPS & AUTOMATION), HPE (XEVYTE TECHNOLOGIES), BENGALURU, KARNATAKA, INDIA; [9]DEPARTMENT OF LABORATORY MEDICINE, MEDANTA HOSPITAL NOIDA, NOIDA, UTTAR PRADESH, INDIA; [10]DEPARTMENT OF ELECTRICAL ENGINEERING, FACULTY OF ENGINEERING, KING ABDULAZIZ UNIVERSITY, JEDDAH, SAUDI ARABIA; [11]SCHOOL OF ELECTRICAL AND MECHANICAL ENGINEERING, UNIVERSITY OF ADELAIDE, ADELAIDE, SA, AUSTRALIA

What you will learn

By the end of this chapter, readers will understand the black-box problem in deep learning (DL)and why explainability is crucial in healthcare, while distinguishing between interpretability and explainability through practical examples. They will be introduced to major XAI methods such as LIME, SHAP, Grad-CAM, and Integrated Gradients, alongside their limitations, and will evaluate how these approaches are applied in imaging, genomics, and multimodal contexts. The chapter also highlights ethical and regulatory challenges, strategies for clinical integration with human-in-the-loop systems, and visualization tools, while offering perspectives on emerging trends including causal explanations, interpretable graph neural networks, and counterfactual

Applying Artificial Intelligence to Computational Biology and Medical Informatics. https://doi.org/10.1016/B978-0-443-44915-4.00003-4

reasoning. Ultimately, readers will gain clinically relevant insights into how explainable AI can support trust, fairness, and broader adoption of DL in medicine.

7.1 Introduction

Artificial intelligence (AI), and in particular DL, has become a transformative force in healthcare. From radiology and pathology to genomics and multimodal patient modeling, DL algorithms are increasingly integrated into clinical research and decision support. Their promise lies in detecting subtle, high-dimensional patterns that are difficult for humans to recognize—for example, automated detection of diabetic retinopathy from retinal fundus images predictive modeling from raw electronic health records (Rajkomar et al., 2018), and improved gene-expression prediction from genomic sequence using transformer-based architectures (Avsec et al., 2021). These developments underpin a shift toward data-driven precision medicine, particularly in neurodegenerative disease research where multimodal models (MRI, CSF biomarkers, genetics) show potential for earlier detection and prognosis (Millar et al., 2023).

However, the rise of DL also introduces one of its most pressing challenges: the black-box problem. Many state-of-the-art models—convolutional neural networks (CNNs), transformers, and multimodal architectures—achieve impressive predictive performance but lack transparency about how predictions are generated. In high-stakes clinical settings this opacity can undermine trust, limit clinical adoption, and even cause harm when outputs are misinterpreted. For instance, DL models trained on radiographs have been shown to exploit spurious dataset artifacts (e.g., portable X-ray markers) rather than pathology, producing deceptively good performance that fails to generalize (Zech et al., 2018). Without interpretable explanations, clinicians may hesitate to act on risk scores for sepsis, cancer, or dementia—especially when interventions have significant consequences (Topol, 2019).

Transparency in healthcare AI is not only a technical desideratum but also an ethical and regulatory requirement. Clinicians are professionally and legally accountable for decisions; an AI system that cannot provide intelligible justification risks being dismissed or misused. Patients and policymakers likewise demand understandability and fairness. Recognizing these stakes, regulators have moved to require greater transparency and human oversight: the European Union's Artificial Intelligence Act (EU AI Act) treats many medical AI systems as "high-risk" and mandates documentation, human oversight, and post-market monitoring (European Commission. (2024). Regulation (EU) 2024/1689 of the European Parliament and of the Council on artificial intelligence (AI Act), 2024), and recent guidance from the U.S. Food and Drug Administration emphasizes transparency and lifecycle governance for AI/ML-enabled medical devices (FDA. (2024). Guiding principles for transparency for machine learning–enabled medical devices. U.S. Food and Drug Administration., no date) (FDA. (2025). Draft guidance:

Artificial intelligence-enabled device software functions — Lifecycle management and marketing submission recommendations. U.S. Food and Drug Administration., no date).

To close the gap between performance and trustworthiness, two complementary approaches have emerged: interpretability (designing models whose internal mechanics are human-readable) and explainability (applying post-hoc methods to make complex models understandable). Intrinsically interpretable models—linear models, decision trees, or generalized additive models—offer direct transparency but may lack the capacity to model high-dimensional, nonlinear biomedical signals. Post-hoc explainability tools (e.g., LIME, SHAP, Grad-CAM, Integrated Gradients) enable interpretation of high-performing DL models by producing feature attributions or spatial saliency maps (Ribeiro et al., 2016; Lundberg and Lee, 2017; Selvaraju et al., 2017; Sundararajan et al., 2017).

Interpretability is especially critical in imaging, genomics, and multimodal integration. In radiology, models that flag a suspicious lesion must indicate the precise region to support clinical validation (Selvaraju et al., 2017). In genomics, models predicting cancer subtype or polygenic risk must show which genes or variants drove the prediction to permit biological interpretation and hypothesis generation (Avsec et al., 2021). In multimodal Alzheimer's prediction—combining MRI, CSF biomarkers, and genotypes—clinicians require clarity on how each modality contributed to the final risk estimate so that model outputs can be meaningfully integrated into trial enrollment or treatment decisions (Millar et al., 2023).

This chapter examines why interpretability and explainability are indispensable for DL in healthcare, emphasizing clinical relevance, ethical safeguards, and regulatory alignment. We outline core XAI concepts and tools, present domain-focused case studies (imaging, genomics, multimodal neurodegeneration), discuss key challenges (bias, overtrust, performance trade-offs, governance), and survey emerging directions such as causal XAI and patient-centric counterfactual explanations. The objective is practical: to show how interpretable DL can be designed, evaluated, and integrated so that AI becomes a trustworthy partner in clinical care rather than an opaque oracle.

Fig. 7.1.

Formula Placeholder:

- Feature attribution in SHAP:

$$\phi_i(f,x) = \sum_{S \subseteqq F \setminus \{i\}} \frac{S \vee !(\vee F \vee - \vee S \vee -1)!}{F \vee !} [f(S \cup \{i\}) - f(S)]$$

Eq. (1). Shapley value attribution for feature i in a predictive model f, where S represents a subset of features from the full feature set F (Lundberg and Lee, 2017).

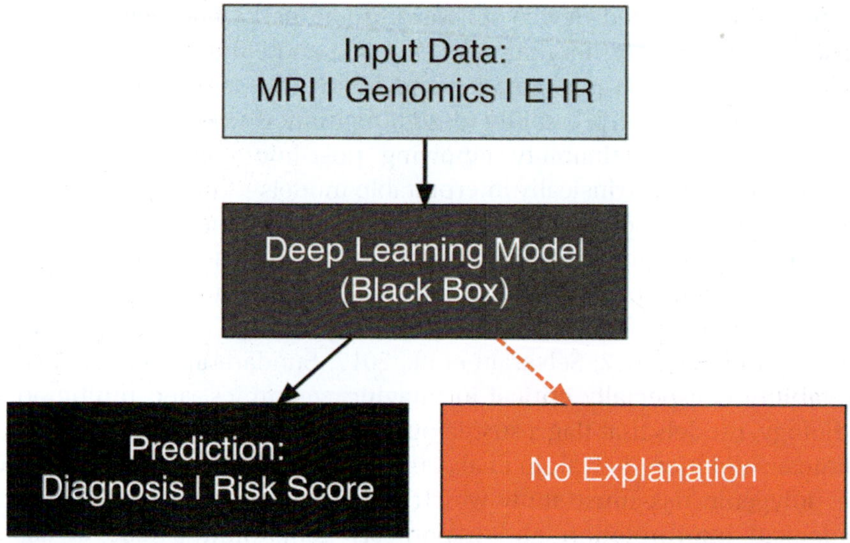

FIGURE 7.1 Conceptual contrast between interpretable models (intrinsically transparent; e.g., GAMs, decision trees) and opaque models + post-hoc explainability (black-box DL models with saliency/attribution overlays). (For final manuscript: Provide a two-panel schematic showing interpretability vs. explainability flows.).

7.2 Core concepts in explainable AI (XAI)

Explainable artificial intelligence (XAI) provides the conceptual and methodological foundation for making DL models transparent and trustworthy. Healthcare requires both interpretability, where models are inherently human-readable, and explainability, where tools are applied to post-hoc analysis of black-box systems (Doshi-Velez and Kim, 2017). This section introduces the major categories of interpretability, the theoretical underpinnings of XAI, and leading tools with their limitations, framed specifically for biomedical data domains.

7.2.1 Global versus local explainability

Global explainability refers to understanding a model's behavior across an entire dataset. In genomics, for example, global explanations can reveal which genes consistently contribute to cancer subtype classification across thousands of patients. Such insights are useful for population-level understanding, but they often miss the nuance needed for individual care.

Local explainability, by contrast, seeks to clarify why a model made a specific prediction for a specific case. In radiology, this could mean visualizing which region of a chest CT scan influenced the classification of a lung nodule as malignant. Local explanations are particularly critical in high-stakes decision-making, where clinicians must justify individual patient care. Both global and local perspectives are needed in healthcare: global to assess model reliability and fairness, and local to support clinical action (Molnar, 2022).

7.2.2 Intrinsic versus post-hoc explainability

- Intrinsic interpretability is built into the model structure. Decision trees, rule-based systems, or generalized additive models (GAMs) are interpretable by design. However, they struggle with high-dimensional, nonlinear biomedical data such as MRI voxels or gene expression matrices.
- Post-hoc explainability applies to complex black-box models after training. Saliency maps, attribution scores, and counterfactual reasoning allow models such as CNNs or transformers to be "explained" even when their internal logic is not directly human-readable.

While intrinsic models promote transparency, post-hoc methods enable the use of state-of-the-art DL architectures without sacrificing interpretability. The choice between them depends on the task: a simple predictive model for readmission risk may suffice with interpretable GAMs, whereas image-based tumor classification often requires CNNs with post-hoc visualization.

7.2.3 Mathematical foundations

A key goal of XAI is to quantify feature importance. SHAP, for example, leverages cooperative game theory, assigning each feature a Shapley value that reflects its contribution to the model output (Lundberg and Lee, 2017):

$$\phi_i(f,x) = \sum_{S \subseteq F \setminus \{i\}} \frac{S \vee !(\vee F \vee - S \vee -1)!}{F \vee !} [f(S \cup \{i\}) - f(S)]$$

where \phi_i is the attribution for feature i, F the set of all features, and f the model. This framework is widely applied in genomics, where SHAP can identify which SNPs drive disease-risk predictions.

Saliency methods for imaging rely on gradient-based attribution. Integrated Gradients (Sundararajan et al., 2017) compute the path integral of gradients between a baseline (e.g., blank image) and input:

$$IG_i(x) = (x_i - x'i) \times \int 0^1 \frac{\partial f(x' + \alpha(x - x'))}{\partial x_i} d\alpha$$

where x′ is a baseline input, x is the sample, and f the model. This method provides faithful attribution while avoiding issues of noise and saturation common in raw saliency maps.

7.2.4 Leading XAI tools in healthcare

- LIME (Local Interpretable Model-Agnostic Explanations): Builds local surrogate models around individual predictions, useful for structured EHR data (Ribeiro et al., 2016). Limitation: unstable results with small perturbations.
- SHAP: Model-agnostic, consistent, and grounded in game theory. Widely used in genomics and EHR predictions. Limitation: computationally expensive for large feature sets.
- Grad-CAM: Produces heatmaps over CNN activations, highlighting clinically relevant regions in imaging studies (Selvaraju et al., 2017). Limitation: resolution is coarse and may miss subtle patterns.
- Integrated Gradients: Axiomatic attribution, providing more stable and faithful saliency maps for both imaging and tabular data. Limitation: requires careful baseline selection (Sundararajan et al., 2017).

7.2.5 Limitations and challenges

Despite their promise, XAI tools face critical challenges:

- Faithfulness: Explanations may not always reflect the true decision boundary of the model.
- Stability: Slightly different inputs can yield very different explanations, reducing reliability.
- Comprehensibility: Visualizations or attributions may not align with clinical reasoning, leading to misinterpretation.
- Scalability: Tools such as SHAP can be computationally prohibitive for large-scale genomic or multimodal datasets.

Therefore, XAI is best understood as decision support rather than a perfect window into DL models. Clinicians must be trained to interpret explanations critically (Fig. 7.2; Table 7.1).

7.3 Case studies and real-world applications

While theoretical frameworks and model-agnostic tools for explainability are essential, the true value of explainable AI (XAI) in healthcare is realized through real-world deployment. In this section, we present three case studies that illustrate the role of explainability in high-stakes clinical decision-making across radiology, genomics, and multimodal disease modeling. Each example highlights how interpretability enhances clinician trust, regulatory compliance, and biological insight—especially in domains such as Alzheimer's disease and cancer where diagnostic uncertainty is high and accountability is paramount.

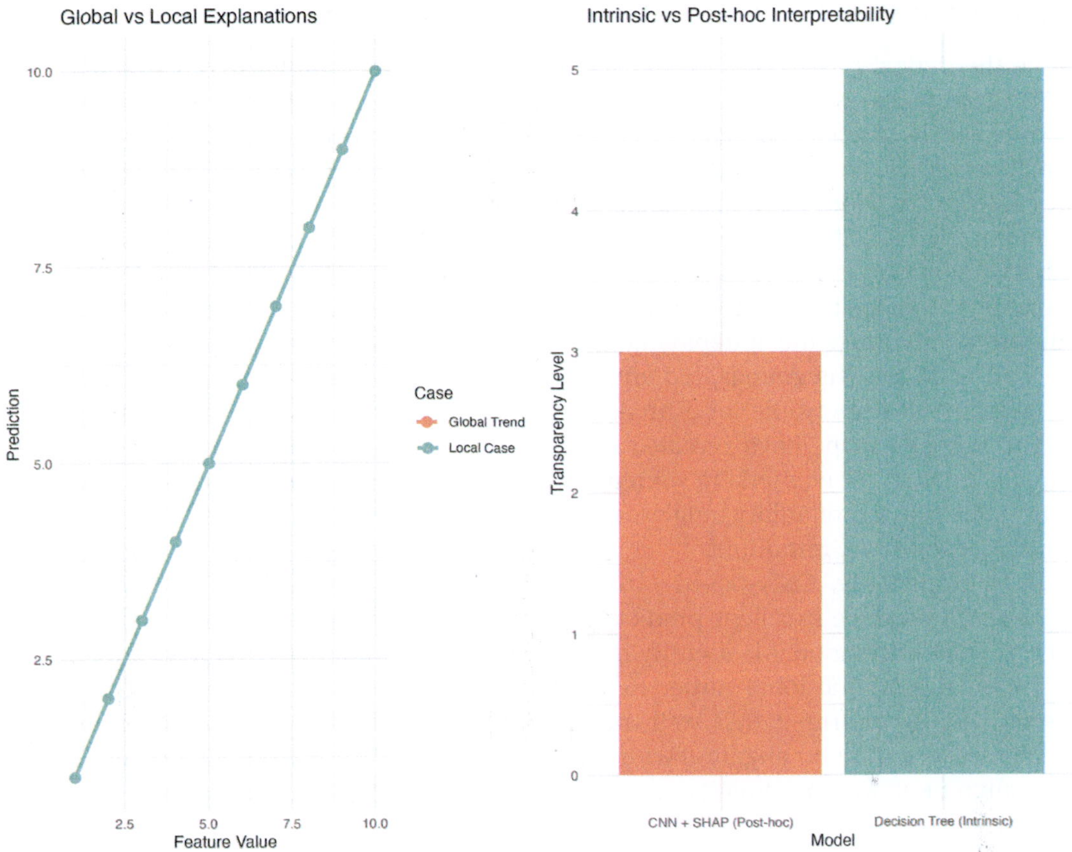

FIGURE 7.2 Overview of global versus local, intrinsic versus post-hoc explainability. Left: Global feature importance in genomic classification. Right: Local saliency map for tumor detection in MRI.

Table 7.1 Comparison of major XAI tools used in healthcare (LIME, SHAP, Grad-CAM, Integrated Gradients) with domains of application, strengths, and limitations.

Tool	Domain of application	Strengths
LIME (local interpretable model-agnostic explanations)	Imaging, EHR, genomics	Model-agnostic, intuitive local explanations
SHAP (SHapley additive exPlanations)	Genomics, multimodal, risk prediction	Consistent, theoretically grounded, patient-level and global insights
Grad-CAM (gradient-weighted class activation mapping)	Radiology, pathology	Visual heatmaps; intuitive for clinicians
Integrated gradients	Imaging, genomics	Theoretically principled attribution; baseline comparisons

7.3.1 Case study 1: interpreting CNNs in radiology

While theoretical frameworks and algorithms provide the scaffolding of explainable AI (XAI), it is in real-world healthcare applications that their value is most evident. The following case studies highlight how interpretability has been leveraged in clinical imaging, genomics, and multimodal prediction models, and why explainability is essential for clinician trust and adoption. Radiology has been one of the earliest and most fertile domains for DL. CNNs trained on large-scale imaging datasets such as chest radiographs, mammography, or brain MRI have achieved radiologist-level performance in detecting abnormalities (McKinney et al., 2020; Ardila et al., 2019). However, without interpretability, the clinical deployment of these models is limited.

Grad-CAM and Integrated Gradients have been widely used to visualize the regions in an image most responsible for a model's decision. For example, in chest X-ray classification, saliency maps have revealed whether predictions of pneumonia are based on lung fields or spurious markers such as embedded text or laterality tokens (Zech et al., 2018). Such interpretability analyses expose hidden confounders, enabling dataset curation and model retraining.

In neuroimaging, CNN-based Alzheimer's disease (AD) classifiers trained on MRI scans often produce excellent predictive accuracy. Yet interpretability tools are necessary to ensure the model is identifying disease-relevant brain regions (e.g., hippocampus, medial temporal lobe) rather than irrelevant structures. Studies using Grad-CAM overlays have confirmed that well-trained networks focus on established biomarkers of neurodegeneration, aligning with neuropathological knowledge (Li et al., 2020). This alignment fosters clinician trust and facilitates integration into early-diagnosis pipelines.

7.3.2 Genomics and omics data interpretation

Genomic and molecular data are highly dimensional and often opaque to clinicians without domain-specific interpretability. DL models such as deep sequence-based predictors or autoencoders can uncover hidden regulatory motifs and disease-associated expression profiles. However, without interpretable outputs, their clinical translation is limited.

SHAP has emerged as a central tool in genomics. For instance, SHAP analyses of DL models trained on single nucleotide polymorphism (SNP) data can highlight which variants most contribute to Alzheimer's disease or cancer risk prediction (Avsec et al., 2021). Importantly, such explanations can be cross-referenced against existing genome-wide association studies (GWAS) to validate biological plausibility.

In transcriptomics, explainability helps differentiate disease-driving pathways from noise. In breast cancer subtyping, saliency and attribution methods have been used to show that specific gene expression signatures (e.g., HER2 pathway genes) drive classification, enabling clinicians to connect predictions with actionable therapies (Koch et al.,

2021). Without such interpretability, the clinical utility of molecular models would remain uncertain.

7.3.3 Multimodal models: Alzheimer's disease and cancer

Healthcare increasingly involves multimodal AI, where imaging, genomics, laboratory values, and structured EHR data are integrated. Multimodal models often achieve higher predictive accuracy than unimodal approaches, but their complexity amplifies the black-box problem.

In Alzheimer's disease, multimodal models combining MRI, cerebrospinal fluid (CSF) biomarkers, and APOE genotype have shown promise for early prediction of progression from mild cognitive impairment to AD (2023). XAI tools are critical here: SHAP can quantify how much of a patient's predicted risk is driven by hippocampal atrophy versus CSF tau levels versus genetic risk. Such explanations allow neurologists to contextualize AI outputs alongside traditional clinical measures.

In oncology, multimodal DL models integrate histopathology slides, genomic alterations, and clinical variables to predict survival or therapy response. Explainability ensures that visual patterns in pathology (highlighted via Grad-CAM) and genomic drivers (revealed via SHAP) are interpretable. For example, in predicting immunotherapy response in melanoma, explainability has clarified how both PD-L1 expression and tumor-infiltrating lymphocytes contribute to predictions, aligning with known biological mechanisms (Fu et al., 2020).

7.3.4 Clinical trust and interpretability

These examples underscore that interpretability is not an academic luxury but a clinical necessity. Without clear explanations, clinicians are unlikely to adopt AI systems, regardless of performance metrics. Importantly, interpretability also reveals model weaknesses: if attribution maps consistently highlight irrelevant features (e.g., imaging artifacts, batch effects in omics), this signals the need for retraining or reevaluation.

Although NLP models applied to clinical text also face explainability challenges. Here, it suffices to note that structured and multimodal datasets pose distinct interpretability challenges—particularly in imaging and molecular data—that cannot be overlooked in translational AI.

Fig. 7.3.

Table 7.2.

7.4 Challenges and ethical considerations

As DL systems grow in sophistication, their integration into healthcare is constrained not only by technical limitations but also by ethical, social, and regulatory challenges. Even with interpretability tools, the clinical use of AI must address fundamental concerns

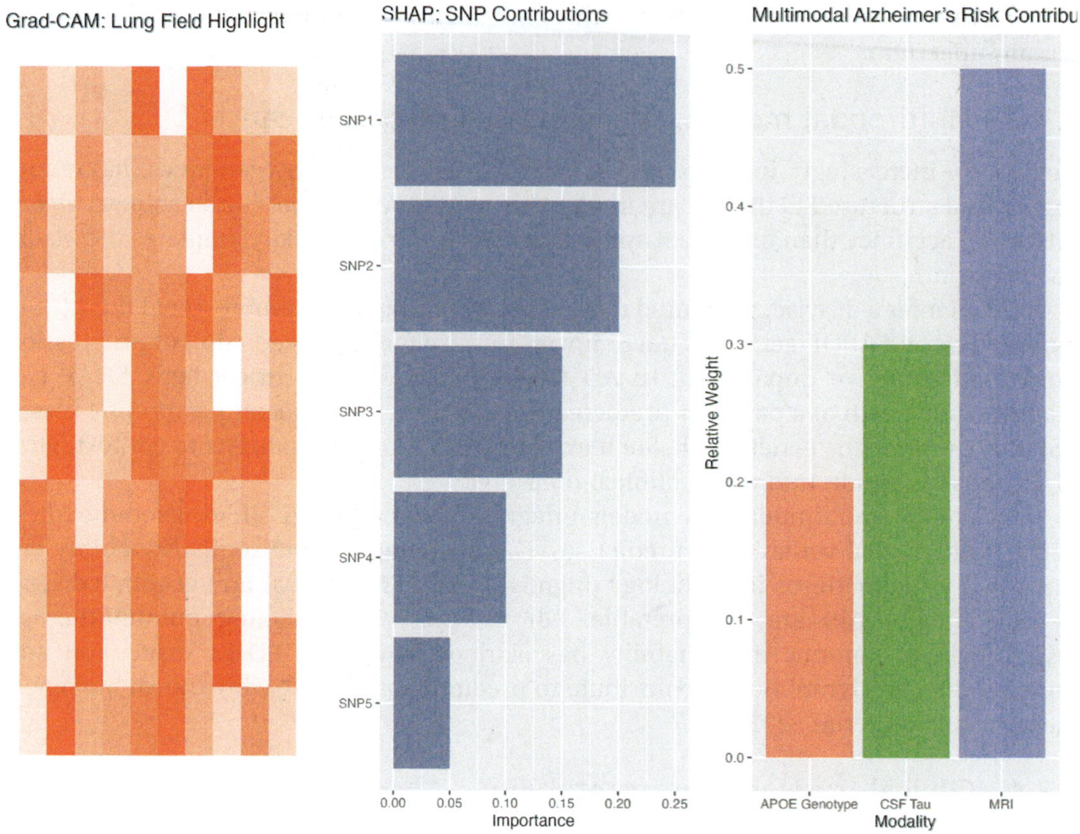

FIGURE 7.3 Case studies of interpretability in healthcare: (A) Grad-CAM highlighting lung fields in pneumonia classification, (B) SHAP plot ranking SNP contributions to Alzheimer's disease risk, (C) multimodal Alzheimer's prediction showing relative contributions of MRI, CSF, and APOE genotype.

Table 7.2 Summary of real-world applications of XAI in healthcare, with domains (imaging, genomics, multimodal, tools used, clinical impact, and key limitations.).

Domain	Tools used	Clinical impact	Key limitations
Imaging (radiology, oncology)	Grad-CAM, integrated gradients	Validates lesion detection, supports radiologist trust	May overfit to artifacts; poor generalization
Genomics (SNPs, expression)	SHAP, LIME	Highlights disease-associated variants and pathways	High-dimensionality complicates explanations
Multimodal (imaging + EHR + genomics)	SHAP, attention, multimodal attribution	Explains relative contributions across data types	Complexity; unstable across datasets

around trust, performance trade-offs, bias, and accountability. In high-stakes medical contexts, these challenges are not optional—they shape whether AI becomes a trusted clinical partner or a source of risk.

7.4.1 Overtrust and undertrust in AI

One ethical dilemma is the balance between overtrust and undertrust in AI systems. Overtrust occurs when clinicians accept AI outputs without sufficient scrutiny. For instance, if a DL system flags a benign lesion as malignant and the clinician defers entirely to the model, unnecessary procedures may result. Conversely, undertrust leads to clinicians ignoring accurate AI predictions, nullifying potential benefits. Both extremes are harmful. XAI aims to mitigate these risks by enabling clinicians to see why a model reached its conclusion. However, explanations themselves must be trustworthy—saliency maps that highlight irrelevant image regions may encourage misplaced confidence or skepticism (Jacobs, 2021).

7.4.2 Explainability versus model performance trade-offs

A recurring challenge is the trade-off between explainability and predictive performance. Intrinsically interpretable models (e.g., decision trees, linear regression) are transparent but often underperform on complex biomedical data such as MRI voxels or multi-omics matrices. Black-box DL models achieve superior accuracy but require post-hoc explanations that may be incomplete or unstable. Clinicians and regulators must navigate this tension: is it acceptable to sacrifice some accuracy for interpretability, or should black-box models be permitted if sufficient post-hoc explanations exist? The answer likely depends on context: a screening tool for diabetic retinopathy may tolerate less interpretability than an algorithm guiding chemotherapy decisions (Topol, 2019).

7.4.3 Dataset bias and hidden confounders

Bias in training datasets remains one of the most pervasive ethical risks. Medical datasets often reflect structural inequities: imaging cohorts may underrepresent minority populations, while genomic datasets such as UK Biobank are skewed toward European ancestry (Bycroft et al., 2018). DL systems trained on such data may inherit or amplify disparities, leading to poorer performance in underrepresented groups. Hidden confounders are equally problematic: models may learn spurious correlations such as hospital-specific imaging artifacts or batch effects in gene expression data (Zech et al., 2018). Explainability tools can expose some of these issues—for example, Grad-CAM revealing that a chest X-ray model focuses on embedded text rather than lung fields—but detecting and correcting bias remains an ongoing challenge.

7.4.4 Regulatory concerns

Healthcare AI operates within strict regulatory frameworks. The EU AI Act (2024) classifies most medical AI as "high-risk" and mandates transparency, documentation of training data, and post-market monitoring (European Commission. (2024). Regulation (EU) 2024/1689 of the European Parliament and of the Council on artificial intelligence (AI Act), 2024). In the United States, the FDA's guidance on AI/ML-enabled medical devices (2024, 2025) emphasizes lifecycle management, transparency, and human oversight (FDA. (2024). Guiding principles for transparency for machine learning–enabled medical devices. U.S. Food and Drug Administration., no date); (FDA. (2025). Draft guidance: Artificial intelligence-enabled device software functions — Lifecycle management and marketing submission recommendations. U.S. Food and Drug Administration., no date). Both frameworks highlight the importance of explainability, not only for safety but also for accountability. If a model misclassifies a tumor or incorrectly predicts Alzheimer's progression, who is responsible—the clinician, the hospital, or the AI developer? Explainability tools can clarify failure modes, but governance structures are still evolving.

7.4.5 Ethical imperatives

Beyond regulation, the ethical imperatives of medicine demand explainable AI. The principle of informed consent implies that patients should understand, at least broadly, how algorithmic predictions inform their care. Opaque models threaten this autonomy. Similarly, the principle of nonmaleficence—to do no harm—requires that AI systems be transparent enough to avoid avoidable errors. In genomics, for example, an uninterpretable risk prediction that leads to unnecessary genetic counseling or invasive testing could violate this principle. Equity is also a central concern: if AI models systematically underperform in disadvantaged groups due to biased data, they risk exacerbating health disparities rather than reducing them (Chen et al., 2021).

Fig. 7.4.

Table 7.3.

7.5 Integration into clinical practice

For explainable AI (XAI) to be impactful in healthcare, it must integrate seamlessly into clinical practice. Beyond methodological innovation, adoption depends on how AI systems are embedded in workflows, interact with clinicians, and support—not replace—human expertise. This section examines integration in structured and multimodal data contexts, explores human-in-the-loop systems, and highlights visualization tools that can bridge the gap between algorithmic output and clinical decision-making.

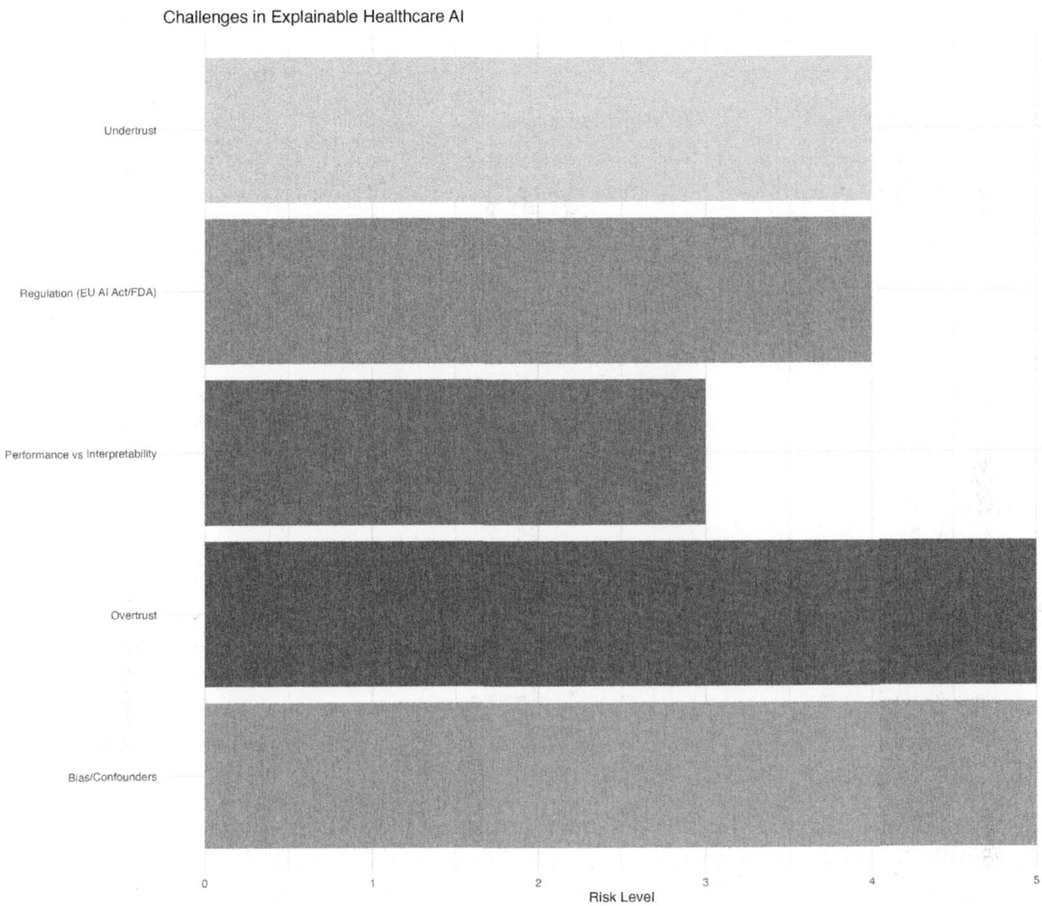

FIGURE 7.4 Ethical and regulatory landscape of XAI in healthcare. Left: Risks of overtrust versus undertrust. Middle: Trade-off between interpretability and performance. Right: Regulatory mandates (EU AI Act, FDA guidance) requiring explainability.

7.5.1 Structured EHR-based predictions

Electronic health records (EHRs) contain structured data such as laboratory test results, medication prescriptions, vital signs, and billing codes. These variables are increasingly used to train DL models for predicting patient outcomes such as hospital readmission, sepsis, or cardiovascular events (Rajkomar et al., 2018). However, clinicians are unlikely to adopt such models unless predictions are interpretable.

Model-agnostic tools such as LIME and SHAP have proven valuable for structured EHR data. For instance, a sepsis prediction model might highlight elevated lactate, tachycardia, and recent antibiotic prescriptions as major contributors to a high-risk score. By surfacing familiar clinical features, explainability aligns algorithmic reasoning with physician intuition, enhancing trust. Without such explanations,

Table 7.3 Challenges in interpretable healthcare AI: description, real-world example, XAI mitigation strategies, and unresolved gaps.

Challenge	Description	Real-world example	XAI mitigation strategy	Unresolved gaps
Overtrust	Clinicians may rely on flawed AI predictions	Pneumonia detection AI misclassifying X-rays	Uncertainty quantification, clinician training	Risk of automation bias remains
Undertrust	Clinicians ignore useful AI due to opacity	Oncologists rejecting risk scores	Visual aids, human-in-the-loop design	Cultural resistance, trust calibration
Dataset bias & confounding	Spurious correlations drive predictions	Skin lesion AI biased to lighter skin tones	Diverse datasets, fairness audits	Ensuring fairness across subgroups
Performance versus interpretability	High accuracy models often less interpretable	Black-box CNN outperforming interpretable models	Hybrid or surrogate models	Balancing trade-offs without loss of clinical utility
Regulation & compliance	Legal need for transparency	EU AI Act, FDA guidance	Model cards, transparent documentation	Standards for evaluation remain evolving

clinicians may view alerts as "black-box alarms" and ignore them, reducing effectiveness (Ghassemi et al., 2021).

7.5.2 Multimodal decision support

Modern medicine increasingly relies on multimodal data: imaging, genomics, and structured EHR variables combined to create a holistic patient profile. Alzheimer's disease research exemplifies this trend: models integrate MRI-derived hippocampal volume, cerebrospinal fluid tau, APOE genotype, and cognitive scores to predict disease progression (2023). Similarly, oncology pipelines combine histopathology slides, somatic mutation data, and treatment history to guide therapy choices.

Integration into clinical practice requires that multimodal AI outputs be disentangled and interpretable. Clinicians need to know not just the risk score, but how much each modality contributed. SHAP values, for instance, can partition contributions into "imaging-driven" versus "genomic-driven" explanations. This is critical for clinical translation: a neurologist may weigh CSF findings more heavily than genetics, while an oncologist may prioritize histopathology over transcriptomics. Without modality-specific explanations, multimodal models risk being inscrutable.

7.5.3 Human-in-the-loop systems

Human-in-the-loop (HITL) frameworks represent a promising path for XAI in practice. Rather than fully automated pipelines, HITL systems allow clinicians to interact with models, adjust parameters, and validate predictions. For example, in radiology, an AI model might propose suspicious regions on a CT scan; the radiologist can then confirm, reject, or annotate these suggestions, which are fed back into model retraining (Kelly et al., 2019). Similarly, in genomics, clinicians could review SHAP-ranked variant contributions, flagging implausible attributions for further investigation.

Such systems foster collaborative intelligence—combining algorithmic pattern recognition with human contextual judgment. Importantly, they also provide continuous feedback loops, ensuring that AI models evolve in alignment with real-world clinical reasoning.

7.5.4 Clinician trust and workflow integration

Successful integration is as much about user experience as technical rigor. Clinicians face time constraints and cognitive burdens; interpretability tools must present explanations concisely and intuitively. Visual dashboards that overlay heatmaps on imaging studies, or SHAP summary plots ranking key predictors, allow rapid comprehension. In contrast, overly technical or verbose explanations may alienate users.

Trust is also built through validation in clinical context. Prospective studies where XAI-augmented predictions demonstrably improve decision-making are essential. For example, a trial demonstrating that interpretable sepsis alerts reduce ICU mortality more effectively than black-box alarms would greatly strengthen adoption (Seymour et al., 2019). Without such real-world evidence, even elegant interpretability methods may remain confined to research.

7.5.5 Distinguishing from NLP applications

Although natural language processing (NLP) models are increasingly used to extract insights from unstructured clinical text, their interpretability challenges differ. Chapter 3 of this volume addresses biomedical NLP in detail. In this chapter, the focus remains on structured and multimodal data integration, where transparency in combining labs, imaging, and genomics is essential for clinical utility.

Fig. 7.5.

Table 7.4.

7.6 Emerging trends and future directions

Explainable AI (XAI) in healthcare continues to evolve rapidly. While current tools such as SHAP, LIME, and Grad-CAM have enabled progress, they remain limited in scope, stability, and alignment with clinical reasoning. Emerging research points to new paradigms that promise deeper interpretability and broader clinical impact.

7.6.1 Causality in XAI

One key direction is integrating causal inference with explainability. Most current XAI tools are correlational—they identify features associated with predictions but cannot

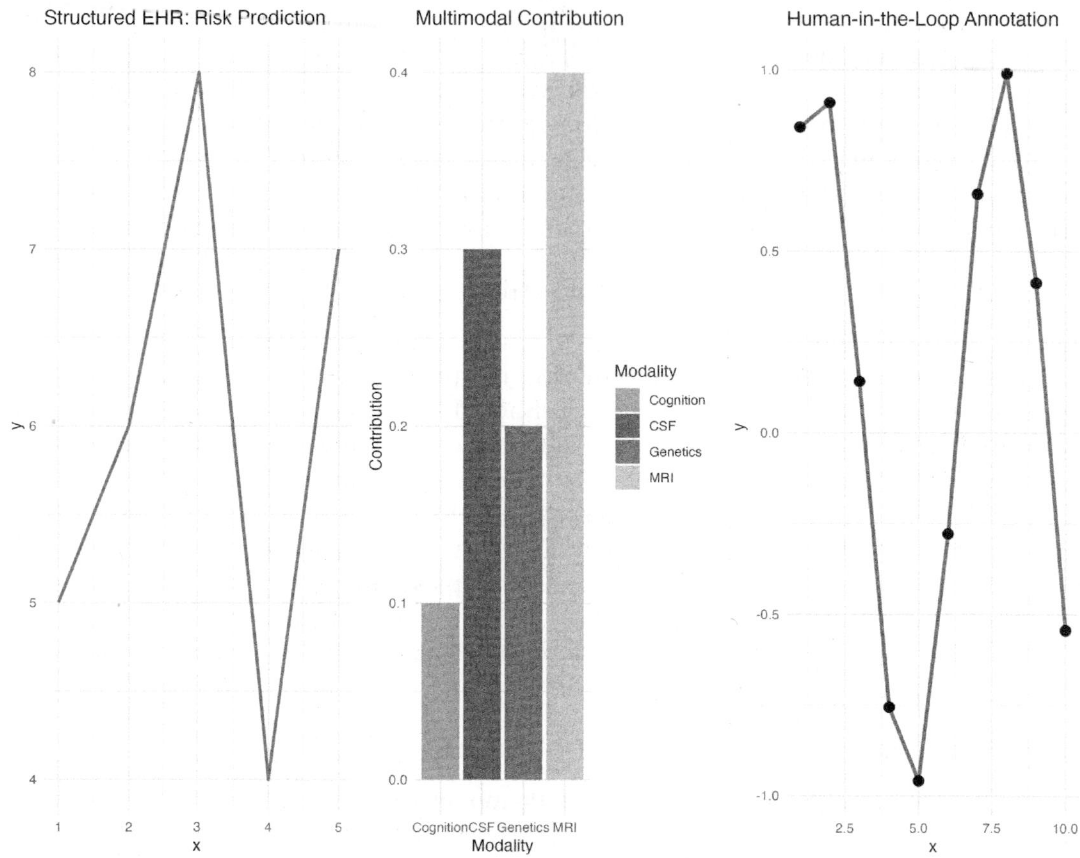

FIGURE 7.5 Integration of explainable AI into clinical practice. Left: SHAP explanation of a structured EHR-based sepsis model. Middle: Modality-specific contribution plot for multimodal Alzheimer's risk prediction. Right: Radiology HITL system showing AI-suggested lesion annotations validated by a clinician.

Table 7.4 Integration strategies for XAI in clinical practice: data modality, interpretability tool, clinician interaction, and example use case.

Data modality	Interpretability tool	Clinician interaction	Example use case
Imaging	Grad-CAM, integrated gradients	Radiologist reviews heatmaps	Chest X-ray AI with lesion highlighting
Genomics	SHAP	Geneticist examines variant contributions	SNP prioritization in cancer risk prediction
EHR (structured)	LIME, SHAP	Clinician reviews lab/vital contributions	Sepsis risk prediction from ICU data
Multimodal	SHAP + attention	Multidisciplinary team interprets contributions	Alzheimer's disease progression prediction

determine whether changing a feature would alter the outcome. In medicine, causality matters: lowering blood pressure should reduce cardiovascular risk, but a purely correlational model may misattribute importance to irrelevant markers. New frameworks that combine structural causal models with DL are being developed to generate causal attributions and counterfactual explanations (Pearl, 2019; Prosperi et al., 2020). These approaches could help clinicians distinguish between biomarkers that merely correlate with disease and those that are actionable.

7.6.2 Interpretable graph neural networks and attention

Another frontier is applying graph neural networks (GNNs) and attention mechanisms in an interpretable way. Biological systems—gene regulatory networks, protein–protein interactions, multimodal patient similarity graphs—are inherently relational. GNNs can model these structures, and explainability methods such as graph attention weights or node importance scoring can reveal which genes, pathways, or patient subgroups drive predictions (Ying et al., 2019). Similarly, attention mechanisms in transformer models offer intuitive explanations by highlighting critical input regions, whether in imaging patches (vision transformers) or multimodal sequences. While not always faithful, attention has proven useful for clinical interpretation.

7.6.3 Counterfactual explanations and visualizations

Counterfactual explanations—statements of the form "if this feature were different, the prediction would change"—are gaining traction in healthcare. For example, a counterfactual explanation in genomics might suggest that if a patient lacked a particular SNP, their predicted risk of Alzheimer's would be reduced. In imaging, counterfactuals can simulate how a tumor margin would need to change for classification to switch from malignant to benign. These approaches are clinically meaningful because they align with "what-if" reasoning familiar to physicians. Visualization techniques, including realistic generative counterfactuals from GANs or diffusion models, are advancing the plausibility of such scenarios (Karimi et al., 2020).

7.6.4 Personalized and patient-centric XAI

Personalized medicine requires individualized explanations. Instead of global feature rankings, clinicians need case-specific rationales that explain why this patient has a high predicted risk and what modifiable factors are involved. Patient-centric XAI also includes designing explanations that are understandable to non-experts. For instance, visualizations that highlight brain atrophy on MRI alongside clinical biomarkers can help patients and families grasp AI-driven dementia predictions, supporting informed decision-making.

7.6.5 Benchmarking and standardization

Fig. 7.6.

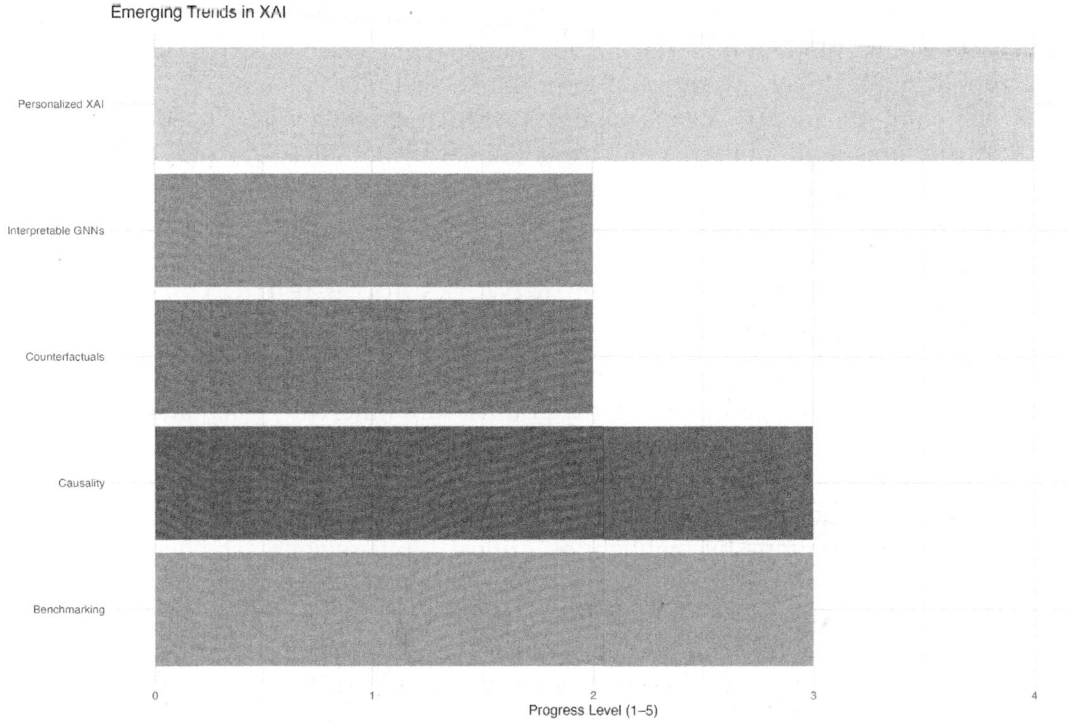

FIGURE 7.6 Emerging XAI paradigms in healthcare: (A) Causal inference with counterfactuals, (B) interpretable GNNs modeling patient networks, (C) personalized explanations for multimodal Alzheimer's prediction.

Finally, the field is moving toward benchmarking and standardization. Current XAI evaluations often rely on subjective human judgment or case studies. Initiatives such as the Explainable AI Consortium and community challenges (e.g., MICCAI XAI Challenge, PhysioNet Challenges) are creating shared benchmarks for evaluating explanation faithfulness, stability, and clinical utility (Holzinger et al., 2022). Standardized metrics will be essential for regulatory approval and for comparing tools across domains such as imaging, genomics, and multimodal AI.

Table 7.5.

7.7 Conclusion

The integration of DL into healthcare has brought unprecedented opportunities for earlier diagnosis, more accurate prognosis, and personalized treatment. Yet, the black-box nature of many high-performing models continues to challenge clinical adoption. This chapter has shown that interpretability and explainability are not optional in medicine—they are central to patient safety, clinician trust, regulatory compliance, and ethical responsibility.

Table 7.5 Summary of emerging trends in explainable healthcare AI: methods, applications, clinical relevance, and open challenges.

Method/Trend	Application	Clinical relevance	Open challenges
Causality in XAI	Identify causal biomarkers, treatment effects	Supports actionable and mechanistic insights	Data requirements, causal discovery under uncertainty
Interpretable graph neural nets	Patient networks, gene/protein interaction graphs	Enables pathway-level reasoning	Scalability, interpretability metrics
Counterfactual explanations	"What-if" reasoning for individual patients	Clinician reasoning, treatment simulation	Ensuring plausibility and clinical safety
Personalized explanations	Tailored outputs at patient level	Aligns with precision medicine	Complexity, validation burden
Benchmarking & standards	Comparing XAI methods across domains	Promotes reproducibility and regulatory adoption	Lack of consensus on evaluation metrics

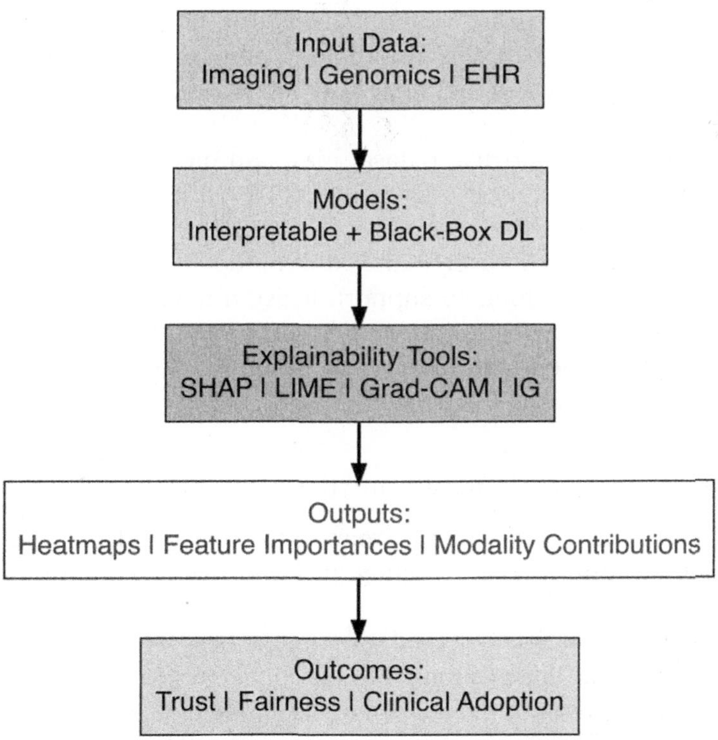

FIGURE 7.7 Roadmap for explainable AI in healthcare. The figure should depict a flowchart: • inputs: Multimodal data (imaging, genomics, EHR).• models: Interpretable (GAMs, decision trees) and black-box DL (CNNs, transformers).• explainability tools: SHAP, LIME, Grad-CAM, integrated gradients, counterfactuals.• outputs: Clinician-facing explanations (heatmaps, feature importance, modality contributions).• outcomes: Trust, adoption, fairness, regulatory compliance, improved patient care.

Table 7.6 Key Takeaways for clinicians using explainable AI in healthcare.

Domain	XAI methods	Clinical value	Pitfalls
Radiology	Grad-CAM, integrated gradients	Highlights lesions and abnormal regions	May highlight irrelevant artifacts
Genomics	SHAP, feature attribution	Identifies key variants and pathways	High dimensionality complicates trust
Multimodal AI	Attention, multimodal SHAP	Explains modality contributions to risk	Complex integration can cause instability
EHR (structured)	LIME, SHAP	Explains labs and meds driving prediction	EHR shifts affect reliability
Ethics/regulation	Documentation, oversight	Ensures fairness, transparency, compliance	Accountability and standardization remain open

Key insights

- Interpretability versus Explainability: Intrinsic interpretability offers clarity but limited scalability, while post-hoc tools extend transparency to black-box models. Both are necessary in practice.
- Healthcare Domains: Case studies in radiology, genomics, and multimodal Alzheimer's/cancer prediction highlight how XAI transforms outputs into clinically actionable insights.
- Challenges: Overtrust, undertrust, dataset bias, and hidden confounders remain pressing obstacles. Regulatory frameworks (EU AI Act, FDA guidelines) mandate explainability as a safeguard.
- Clinical Integration: Structured EHR and multimodal models require HITL systems and intuitive visualization tools to support decision-making.
- Emerging Trends: Causality, counterfactual reasoning, interpretable GNNs, and patient-centric explanations are shaping the future of trustworthy AI.

The path forward

For XAI to achieve its promise in healthcare, three priorities stand out:

1. Clinical alignment: Explanations must map onto concepts that clinicians use in practice—brain atrophy in neuroimaging, pathway dysregulation in genomics, or lab abnormalities in EHR.
2. Evaluation and benchmarks: The field must move beyond technical metrics and demonstrate real-world improvements in outcomes, trust, and equity.
3. Interdisciplinary collaboration: Data scientists, clinicians, ethicists, and regulators must co-design models and explanations to ensure utility, fairness, and accountability.

In neurodegenerative disease research, oncology, and beyond, interpretable AI can shift from opaque prediction to transparent partnership. The future of AI in medicine will not be measured solely by accuracy, but by whether patients and clinicians can understand, trust, and act upon its outputs (Fig. 7.7; Table 7.6).

References

Ardila, D., Kiraly, A.P., Bharadwaj, S., Choi, B., Reicher, J.J., Peng, L., Tse, D., Etemadi, M., Ye, W., Corrado, G., Naidich, D.P., Shetty, S., 2019. End-to-end lung cancer screening with three-dimensional deep learning on low-dose chest computed tomography. Nat. Med. 25 (6), 954–961. https://doi.org/10.1038/s41591-019-0447-x.

Avsec, Ž., Agarwal, V., Visentin, D., Ledsam, J.R., Grabska-Barwinska, A., Taylor, K.R., Assael, Y., Jumper, J., Kohli, P., Kelley, D.R., 2021. Effective gene expression prediction from sequence by integrating long-range interactions. Nat. Methods 18 (10), 1196–1203. https://doi.org/10.1038/s41592-021-01252-x.

Bycroft, C., Freeman, C., Petkova, D., Band, G., Elliott, L.T., Sharp, K., Motyer, A., Vukcevic, D., Delaneau, O., O'Connell, J., Cortes, A., Welsh, S., Young, A., Effingham, M., McVean, G., Leslie, S., Allen, N., Donnelly, P., Marchini, J., 2018. The UK Biobank resource with deep phenotyping and genomic data. Nature 562 (7726), 203–209. https://doi.org/10.1038/s41586-018-0579-z.

Chen, I.Y., Joshi, S., Ghassemi, M., 2021. Treating health disparities with artificial intelligence. Nat. Med. 27, 1905–1906. https://doi.org/10.1038/s41591-021-01535-0.

F. Doshi-Velez, B. Kim, Towards a rigorous science of interpretable machine learning. (2017), Available from: https://arxiv.org/abs/1702.08608.

European Commission. (2024). Regulation (EU) 2024/1689 of the European parliament and of the council on artificial intelligence (AI Act). (2024), Available from: https://eur-lex.europa.eu.

FDA, 2024. Guiding Principles for Transparency for Machine Learning–Enabled Medical Devices. U.S. Food and Drug Administration. Available from. https://www.fda.gov.

FDA, 2025. Draft Guidance: Artificial Intelligence-Enabled Device Software Functions — Lifecycle Management and Marketing Submission Recommendations. U.S. Food and Drug Administration. Available from. https://www.fda.gov.

Fu, J., Li, K., Zhang, W., Wan, C., Zhang, J., Jiang, P., Liu, X.S., 2020. Large-scale public data reuse to model immunotherapy response and resistance. Genome Med. 12 (1). https://doi.org/10.1186/s13073-020-0721-z.

Ghassemi, M., Oakden-Rayner, L., Beam, A.L., 2021. The false hope of current approaches to explainable artificial intelligence in health care. Lancet Digit. Health 3 (11), e745. https://doi.org/10.1016/S2589-7500(21)00208-9.

Holzinger, A., Goebel, R., Ferri, C., Palade, V., Müller, H., 2022. Towards the evaluation of explainable artificial intelligence (XAI) in healthcare. npj Digit. Med. 5, 93. https://doi.org/10.1038/s41746-022-00676-5.

Jacobs, A.Z., 2021. Measurement and fairness. In: FAccT 2021 - Proceedings of the 2021 ACM Conference on Fairness, Accountability, and Transparency. Association for Computing Machinery, Inc, United States, pp. 375–385. https://doi.org/10.1145/3442188.3445901. http://dl.acm.org/citation.cfm?id=3442188.

Karimi, A.H., Barthe, G., Balle, B., Valera, I., 2020. Model-Agnostic counterfactual explanations for consequential decisions. Proceedings of Machine Learning Research 108, 895–905. In: https://proceedings.mlr.press/.

Kelly, C.J., Karthikesalingam, A., Suleyman, M., Corrado, G., King, D., 2019. Key challenges for delivering clinical impact with artificial intelligence. BMC Med. 17 (1). https://doi.org/10.1186/s12916-019-1426-2.

Koch, L.M., Kampichler, J., Fehringer, G., Krall, A., Fuchs, S., Mair, A., 2021. Explainable deep learning reveals novel gene expression signatures in breast cancer subtypes. BMC Bioinf. 22 (1), 341. https://doi.org/10.1186/s12859-021-04239-6.

Li, H., Habes, M., Wolk, D.A., Fan, Y., 2020. A deep learning model for early prediction of Alzheimer's disease dementia based on hippocampal MRI. Alzheimer's Dement. 16 (11), 1581–1592. https://doi.org/10.1002/alz.12142.

Lundberg, S.M., Lee, S.I., 2017. A unified approach to interpreting model predictions. Adv. Neural Inf. Process. Syst. 2017, 4766–4775.

McKinney, S.M., Sieniek, M., Godbole, V., Godwin, J., Antropova, N., Ashrafian, H., Back, T., Chesus, M., Corrado, G.C., Darzi, A., Etemadi, M., Garcia-Vicente, F., Gilbert, F.J., Halling-Brown, M., Hassabis, D., Jansen, S., Karthikesalingam, A., Kelly, C.J., King, D., Ledsam, J.R., Melnick, D., Mostofi, H., Peng, L., Reicher, J.J., Romera-Paredes, B., Sidebottom, R., Suleyman, M., Tse, D., Young, K.C., De Fauw, J., Shetty, S., 2020. International evaluation of an AI system for breast cancer screening. Nature 577 (7788), 89–94. https://doi.org/10.1038/s41586-019-1799-6.

Millar, P.R., Vogel, J.W., Thomas, R.A., Insel, P.S., Hansson, O., Alzheimer's Disease Neuroimaging Initiative, 2023. Multimodal brain age estimates relate to Alzheimer's disease biomarkers and cognitive decline. Brain 146 (2), 706–718. https://doi.org/10.1093/brain/awac460.

C. Molnar, Interpretable machine learning: a guide for making black box models explainable . (2022).

J. Pearl, The book of why: the new science of cause and effect . (2019).

Prosperi, M., Guo, Y., Sperrin, M., Koopman, J.S., Min, J.S., He, X., Rich, S., Wang, M., Buchan, I.E., Bian, J., 2020. Causal inference and counterfactual prediction in machine learning for actionable healthcare. Nat. Mach. Intell. 2 (7), 369–375. https://doi.org/10.1038/s42256-020-0197-y.

Rajkomar, A., Oren, E., Chen, K., Dai, A.M., Hajaj, N., Hardt, M., Liu, P.J., Liu, X., Marcus, J., Sun, M., Sundberg, P., Yee, H., Zhang, K., Zhang, Y., Flores, G., Duggan, G.E., Irvine, J., Le, Q., Litsch, K., Mossin, A., Tansuwan, J., Wang, D., Wexler, J., Wilson, J., Ludwig, D., Volchenboum, S.L., Chou, K., Pearson, M., Madabushi, S., Shah, N.H., Butte, A.J., Howell, M.D., Cui, C., Corrado, G.S., Dean, J., 2018. Scalable and accurate deep learning with electronic health records. npj Digit. Med. 1 (1). https://doi.org/10.1038/s41746-018-0029-1.

Ribeiro, M.T., Singh, S., Guestrin, C., 2016. "Why should i trust you?" Explaining the predictions of any classifier. Proceedings of the ACM SIGKDD International Conference on Knowledge Discovery and Data Mining 13–17, 1135–1144. https://doi.org/10.1145/2939672.2939778.

Selvaraju, R.R., Cogswell, M., Das, A., Batra D., V.R.P.D., 2017. Grad-CAM: visual explanations from deep networks via gradient-based localization. Proceedings of the IEEE International Conference on Computer Vision 2017, 618–626. https://doi.org/10.1109/ICCV.2017.74.

Seymour, C.W., Kennedy, J.N., Wang, S., Chang, C.C.H., Elliott, C.F., Xu, Z.B.S.C.G., Cooper, G., Gomez, H., Huang, D.T., Kellum, J.A., Mi, Q., Opal, S.M., Talisa, V., Van Der Poll, T., Visweswaran, S., Vodovotz, Y., Weiss, J.C., Yealy, D.M., Yende, S., Angus, D.C., 2019. Validation, and potential treatment implications of novel clinical phenotypes for sepsis. JAMA, J. Am. Med. Assoc. 321 (20), 2003–2017. https://doi.org/10.1001/jama.2019.5791.

Sundararajan, M., Taly, A., Yan, Q., 2017. Axiomatic attribution for deep networks. In: 34th International Conference on Machine Learning, ICML 2017. International Machine Learning Society (IMLS), United States, pp. 5109–5118.

E.J. Topol, Deep medicine: how artificial intelligence can make healthcare human again. (2019).

Ying, R., Bourgeois, D., You, J., Zitnik, M., Leskovec, J., 2019. Neural information processing systems foundation United States GNNExplainer: generating explanations for graph neural networks. Adv. Neural Inf. Process. Syst. 32. https://papers.nips.cc/.

Zech, J.R., Badgeley, M.A., Liu, M., Costa, A.B., Titano, J.J., Oermann, E.K., 2018. Variable generalization performance of a deep learning model to detect pneumonia in chest radiographs: a cross-sectional study. PLoS Med. 15 (11). https://doi.org/10.1371/journal.pmed.1002683.

8

Using hybrid models in healthcare data

Priyanka Mishra[1] and Siddhant Panigrahi[2]

[1]CENTRE OF EXCELLENCE FOR MICROBIOME RESEARCH (CCMR) LAB, DEPARTMENT OF BIOCHEMISTRY, ALL INDIA INSTITUTE OF MEDICAL SCIENCE, BHUBANESWAR, ODISHA, INDIA; [2]ROBOTICS LAB, DEPARTMENT OF ENGINEERING DESIGN, INDIAN INSTITUTE OF TECHNOLOGY MADRAS, CHENNAI, TAMIL NADU, INDIA

What you will learn

Understand the foundational role of descriptive and inferential statistics as well as their limitations in handling high-dimensional data necessitating the move toward machine learning.

Gain a working knowledge of supervised and unsupervised learning techniques, including Support Vector Machine for classification and K-Means for learning in unlabeled datasets.

Learn how to design and interpret hybrid models that fuse disparate data types, such as combining multi-omics data with medical imaging and real-time physiological signals, to improve diagnostic accuracy.

Identify specific roles of various neural networks (CNN, RNN, LSTMs, and GNN).

Critically analyze the performance trade-offs between Classical, ML, and Deep Learning models.

Explore emerging shifts in the biomedical landscape and how hybrid models are shaping the future of personalized healthcare and predictive analytics.

8.1 Overview of existing methods

Biomedical engineering combines quantitative reasoning with biological complexity to develop diagnostic, therapeutic, and analytical systems that enhance health outcomes. Historically, inferential statistics, such as regression, analysis of variance (ANOVA), and hypothesis testing, have formed the analytical backbone of biomedical research (Montgomery and Runger, 2018). Yet as data volume and dimensionality expanded through high-throughput sequencing, advanced imaging, and sensor networks, the linear assumptions of these models proved insufficient (Hastie et al., 2017). To visualize the shift,

Applying Artificial Intelligence to Computational Biology and Medical Informatics. https://doi.org/10.1016/B978-0-443-44915-4.00008-3

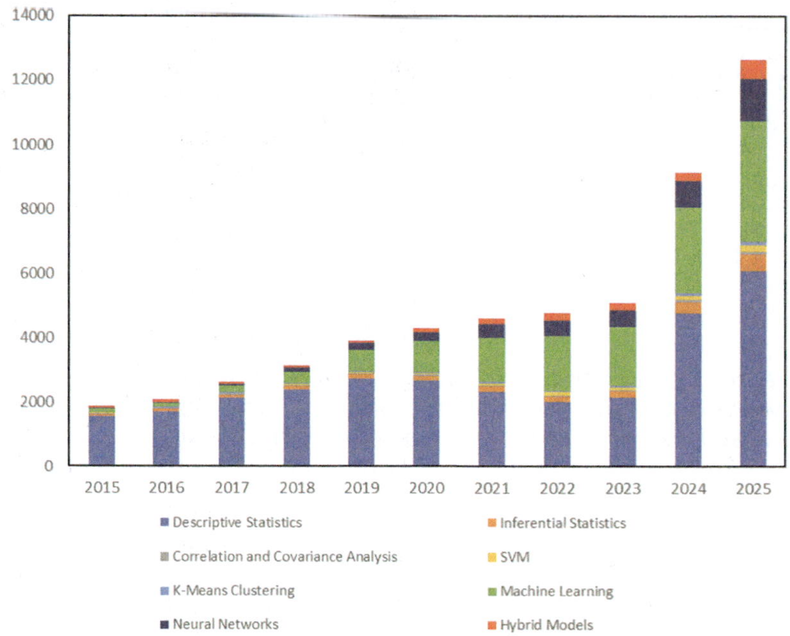

FIGURE 8.1 Trends in methodological application (2015–25).

the number of research studies utilizing various analytical approaches is quantified and plotted in the accompanying stacked bar chart (Fig. 8.1). This figure highlights a transition from traditional statistical methods to advanced computational modeling over the last decade. While descriptive statistics consistently serve as the foundational methodology, maintaining the highest volume throughout the decade, the most striking trend is the exponential rise in Machine Learning and Neural Networks starting from 2021. This shift led to the rise of ML-enabled algorithms to recognize nonlinear patterns and interactions in complex data. Methods such as Support Vector Machines (SVMs) and K-Means Clustering bridged the gap between statistical inference and pattern recognition. While the total frequency of applied methods grew steadily between 2013 and 2023, 2024 and 2025 mark a significant inflexion point, with total counts more than doubling. This surge is primarily driven by the machine learning and descriptive statistics segments, which together dominate the 2025 peak of over 12,000 instances. In contrast, specialized techniques such as SVMs and K-Means Clustering show stable, marginal growth. Later, deep neural networks extended this capability to understand unstructured data, including images, audio, and free text records (Topol, 2019).

Today's hybrid models, which combine statistical, mechanistic, and data-driven components, represent the next step, uniting interpretability with predictive power (Panch et al., 2018). This chapter surveys these developments, their biomedical applications, and the challenges that remain in making hybrid analytics contribute to routine clinical tools.

8.2 Statistical methods used in biomedical engineering

Statistical methods have been integral to biomedical engineering since the discipline's inception. They allow researchers to quantify biological variability, establish cause-and-effect relationships, and test hypotheses about health outcomes. From clinical trial analysis to bio-signal processing, statistical reasoning ensures that observed effects are significant and reproducible (Fig. 8.2).

8.2.1 Descriptive statistics

The first step in biomedical data analysis involves summarizing datasets using measures such as mean, median, standard deviation, variance, and interquartile range. For example, in analyzing blood glucose levels across patient cohorts, descriptive statistics quickly convey central tendencies and dispersion, which are crucial for quality control and cohort matching.

8.2.2 Inferential statistics

Inferential statistics form the foundation for concluding a larger population based on observations from a limited sample. The primary goal is to determine if observed patterns (differences or associations) are statistically significant, and whether they are likely to reflect an actual effect in the population or are merely due to random chance or sampling variability, often quantified by a *P*-value (Rosner, 2016).

8.2.2.1 T-tests

The *t*-test is used to compare the means of exactly two groups, typically under the assumption that the data are approximately normally distributed. It is calculated as t-statistic by dividing the difference between the group means by the standard error of the difference. A larger *t*-statistic suggests a more negligible probability that the observed difference is due to chance.

Application: In clinical research, this is critical for assessing the efficacy of an intervention.

Example: A paired *t*-test would compare a patient's blood pressure measured before drug administration with the same patient's blood pressure after administration (pre- and post-intervention) to depict the drug interaction and effects on the patient.

8.2.2.2 Analysis of variance (ANOVA)

ANOVA extends the *t*-test to compare the means of three or more independent groups. This is performed by computing the F-statistic which evaluates a ratio of the variance between the groups to the variance within the groups. A higher F-statistic suggests that the groups are genuinely different from each other, rather than just varying by random chance.

Application: It is frequently used to determine whether a single variable (for instance, biomarker concentration) varies significantly across different stages or categories of a disease.

Example: Assessing if the average level of a specific inflammatory cytokine differs across early stage cancer, late-stage cancer, and healthy controls.

8.2.2.3 Chi-square tests

Chi-square test assesses the association (dependence) between two or more categorical variables. It compares the observed frequencies in a contingency table with the expected frequencies (the frequencies that would occur if there were no association between the variables). A significant chi-square value indicates that the observed data deviates significantly from the expected data, suggesting a statistically significant relationship.

Application: This is a fundamental tool in genetic epidemiology and risk assessment.

Example: The Chi-square test can be used to determine whether a specific gene polymorphism is associated with a clinical outcome.

8.2.2.4 Regression models

Regression analysis is used to model the predictive relationship between one or more independent variables (predictors or features, X) and a dependent variable (outcome, Y). It is used to express the analytical relationship between the cause and the effect to understand the critical input variables resulting in a linear or nonlinear formulation on the outcome.

8.2.2.4.1 Linear regression

Linear regression models the continuous outcome as a linear function of the predictors. The model finds the best-fitting straight line that minimizes the Sum of Squared Errors (SSE) between the predictive values and the actual observed values. The resulting coefficients indicate the magnitude and direction of the change in Y (independent variable) associated with a one-unit change in X (dependent variable).

Application: Used for quantifying the influence of multiple clinical factors on a quantitative outcome.

Example: Modeling systolic blood pressure (continuous outcome) as a function of Body Mass Index (BMI), age, and salt intake (as independent variables).

8.2.2.4.2 Logistic regression

Logistic regression is the primary statistical model used when the outcome variable is binary. Instead of predicting the outcome directly, it predicts the log odds (the logarithm of the odds of the event occurring) as a linear combination of the predictors. This log-odd is then transformed using the logistic function (or sigmoid function) to yield a probability that always falls between 0 and 1 (Rajkomar et al., 2019).

Application: Essential for predicting the probability of a clinical event or diagnosis.

Example: Modeling the presence/absence of diabetes (binary outcome) based on age, BMI, and fasting glucose levels.

8.2.3 Correlation and covariance analysis

Correlation analysis measures the strength and direction of a linear relationship between two random variables. A high correlation coefficient (close to $+1$ or -1) indicates that the signals are strongly related. A high positive correlation in the EEG (electroencephalogram) brain signals between two distant brain regions implies functional connectivity; which implies that these regions are likely part of a shared neural network working in tandem.

Application in Neuroimaging: In functional magnetic resonance imaging (fMRI) or electroencephalography (EEG) studies, the brain is partitioned into multiple regions of interest (ROIs). Correlation matrices are constructed by calculating the pairwise Pearson correlation coefficient between the activities (time series signal) of every pair of ROIs (Friston, 2011).

8.2.4 Principal Component Analysis

Principal Component Analysis (PCA) is a classic, unsupervised statistical technique for dimensionality reduction. It is often the first step in analyzing complex biological data. Data from modern omics technologies can involve tens and thousands of features (genes or metabolites) that are often highly correlated as well as redundant. Training models on this high-dimensional, noisy data leads to overfitting and computational intractability. PCA works by finding a new set of orthogonal (uncorrelated) variables called Principal Components (PCs). The first PC is the single direction in the original data space that accounts for the maximum possible variance. The second PC accounts for the maximum remaining variance, being orthogonal to the first PC, and so on.

Application: PCA allows researchers to compress 20,000 gene expression features into a handful of PCs that retain 90% or more of the original biological variability, effectively summarizing the data's underlying structure (Jolliffe and Cadima, 2016). These PCs can then be used as input for subsequent machine learning models, greatly enhancing their stability and efficiency.

8.2.5 Limitations of classical statistics

Traditional models assume linear relationships, normal data distributions, and independence between predictors. However, biological systems exhibit nonlinearity (e.g., enzyme saturation kinetics), multicollinearity (e.g., gene coexpression), and feedback loops (e.g., hormone regulation). Furthermore, real-world biomedical data are noisy, incomplete, and multidimensional (Bishop, 2006). These constraints have led to the rise of ML techniques capable of capturing complex, nonlinear dependencies.

8.2.6 Limitations of statistical methods

8.2.6.1 *Why do classical statistics struggle?*

Classical statistical methods (such as linear regression and ANOVA) operate under the assumption of a parametric model, where the underlying data generation process can be described by a known functional form with a finite number of parameters (e.g., a straight line or a Gaussian distribution). When biomedical reality deviates from these assumptions, these methods struggle to provide accurate and generalizable results (Lu et al., 2021).

Nonlinearity: The majority of biological phenomena are nonlinear. The relationship between physiological variables or molecular concentrations is rarely additive or proportional. Classical linear models assume that the effect on an outcome Y is a simple sum of the impacts of predictors X_i, as expressed in Eq. (8.1) (Montgomery and Runger, 2018):

$$Y = \beta_0 + \sum \beta_i X_i \tag{8.1}$$

It represents a multiple linear regression model. In this framework, Y denotes the dependent variable or clinical outcome, and β_0 represents the y-intercept, or the baseline value when all predictors are zero. The term $\sum \beta_i X_i$ signifies the summation of each independent predictor X_i multiplied by its respective coefficient β_i, which quantifies the individual contribution of that variable to the final result. While this model offers high interpretability, its additive nature inherently struggles to capture the complex, nonlinear interactions typically found in high-dimensional biological datasets.

For example, the dose-response curve of a drug or the relationship between the heart rate variability and stress level often follows a sigmoid or exponential curve, rendering the constant beta parameter inaccurate for prediction. ML models, such as SVMs, are designed to implicitly model these complex, nonlinear dependencies, allowing for a more accurate representation of biological systems.

High Dimensionality: This is considered one of the critical challenges in modern omics and biomedical data, where the number of features or variables (p) vastly exceeds the number of observations or samples (n). In genomics, p (e.g., 20,000 genes) is much greater than n (e.g., 100 patient samples). Classical parametric models, such as linear regression, become unidentifiable and highly prone to overfitting. Overfitting occurs because the model finds an overly complex solution that fits the noise and neglects the majority of the data, focusing instead on a sample dataset or outliers. This results in zero generalization when tested on unseen data. ML methods are inherently built to handle high feature spaces, often through sophisticated regularization techniques that penalize model complexity.

Interaction Effects: Biological effects often result from the multiplicative interaction of multiple factors, rather than the sum of their individual contributions. Consider two genes, for instance, G_1 and G_2. Classical statistics may assume that their effects are additive. However, a disease risk might only manifest when a patient has a specific G_1

allele and is exposed to a particular environmental factor E. This is a high-order, multiplicative interaction. Parametric models often require these interactions to be explicitly defined and added ($\beta_{12}G_1G_2$), which is infeasible given the vast number of potential interactions. Nonparametric models with intensive learning can automatically discover these complex, implicit interaction terms.

Generalization: The reliance on a fixed, parametric structure severely limits classical models' ability to adapt to patient heterogeneity. Traditional models assume that the underlying relationships are constant across the population. When applied to highly diverse clinical data, such as data spanning across different demographics or disease subtypes (population heterogeneity), the model performs poorly because its fixed structure lacks the adaptability required to capture the varying decision boundaries across subgroups (Lu et al., 2021). Nonparametric models are, by definition, more flexible, allowing them to generalize better to unseen data that differs slightly from the training set.

8.2.7 Scope of SVM and K-means clustering

8.2.7.1 Support vector machines

SVMs are supervised learning models that overcome the limitations of linearity and high dimensionality through a unique approach rooted in geometric optimization and the kernel trick (Noble, 2006).

Working Mechanism:

The fundamental objective of an SVM is to find the optimal separating hyperplane that separates data points into distinct classes.

1. The Hyperplane: For N features, the data is visualized in an N-dimensional space. The hyperplane is a decision boundary that best divides the classes.
2. Support Vectors: The SVM identifies the data closest to this hyperplane. These are called the support vectors. They are the most crucial, difficult-to-classify samples, and they alone define the position and orientation of the optimal hyperplane.
3. Maximum Margin: Here, SVM is engineered to choose the hyperplane that maximizes the margin, the perpendicular distance between the hyperplane and the nearest support vectors of each class. A wider margin directly translates into superior generalization and a reduced risk of overfitting, making SVMs particularly effective in the $p >> n$ scenarios common in genomics, where p denotes the number of parameters and n denotes the number of samples (Guyon et al., 2002).
4. Regularization: The degree of overfitting is controlled by a regularization parameter (C). A more minor C allows for a wider margin but results in more training errors, thereby improving robustness to noise.

The Kernel Trick and Nonlinearity

SVM's most significant power lies in its ability to handle nonlinear separation without explicitly performing a computationally expensive transformation of the entire dataset.

Kernels (for instance, Radial Basis Function Polynomial (RBF) implicitly map the original input features into a much higher-dimensional space (the Kernel Space). In this higher space, data points that were nonlinearly separable in the original space become linearly separable by the hyperplane. This technique, known as the Kernel trick, allows the SVM to find complex boundaries while maintaining computational efficiency (Noble, 2006).

Biomedical Applications:

Methods like SVMs excel in domains requiring robust classification with limited sample sizes.

1. Cancer Genomics: SVMs have been historically crucial for building diagnostic classifiers, often outperforming logistic regression, by using their regularization capacity to select a small, highly predictive set of gene-expression signatures for cancer type or prognosis (Guyon et al., 2002).
2. EEG/Brain Computer Interfaces (BCIs): SVMs are fast and effective at distinguishing between patterns of electrical brain activity, such as those corresponding to a user's intent to move a limb, making them a cornerstone in the signal decoding pipeline for BCIs (Lotte et al., 2018).
3. Histopathology and Cytology: They are used to classify tissue sections or cell images by extracting morphological features (such as texture, shape) and using the SVM to distinguish between normal, benign, and malignant samples accurately.

8.2.7.2 K-Means clustering

K-Means is a foundational algorithm in unsupervised learning, where the goal is to discover inherent structure within unlabeled data by grouping similar data points.

Working Mechanism: The K-Means algorithm partitions N data points into a pre-specified number of k clusters, aiming to make the clusters as distinct and compact as possible.

1. Centroid Initialization: The algorithm begins by randomly selecting k initial data points as cluster centers (centroids).
2. Assignment Step: Each data point is assigned to the nearest centroid, based on the Euclidean distance.
3. Update Step: The centroids are recalculated as the mean vector of all data points assigned to that cluster.
4. Iteration: Steps 2 and 3 are repeated until the assignments no longer change, or the Within-Cluster Sum of Squares (WCSS) is minimized. WCSS is the sum of the squared distances between each data point and its assigned cluster centroid (MacQueen, 1967).

Biomedical uses:

K-Means is widely used to discover patterns and segment heterogeneous populations.

1. Patient Phenotyping: In Electronic Health Records (EHRs), K-Means can be applied to patient vitals, lab results, and demographic data to identify previously unknown subtypes of a disease (e.g., clustering diabetes patients into distinct groups based on their complication profiles), allowing for tailored treatment strategies.
2. Microarray Analysis: This technique is used to cluster genes based on their expression profiles across various experimental conditions. Genes clustered together are often functionally related or involved in the same biological pathway, which aids in hypothesis generation in molecular biology (Eisen et al., 1998).
3. Image Segmentation: In basic medical image processing, K-Means can segment an image by clustering pixels based on intensity or color, helping to isolate regions of interest, such as lesions or organs.

8.3 Machine learning and different models used in biomedical studies

The transition from classical statistics to machine learning is driven by a data revolution in healthcare, where the need to handle high-dimensional, nonlinear biological data, such as found in electronic medical records and omics datasets, has become paramount. These models are generally categorized by their learning paradigm, supervised methods for predictive diagnostics and unsupervised methods for discovery and pattern recognition.

8.3.1 Supervised learning

Supervised learning requires labeled datasets to train predictive models. Here, the dataset is split into training and validation sets, with the training set annotated for features to ensure the model generalizes well to unseen patient data.

Logic-Based Models: Decision Trees and Random Forests utilize hierarchical rules to split data based on thresholds for clinical variables. While a single tree is easy to interpret, Random Forests aggregate multiple trees to improve robustness and reduce the variance often found in noisy clinical records. These are frequently employed to predict mortality risk or heart disease from electronic health record data (Ghassemi et al., 2015).

Boundary and Probabilistic Models: For high-dimensional data where the number of features often exceeds the number of samples, SVMs are highly effective at finding the optimal separating hyperplane. In contrast, Naive Bayes offers a probabilistic approach based on Bayes' theorem, making it well-suited for classifying complex gene-expression patterns.

Baseline Interpretability: Despite the rise of complex algorithms, Logistic Regression remains a vital baseline. It provides a transparent, interpretable model across many clinical studies, enabling clinicians to understand the specific impact of each predictor.

8.3.2 Unsupervised learning

Unsupervised algorithms operate on unlabeled data to uncover hidden structures, which is essential for defining new disease subtypes or identifying biomarkers.

Clustering Techniques: K-Means and Hierarchical Clustering are primary tools for identifying biological subtypes or molecular clusters in omics data. By grouping patients based on similar profiles, these methods allow for the discovery of distinct disease endotypes but can fail on unseen datatypes.

Dimensionality Reduction and Visualization: Because biomedical data, such as cell nuclei variables or pixel intensity vectors, are often overwhelming in its complexity, traditional linear techniques like PCA are often used to filter noise. However, linear mapping cannot keep similar data points close together when the data lie on a nonlinear manifold. To address this, t-SNE (t-Distributed Stochastic Neighbor Embedding) uses a heavy-tailed Student-t distribution to preserve local structure and reveal clusters at multiple scales without the crowding problem typical of other methods (Van der Maaten and Hinton, 2008).

8.3.3 Hybrid and ensemble methods

Modern biomedical data are heterogeneous and multifaceted, encompassing molecular, physiological, imaging, and clinical domains. Integrating these data processing sources can provide comprehensive insights into disease mechanisms.

Hybrid methods combine multiple algorithms for improved generalization and stability.

Random Forest and Neural Network: Hybrid architectures leverage the feature-selection strengths of models like Random Forests to identify critical biomarkers, while Deep Neural Networks perform hierarchical pattern extraction to predict patient outcomes. For instance, adding genomic markers to image-based convolutional neural networks (CNNs) has been shown to improve the concordance index (widely used to rank the predictive accuracy of models) in survival analysis significantly (Zhu et al., 2020).

SVM + Genetic Algorithm: The GALORIS model combines Genetic Algorithms (GA) with Logistic Regression (LoR) to optimize feature selection from complex EEG signals. This hybrid approach identifies critical frequency bands and structures reduced datasets that enhance the precision of SVMs to achieve 90% accuracy in recognizing high and low cognitive workloads (Becerra-Sánchez et al., 2020).

K-Means + Autoencoders: To address the curse of dimensionality in multi-omics data, unsupervised Autoencoders are combined with clustering techniques like K-Means. This integration enables effective dimensionality reduction and feature transformation, enabling the discovery of distinct biological endotypes or tumor tissue subtypes that traditional linear models may overlook.

Emergent LLM-Augmented Hybrid Agents: The latest paradigm involves Large Language Model (LLM) agents that coordinate specialized toolchains for tasks like clinical trial matching or pharmacovigilance. These agents utilize Retrieval-Augmented Generation (RAG) and function calling to bridge the gap between static linguistic reasoning and real-world task execution (Xu and Sankar, 2025).

8.3.3.1 Multi-omics integration

Combining genomics (complete genetic blueprint), proteomics (cellular protein information), metabolomics (cellular byproduct state), and microbiome (internal microbial ecosystem) profiles reveals molecular networks underlying disease. For example, multi-omics integration can identify host-microbe interactions influencing metabolic syndrome (Hasin et al., 2017).

8.3.3.2 Medical imaging and clinical data fusion

Linking radiological features (e.g., CT, MRI) with patient demographics, vital signs, and lab values yields hybrid predictors of disease severity. Example: Radio genomic models correlate imaging phenotypes with genomic alterations in glioblastoma.

8.3.3.3 Physiological signal integration

Wearable devices continuously record heart rate, motion, and temperature. Coupled with machine learning, these signals can predict health deterioration in chronic diseases or even in daily diagonostic settings.

8.3.4 Scope and applications

Early Disease Detection and Progression Modeling: Advanced analytic strategies allow for the identification of effective models that predict phenotypic outcomes, facilitating the discovery of early biomarkers for complex traits. These models help elucidate the genetic underpinnings of heritability, allowing researchers to track how small molecular changes escalate into a clinical disease state.

Drug Response Prediction and Precision Therapeutics: By integrating disparate data types, such as genotype-phenotype interactions that dictate how an individual will respond to a particular pharmacological intervention. This granularity is essential for developing precision therapeutics that account for the non-Mendelian inheritance patterns typical of complex diseases.

Systems-Level Understanding of Health and Disease: Rather than analyzing molecular factors in isolation, integration strategies aim to uncover the collective interactions among multiple factors, including environmental and behavioral influences. This systems-level perspective is critical for reducing false associations and identifying proper biological drivers within comprehensive, high-throughput datasets (Ritchie et al., 2015).

8.4 Neural networks and their usage in biomedical data

Neural networks (NNs), are often referred to as combinations ofe sophisticated, nonlinear function approximators inspired by the structure of the brain. They consist of a layered structure of interconnected nodes (neurons) that process inputs by applying weights and biases, followed by a nonlinear activation function (e.g., ReLU or Sigmoid). This architecture allows them to automatically learn complex, hierarchical representations from raw data, a capability that distinguishes them from traditional ML methods (Schmidhuber, 2015).

8.4.1 Artificial neural network

The Artificial Neural network (ANN), or Multilayer Perceptron (MLP), is the most basic form of a feedforward NN. Data flows in one direction, from the input layer, through one or more hidden layers, to the output layer.

Mechanism: Each neuron in one layer is fully connected to every neuron in the next layer. During training (via backpropagation), the network learns the optimal weights that minimize the difference between the predicted output and the actual label. The hidden layers enable the network to learn progressively more abstract features.

Biomedical Applications: ANNs are well-suited for tabular data and simpler classification tasks.

1. Classification of Physiological Signals: Used to classify one-dimensional time series data, such as distinguishing normal from abnormal ECG/EMG patterns or detecting phases like muscle fatigue based on statistical features extracted from raw signals.
2. Risk Prediction: Calculating a patient's risk score (e.g., for cardiovascular disease) based on a vector of clinical variables (age, cholesterol, blood pressure).

8.4.2 Convolutional neural networks

CNNs are the dominant architecture for analyzing data with a known grid-like topology, such as images (2D) or time series (1D). They are designed to exploit the spatial or temporal structure of data.

Mechanism: CNNs use convolutional filters (or kernels) that slide across the input data. Each filter learns to extract a specific, localized feature (e.g., edges, corners, textures in an image). These features are then progressively combined in deeper layers to form hierarchical representations (e.g., from edges to shapes to organs). This process, known as parameter sharing, drastically reduces the number of parameters, making CNNs efficient and robust to variations in position.

Biomedical Applications: CNNs have revolutionized medical imaging analysis (Litjens et al., 2017).

1. Detection and Diagnosis in Imaging: Used to detect subtle anomalies, such as lung nodules in CT scans or microaneurysms indicative of diabetic retinopathy from retinal fundus images.

2. Histopathology Analysis: Analyzing high-resolution digital slide images to perform tasks like cancer grading (e.g., classifying breast or prostate cancer severity) and segmenting cell nuclei.
3. Medical Signal Processing: Applied in 1D form to ECG or EEG signals to automatically extract and classify temporal features.

8.4.3 Recurrent neural networks and long short-term memory

RNNs are specialized for processing sequential data where the order of observations matters (e.g., time-series data or clinical notes). Unlike ANNs, RNNs have loops in their connections, allowing information from a previous step to influence the current step.

RNN Limitation (Vanishing Gradient): Standard RNNs struggle with long-term dependencies. Information from the beginning of a long sequence tends to fade, making it challenging to link distant events which might affect each others occurence.

LSTM Solution: Long Short-Term Memory (LSTM) networks were designed to solve the vanishing gradient problem. LSTMs contain specialized internal structures called memory cells or gates (input, forget, and output gates) that explicitly control the flow of information. The forget gate determines which data to discard, and the input gate decides what new information to store, allowing the model to retain critical context over extremely long sequences (Mienye et al., 2024).

Biomedical Applications:

1. Time-Series Prediction: Predicting the onset of cardiac arrhythmias from continuous ECG waveforms or forecasting patient deterioration in the ICU.
2. Natural Language Processing (NLP): Used for tasks like extracting information from unstructured clinical notes (e.g., identifying medication dosages or adverse events).

8.4.4 Graph neural networks

Graph neural networks (GNNs) are designed for data naturally represented as graphs, where entities (nodes) are connected by relationships (edges). This architecture is fundamentally aligned with the structure of many biological systems.

Mechanism: GNNs operate by iteratively performing message passing (or aggregation), where each node accumulates information from its immediate neighbors. By repeating this process across multiple layers, a node's final representation is an aggregated summary of its local and extended neighborhood in the graph. This allows the model to learn structure-aware node representations.

Biomedical Applications:

1. Biological Network Analysis: Analyzing complex interactions between molecules, and for example, predicting drug-target interactions by modeling drugs and proteins as nodes in a graph.
2. Gene Regulatory Network Reconstruction: Mapping complex connections between genes to understand how they influence each other (Zitnik et al., 2018).

3. Disease Similarity: Modeling diseases as nodes and their relationships (e.g., shared symptoms) as edges to predict new disease associations.

8.4.5 Transformer and attention models

The Transformer architecture, introduced in 2017, relies solely on a mechanism called self-attention and has become the new state-of-the-art for sequence processing, and increasingly for vision tasks.

Mechanism (Attention): The core innovation is the attention mechanism, which dynamically weighs the relevance of every element in the input sequence to every other element when making a prediction. Unlike RNNs, which process data sequentially, Transformers process all elements simultaneously, making them highly efficient and capable of capturing global dependencies (long-range context) immediately.

Biomedical Applications:

1. Biomedical NLP (e.g., BioBERT): Models like BioBERT are pretrained on a massive corpus of biomedical text (e.g., PubMed abstracts) (Lee et al., 2020). They use the attention mechanism to accurately interrupt unstructured clinical notes, extract key entities, and answer questions by understanding the context across long documents.
2. Vision Transformers (ViTs): Recently adapted to medical imaging, ViTs treat image patches as "tokens" in a sequence. By applying self-attention to these patches, they excel at classifying medical images. They are competitive with, or even superior to, CNNs, especially when large amounts of training data are available.

8.5 Case studies

This section discusses the application of hybrid models in healthcare. Each of these case studies follows a unified methodology that comprises data collection, data processing, and feature extraction. These steps ensure the data comprises distinct samples and remove outliers, which can then be fed into the learning pipeline and validated with unseen data to assess their inference in biomedical studies. Hence, this section discusses multiple such methods applied to various biomedical conditions to understand their inherent advantages and challenges. The unified methodology adopted across these applications is visually represented in the provided figure (Fig. 8.2), which illustrates a hierarchical pipeline from raw data acquisition to clinical insight. The process begins with patient data, such as blood and DNA, and biomedical data, including imaging and omics, which undergo rigorous data preprocessing via noise removal and normalization to ensure sample quality. This leads to Feature Engineering and Representation Learning, where statistical or learned characteristics are extracted to feed into the hybrid model core, combining the strengths of machine learning, deep learning, and traditional statistics. The final stages involve decision and inference for tasks such as classification or prediction, followed by model validation to evaluate performance, often using metrics like ROC-AUC and Accuracy, ensuring the outcomes have biomedical interpretation and clinical significance.

Note: Accuracy (Acc) represents the most straightforward performance metric, calculating the percentage of total correct predictions made by a model. However, to better account for clinical complexities, researchers rely on the **ROC** curve (Receiver Operating Characterstics), a graphical plot that illustrates the trade-off between sensitivity and specificity across various diagnostic thresholds. The **AUC** (Area Under Curve) summarizes this performance into a single value between 0.5 and 1.0, providing a robust measure of a model's fundamental ability to discriminate between disease and healthy states. These metrics are essential for validating the ML and Neural Network approaches that have seen an exponential rise in recent years.

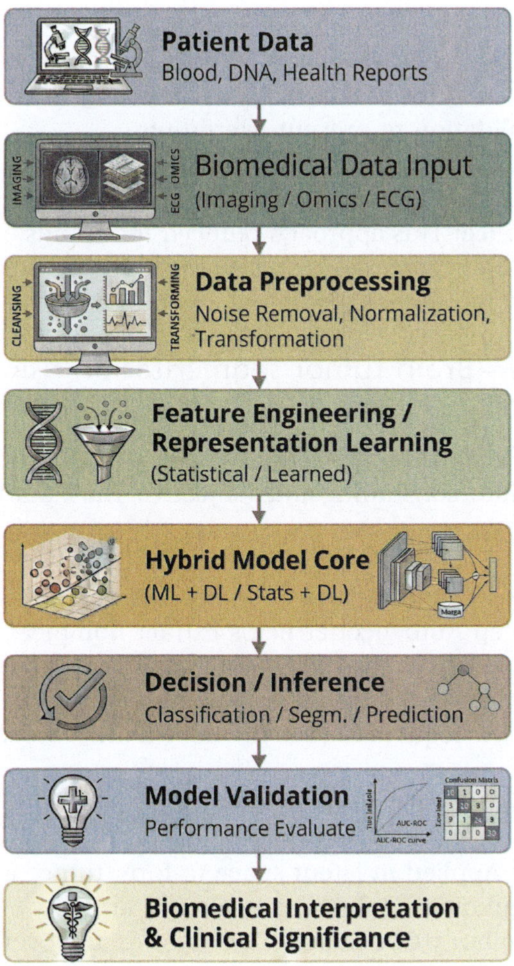

FIGURE 8.2 Generalized flowchart for hybrid models in the biomedical analysis.

8.5.1 Case study I—Breast cancer classification using SVM and gene expression

Breast Cancer is a heterogeneous disease, and its molecular profiling through high-throughput microarray experiments has revolutionized diagnosis and prognosis. Early studies typically relied on differential expression statistics; however, the high dimensionality of microarray data (often exceeding 10,000 genes per sample) posed challenges for traditional regression and clustering methods, leading to overfitting and noise.

Approach: Guyon et al. introduced the use of SVMs with Recursive Feature Elimination (RFE) to identify the most discriminative genes for classifying malignant and benign samples (Guyon et al., 2002). RFE iteratively removed genes contributing least to classification accuracy, producing a compact gene signature.

Workflow:

Data Acquisition: Breast-tissue microarray datasets (e.g., BRCA cohort).

Preprocessing: Log transformation and normalization to adjust for hybridization variance.

Feature Selection: RFE + SVM to select ~30 informative genes.

Classification: RBF kernel for nonlinear decision boundaries.

Validation: Cross-validation to prevent overfitting.

Results: The hybrid SVM-RFE model achieved >90% accuracy, identifying key oncogenes (BRCA1, ERBB2, and ESR1).

Biomedical significance: This approach reduced thousands of redundant features to a biologically interpretable gene set, establishing SVM-based feature selection as a standard in modern bioinformatics pipelines.

8.5.2 Case study II—Brain tumor segmentation using K-Means and deep autoencoder

Brain tumor detection from MRI scans requires precise localization of lesion boundaries. Manual annotation is time-consuming and subjective; thus, automated segmentation is critical in neuro-oncology.

Approach:

Traditional K-Means clustering struggles with intensity inhomogeneity and noise. Integrating it with a Deep Autoencoder helps extract nonlinear representations before clustering (Pereira et al., 2016).

Workflow:

Input: Multimodal MRI sequences (T1, T2, and FLAIR protocols to capture tissues of varying contrast).

Autoencoder Encoding: Dimensionality reduction to latent features, capturing high-level structure.

K-Means Clustering: Applied in latent space to form tumor and nontumor clusters.

Postprocessing: Morphological filtering to remove artifacts.

Evaluation: Dice Similarity Coefficient (DSC) to measure overlap between two sets of data.

Results: DSC = 0.92, significantly outperforming traditional K-Means (~0.83). **Biomedical significance:** This hybrid architecture preserves spatial context, enhances accuracy in low contrast regions, and accelerates radiological workflows by providing pre-segmented tumor masks for expert review.

8.5.3 Case study III—ECG signal classification using hybrid SVM neural network

Cardiac arrhythmia diagnosis relies on interpreting ECG waveforms that vary across patients and noise levels. Manual interpretation is prone to variability.

Approach:

A hybrid SVM-Artificial Neural Network (ANN) pipeline exploits both linear margin-based discrimination (SVM) and nonlinear hierarchical feature learning (ANN) (Hybrid SVM-ANN classifier is used for heart disease prediction system, 2019).

Workflow:

Data Input: ECG signals from the MIT-BIH Arrhythmia Database.

Preprocessing: Wavelet transform for time-frequency decomposition.

Feature Extraction: Heart-rate variability, QRS complex width, and spectral energy of the ECG signal.

Stage 1 (SVM): Binary classification (standard vs. abnormal).

Stage 2 (ANN): Multi-class classification of arrhythmia types.

Validation: 10-fold cross-validation.

Results: 98.1% accuracy, outperforming SVM-only (95%) and ANN-only (96%) models.

Biomedical significance: Combining explicit statistical features with learned nonlinear representations enables real-time arrhythmia detection for wearable ECG devices.

8.5.4 Case study IV—Retinal disease detection using CNN with statistical preprocessing

Diabetic retinopathy (DR) is a leading cause of vision loss. Detecting early lesions such as microaneurysms in fundus images is vital for prevention and early diagnostic care.

Approach:

Combining statistical preprocessing with a CNN improves feature contrast and model robustness (Gulshan et al., 2016).

Workflow:

Dataset: 12,000 retinal images from EyePACS and MESSIDOR datasets.

Preprocessing: Histogram equalization + illumination normalization to correct lighting variations and reduce Gaussian noise.

CNN Architecture: Modified Inception-V3 network fine-tuned for DR grading.

Evaluation: Receiver Operating Characteristic (ROC) and Precision-Recall curves.

Results: AUC = 0.99 for detecting referable DR with a sensitivity >94%.

Biomedical significance: This hybrid pipeline demonstrates how basic statistical normalization steps enhance sustainability and improve deep-learning performance in medical imaging, paving the way for tele-ophthalmology screening tools.

8.5.5 Synthesis of case studies

The presented case studies illustrate a generalized workflow for biomedical data analysis across applications such as medical imaging, physiological signal analysis (ECG/EEG), and clinical diagnosis. Each study begins with patient data acquisition, followed by preprocessing to normalize the data and extract meaningful features. Further in this vertical pipeline, these processed data undergo training to understand meaningful insights on this data which is ranked by their performance on the validation dataset. These findings emphasize that selecting algorithms tailored to the data modality as well as clinical objective can enhance diagnostic accuracy, biomarker discovery, and accurate disease detection. Apart from highlighting the efficacy of integrating hybrid models in healthcare, they also present a generalized framework that can be readily adapted to tackle any multi-omics dataset to understand them in detailed manner helping to classify, categorize and predict anomalies amongst these datapoints.

8.6 Comparative analysis of different models

Hybrid frameworks derive their power from combining complementary algorithms. A comparative evaluation clarifies where each method class performs best (Table 8.1).

Table 8.1 Summary of hybrid model case studies across biomedical domains.

Case	Model	Data type	Objective	Key outcome	Metric	References
I	SVM + RFE	Microarray	Gene-signature discovery	30 gene biomarker panel	Acc >90%	Guyon et al. (2002)
II	K-Means+ Autoencoder	MRI	Tumor segmentation	Accurate region mapping	Dice 0.92	Pereira et al. (2016)
III	SVM + ANN	ECG	Arrhythmia classification	Real-time detection	Acc >98%	Hybrid SVM-ANN classifier is used for heart disease prediction system, (2019)
IV	Network autoencoder	Multi-omics	Metabolic risk prediction	Cross-omics integration	AUC 0.91	Pereira et al. (2016)
V	Stats + CNN	Fundus images	DR screening	Clinical-grade accuracy	AUC 0.99	Gulshan et al. (2016)

8.6.1 Classical statistical models

Statistical models remain essential when interpretability is critical (e.g., clinical guidelines) or when sample sizes are small. However, they struggle to capture nonlinear interactions or contextual dependencies among biomarkers (Table 8.2).

8.6.2 Machine-learning model

Machine learning algorithms outperform traditional statistics in handling nonlinearities and high-dimensionality, but may lack transparency. They often serve as the computational core within hybrid systems (Table 8.3).

8.6.3 Deep and hybrid learning models

Hybrid models consistently outperform single-method approaches by exploiting the complementary strengths of each component, statistical rigor for interpretability and deep learning for representation power (Table 8.4).

Table 8.2 Comparative analysis of classical statistical methods.

Model	Category	Strengths	Limitations	Biomedical use	References
Regression/ ANOVA	Statistical	Simple, interpretable, fast, baseline inference	Linear assumptions, limited nonlinearity	Clinical risk modeling, trial analysis	Rajkomar et al. (2019) ; Bishop (2006)
PCA	Statistical	Dimensionality reduction, noise filtering	Loss of biological interpretability	Genomics, fMRI	Jolliffe and Cadima (2016)

Table 8.3 Comparative analysis of core machine learning models.

Model	Type	Strengths	Limitations	Biomedical use	References
SVM	ML (supervised)	Handles nonlinearity, robust in high dimensions	Kernel selection is sensitive to computational cost	Cancer, EEG	Guyon et al. (2002); Lotte et al. (2018)
K-means	ML (unsupervised)	Fast pattern discovery	Initialization bias, spherical assumption	Imaging, gene clustering	MacQueen (1967); Xu et al. (2024)
Random forest	Ensemble	Reduces overfitting, estimates feature importance	Interpretability is lower than that of linear models	Clinical decision support	Ghassemi et al. (2015); Zhu et al. (2020)

Table 8.4 Comparative analysis of deep and hybrid learning models.

Model	Category	Strengths	Limitations	Biomedical use	References
CNN	Deep	Learn spatial hierarchies, state-of-the-art image performance	Data-hungry, black box	Radiology, pathology	Litjens et al. (2017); [33]
RNN/LSTM	Deep	Captures temporal patterns in signals	Vanishing gradient, long training	ECG, EEG	Mienye et al. (2024)
GNN	Deep	Exploits network structure, models biological interactions	Complex architecture and tuning	Drug discovery, omics	Zitnik et al. (2018)
Autoencoder + K-Means	Hybrid	Combines representational learning and clustering	Sensitive to the latent-space dimension	MRI segmentation	Pereira et al. (2016)
SVM + ANN	Hybrid	Balances interpretability and nonlinearity	Requires two-stage training	ECG classification	Hybrid SVM-ANN classifier is used for heart disease prediction system, (2019)
Stats + CNN	Hybrid	Statistical preprocessing enhances image features	Additional computational steps	Retinal imaging	Gulshan et al. (2016)

8.7 Emerging trends and future outlook

1. Federated Learning: Enabling collaborative model training across hospitals while preserving patient privacy.
2. Edge Computing: Real-time inference on wearable devices for continuous monitoring.
3. Integration with Computational Biology: Linking predictive AI(Artificial Intelligence) outputs to biological mechanisms such as pathway activation or microbial gene expression.

These directions bridge machine intelligence and biomedical insight, driving precision health toward clinical translation.

8.8 Conclusion

The field of biomedical engineering has undergone a fundamental transformation in its analytical approach to managing the complexity of multidimensional healthcare data. The journey has progressed from foundational classical statistical methods (like regression and ANOVA), which struggle with biological system nonlinearity and high dimensionality, to sophisticated learning techniques.

The current gold standard is the hybrid model architecture. These models strategically combine complementary components, such as fusing the geometric optimization of SVM or the unsupervised clustering of K-Means with the deep feature learning of

Neural Networks (e.g., Autoencoder + K-Means for MRI segmentation, SVM + ANN for ECG classification). This fusion successfully leverages the interpretability and rigor of statistics with the predictive power of deep learning to achieve superior performance, as demonstrated across diverse applications like gene expression classification, MRI segmentation, and retinal disease screening.

The adoption of hybrid models is crucial for advancing precision health and necessitates a continued focus on challenges such as interpretability, reproducibility, and ensuring clinical translation. Emerging trends, including federated learning and edge computing, are poised to further drive these powerful analytical tools into routine clinical practice as shown in Fig 8.3.

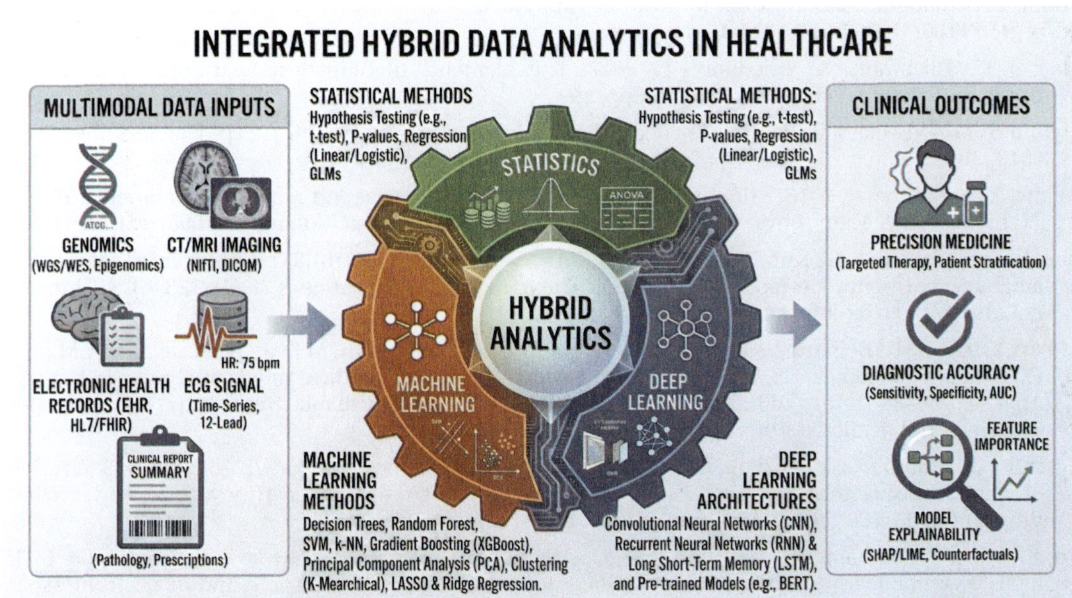

References

Becerra-Sánchez, P., Reyes-Munoz, A., Guerrero-Ibañez, A., 2020. Feature selection model based on EEG signals for assessing the cognitive workload in drivers. Sensors 20 (20), 5881. https://doi.org/10.3390/s20205881.

Bishop, C.M., 2006. Pattern Recognition and Machine Learning. Springer.

Eisen, M.B., Spellman, P.T., Brown, P.O., Botstein, D., 1998. Cluster analysis and display of genome-wide expression patterns. Proc. Nat. Acad. Sci. 95 (25), 14863–14868. https://doi.org/10.1073/pnas.95.25.14863.

Friston, K.J., 2011. Functional and effective connectivity: a review. Brain Connect. 1 (1), 13–36. https://doi.org/10.1089/brain.2011.0008.

Ghassemi, M., Celi, L.A., Stone, D.J., 2015. State of the art review: the data revolution in critical care. Crit. Care 19 (1). https://doi.org/10.1186/s13054-015-0801-4. Available from: http://ccforum.com/content/17.

Gulshan, V., Peng, L., Coram, M., Stumpe, M.C., Wu, D., Narayanaswamy, A., Venugopalan, S., Widner, K., Madams, T., Cuadros, J., Kim, R., Raman, R., Nelson, P.C., Mega, J.L., Webster, D.R., 2016. Development and validation of a deep learning algorithm for detection of diabetic retinopathy in retinal fundus photographs. JAMA 316 (22), 2402. https://doi.org/10.1001/jama.2016.17216.

Guyon, I., Weston, J., Barnhill, S., Vapnik, V., 2002. Gene selection for cancer classification using support vector machines. Mach. Learn. 46 (1–3), 389–422. https://doi.org/10.1023/a:1012487302797.

Hasin, Y., Seldin, M., Lusis, A., 2017. Multi-omics approaches to disease. Genome Biol. 18 (1). https://doi.org/10.1186/s13059-017-1215-1.

Hastie, T., Tibshirani, R., Friedman, J., 2017. The Elements of Statistical Learning: Data Mining, Inference, and Prediction, second ed. Springer.

Hybrid SVM-ANN classifier is used for heart disease prediction system. Int. J. Exp. Diabetes Res. 7 (3), 2019, 365–372.

Jolliffe, I.T., Cadima, J., 2016. Principal component analysis: a review and recent developments. Philos. Trans. R. Soc. A Math. Phys. Eng. Sci. 374 (2065), 20150202. https://doi.org/10.1098/rsta.2015.0202.

Lee, J., Yoon, W., Kim, S., Kim, D., Kim, S., So, C.H., Kang, J., 2020. BioBERT: a pre-trained biomedical language representation model for biomedical text mining. Bioinformatics 36 (4), 1234–1240. https://doi.org/10.1093/bioinformatics/btz682.

Litjens, G., Kooi, T., Bejnordi, B.E., Setio, A.A.A., Ciompi, F., Ghafoorian, M., van der Laak, J.A.W.M., van Ginneken, B., Sánchez, C.I., 2017. A survey on deep learning in medical image analysis. Med. Image Anal. 42, 60–88. https://doi.org/10.1016/j.media.2017.07.005. Available from: http://www.elsevier.com/inca/publications/store/6/2/0/9/8/3/index.htt.

Lotte, F., Bougrain, L., Cichocki, A., Clerc, M., Congedo, M., Rakotomamonjy, A., Yger, F., 2018. A review of classification algorithms for EEG-based brain–computer interfaces: a 10 year update. J. Neural. Eng. 15 (3), 031005. https://doi.org/10.1088/1741-2552/aab2f2.

Lu, J., Wang, L., Bennamoun, M., Ward, I., An, S., Sohel, F., Chow, B.J.W., Dwivedi, G., Sanfilippo, F.M., 2021. Machine learning risk prediction model for acute coronary syndrome and death from use of non-steroidal anti-inflammatory drugs in administrative data. Scient. Rep. 11 (1). https://doi.org/10.1038/s41598-021-97643-3. Available from: www.nature.com/srep/index.html.

MacQueen, J., 1967. Some methods for classification and analysis of multivariate observations. In: Proceedings of the Fifth Berkeley Symposium on Mathematical Statistics and Probability, Vol 1, pp. 281–297.

Mienye, I.D., Swart, T.G., Obaido, G., 2024. Recurrent neural networks: a comprehensive review of architectures, variants, and applications. Information 15 (9), 517. https://doi.org/10.3390/info15090517.

Montgomery, D.C., Runger, G.C., 2018. Applied Statistics and Probability for Engineers, seventh ed. Wiley.

Noble, W.S., 2006. What is a support vector machine? Nat. Biotechnol. 24 (12), 1565–1567. https://doi.org/10.1038/nbt1206-1565.

Panch, T., Szolovits, P., Atun, R., 2018. Artificial intelligence, machine learning and health systems. J. Glob. Health 8 (2). https://doi.org/10.7189/jogh.08.020303.

Pereira, S., Pinto, A., Alves, V., Silva, C.A., 2016. Brain tumor segmentation using convolutional neural networks in MRI images. IEEE Trans. Med. Imag. 35 (5), 1240–1251. https://doi.org/10.1109/TMI.2016.2538465.

Rajkomar, A., Dean, J., Kohane, I., 2019. Machine learning in medicine. N. Engl. J. Med. 380 (14), 1347–1358. https://doi.org/10.1056/nejmra1814259.

Ritchie, M.D., Holzinger, E.R., Li, R., Pendergrass, S.A., Kim, D., 2015. Methods of integrating data to uncover genotype-phenotype interactions. Nat. Rev.Genet. 16 (2), 85–97. https://doi.org/10.1038/nrg3868.

Rosner, B., 2016. Fundamentals of Biostatistics, eighth ed. Cengage Learning.

Schmidhuber, J., 2015. Deep learning in neural networks: an overview. Neural Netw. 61, 85–117. https://doi.org/10.1016/j.neunet.2014.09.003.

Topol, E.J., 2019. Deep Medicine: How Artificial Intelligence Can Make Healthcare Human Again. Basic Books.

Van der Maaten, L., Hinton, G., 2008. Visualising data using t-SNE. J. Mach. Learn. Res. 9 (11), 2579–2605.

Xu, X., Sankar, R., 2025. Large Language model agents for biomedicine: a comprehensive review of methods, evaluations, challenges, and future directions. Information (Switzerland) 16 (10), 20782489. https://doi.org/10.3390/info16100894.

Xu, Y., Quan, R., Xu, W., Huang, Y., Chen, X., Liu, F., 2024. Advances in medical image segmentation: a comprehensive review of traditional, deep learning and hybrid approaches. Bioengineering 11 (10), 1034. https://doi.org/10.3390/bioengineering11101034.

Zhu, W., Xie, L., Han, J., Guo, X., 2020. The application of deep learning in cancer prognosis prediction. Cancers 12 (3), 603. https://doi.org/10.3390/cancers12030603.

Zitnik, M., Agrawal, M., Leskovec, J., 2018. Modeling polypharmacy side effects with graph convolutional networks. Bioinformatics 34 (13), i457. https://doi.org/10.1093/bioinformatics/bty294.

9

Improving disease prediction by integrating multiple ML and optimization techniques

Sara Nadeem[1], Mohammad Subhan Ali[2], Syed Ansar Ahmad[3], Anima Sushim Mathur[4, 5, 6], Md Zaid Hussain Siddique[7], Aisha Idris[7] and Mohammad "Sufian" Badar[8, 9, 10, 11]

[1]DEPARTMENT OF COMPUTER SCIENCE AND ENGINEERING, SCHOOL OF ENGINEERING SCIENCES AND TECHNOLOGY (SEST), JAMIA HAMDARD UNIVERSITY, NEW DELHI, INDIA; [2]WHO, KATIHAR, BIHAR, INDIA; [3]DEPARTMENT OF MAXILLOFACIAL SURGERY, FACULTY OF DENTISTRY, JAMIA MILLIA ISLAMIA, NEW DELHI, INDIA; [4]DESERT MEDICINE RESEARCH CENTRE (INDIAN COUNCIL OF MEDICAL RESEARCH), WORLD HEALTH ORGANIZATION PROJECT, JODHPUR, RAJASTHAN, INDIA; [5]MANAV RACHNA INTERNATIONAL INSTITUTE OF RESEARCH AND STUDIES, FARIDABAD, HARYANA, INDIA; [6]SCIENTIFIC HUB RESEARCH WRITING SERVICES, JODHPUR, RAJASTHAN, INDIA; [7]DEPARTMENT OF BIOTECHNOLOGY, JAMIA HAMDARD, NEW DELHI, INDIA; [8]DEPARTMENT OF BIOENGINEERING, UNIVERSITY OF CALIFORNIA, RIVERSIDE, CA, UNITED STATES; [9]UNIVERSAL SCIENTIFIC EDUCATION AND RESEARCH NETWORK (USERN), TEHRAN, IRAN; [10]SPI DARBHANGA, DARBHANGA, BIHAR, INDIA; [11]DEPARTMENT OF COMPUTER SCIENCE AND ENGINEERING (BIOINFORMATICS), SCHOOL OF ENGINEERING SCIENCES AND TECHNOLOGY (SEST), JAMIA HAMDARD UNIVERSITY, NEW DELHI, INDIA

What you will learn

This chapter discusses how machine learning and optimization can assist medical professionals and researchers in making more accurate predictions regarding the likely occurrence of a heart attack in a patient. This article talks about how using synchronized resources can help make predictions more accurate and reliable. Some of the machine learning methods that will be talked about at the start of the talk are Support Vector Machines (SVMs), Random Forest, Logistic Regression, K-Nearest Neighbors (KNN), and XGBoost. Each strategy uses a different method to analyze heart disease data, and each has its own strengths when it comes to finding patterns in patient data. This chapter examines the identification of the most critical elements from the data of the patient. This process is referred to as feature selection. It uses statistical tests like Chi-square and ANOVA, as well as optimization methods like genetic algorithms, to find the data points

Applying Artificial Intelligence to Computational Biology and Medical Informatics. https://doi.org/10.1016/B978-0-443-44915-4.00009-5

that are most useful for figuring out how likely it is that someone will have a heart attack. The author talks about how to use optimization methods like grid search, random search, and Bayesian optimization to make models better in the next part of the chapter. These help you find the best combinations for each model so that it works better.

This means using SMOTE to make the dataset more even, filling in missing values, getting rid of outliers, and making sure that the scales are the same. These methods make sure that the model gets data that is both correct and fair, which makes the predictions more likely to be correct. Then it talks about ensemble learning, which is when you use more than one model together.

Bagging, boosting, stacking, and voting methods use the best parts of several models to make predictions more accurate.

Finally, you'll learn about the bigger issues, real-life problems, and future possibilities of using machine learning and optimization to figure out when someone is going to have a heart attack. This will prepare you to apply this extensive knowledge in both scholarly research and practical healthcare environments.

9.1 Introduction

Cardiovascular diseases, especially myocardial infarctions, remain the leading cause of morbidity and mortality globally, requiring the advancement of innovative predictive methodologies that exceed traditional clinical frameworks. This chapter presents a hybrid predictive modeling methodology that integrates the advantages of various machine learning models and optimization techniques to tackle the increasing complexity, diversity, and volume of modern healthcare data.

9.1.1 The goal and scope of hybrid predictive modeling

Heart disease is getting worse and more common all over the world. This makes it harder for doctors to understand how all of these things—genetic, lifestyle, environmental, and clinical—affect each other. A lot of the time, old-fashioned prediction systems don't think about everything that could cause this disease. This can lead to wrong classifications, delays in treatment, or a lack of personalization. AI and machine learning are now very good at filling in these gaps. But not all types of predictions can be made with the same algorithm. This is why it is so important to use hybrid predictive modeling. It uses different ML models, ways to pick features, and ways to make things work better. This type of integration not only combines the best parts of different algorithms, such as statistical, deep learning, and tree-based algorithms, but it also makes risk scores that are more accurate, reliable, and generalizable. It is important to do personalized, data-driven risk assessments, and the fact that there is so much multi-modal digital health data, such as electronic health records, medical imaging, genomics, and wearable patient monitoring devices, means that this hybrid approach is needed. Techniques like stacking, bagging, and boosting, as well as feature engineering (like principal component analysis and rough set theory) and hyperparameter optimization (like grid search and particle swarm optimization), can make data that is unbalanced, has a lot of dimensions, or is noisy work much better.

9.1.2 Benefits of combining healthcare ML and optimization

There are many ways that machine learning and optimization change how well we can guess when people will get sick. There are many ways to make machine learning models more accurate and clear. These include ways to choose features, change hyperparameters, and make choices based on more than one thing. When it comes to important clinical tasks that need both sensitivity and specificity, ensemble learning solutions like voting systems and advanced stacking always work better than single models. When doctors and nurses use new explainable AI tools like SHAP and LIME to show how different features affect model predictions, they can trust each other more and talk to each other more easily. Health systems and practitioners can use these integrated models to work with different diseases and datasets, as well as data that is incomplete or is in real time. Research has shown that hybrid models are more accurate, have a higher F1-score, and a larger area under the curve (AUC) than both traditional and deep learning models when looking at different models on standard clinical datasets. They render computers more affordable and user-friendly.

9.1.3 Examination of the chapter's structure

This section extensively discusses the concept, framework, and practical applications of machine learning and optimization systems that collaboratively predict the occurrence of a heart attack:

- Initially, it elaborates on the supervised machine learning algorithms, optimization techniques, and ensemble learning methods employed to assess cardiovascular risk.
- Subsequently, it discusses the advantages and disadvantages of utilizing hybrid methods and innovative approaches. It shows how these methods could make estimates more accurate, easier to understand, and more useful in clinical settings.
- We show that you can use both real-world and standard datasets to make sure that the ways you collect data, prepare it, build models, change hyperparameters, and test these models are all the same.
- It begins by examining the supervised machine learning algorithms, optimization techniques, and ensemble learning approaches employed to assess an individual's likelihood of developing cardiac issues. Next, it talks about why the hybrid approach is better, how to do it, and how it can help in the clinic. It shows how it can make predictions more accurate, how easy it is to understand, and how useful it is. We show how to use both real-world and benchmark datasets to set up

standard methods for collecting, cleaning, building models, optimizing hyper-parameters, and testing data.

The final part of the chapter talks about the problems, limitations, and new trends that are always present in the field. It offers a conceptual framework for future research on hybrid machine learning and optimization in the healthcare system. Using hybrid predictive modeling can help find heart attacks and other heart problems early, find people who are at higher risk, and make predictions about what might happen. This part has all the information you need to get started. It does this by using new research, case studies from the clinic, and stories from real life.

9.2 Machine learning algorithms in heart attack prediction

For the purpose of determining the likelihood that an individual will experience heart problems, machine learning is becoming an increasingly popular method. It is possible that it examined a wide variety of data, including medical records, test results, genomic markers, and medical pictures, among other types of information. As demonstrated by the studies that you provided, a significant number of individuals employ deep learning architectures in conjunction with conventional supervised algorithms in order to detect heart problems, such as heart attacks, at an earlier stage. Each individual executes this in a distinct manner. This section discusses their profession, their level of proficiency, and their limitations (Fig. 9.1).

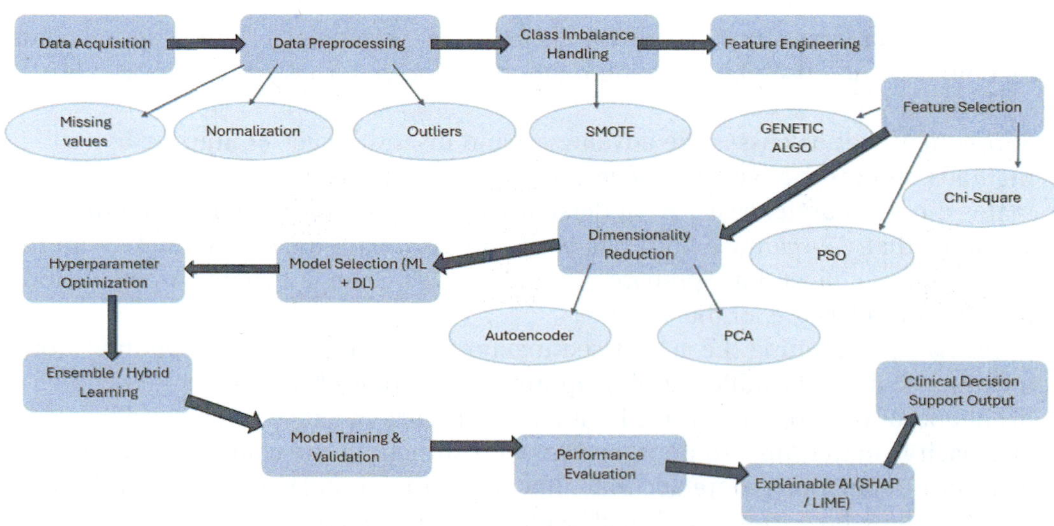

FIGURE 9.1 Flowchart of the integrated machine learning and optimization framework for heart attack prediction.

9.2.1 Supervised machine learning algorithms

9.2.1.1 Decision tree

Some of the first supervised models used to predict heart disease were Decision Trees (DT). They are helpful because they are easy to understand and have a clear, rule-based structure. A lot of research shows that DTs can be very accurate, but they often have trouble with overfitting, especially when they are used with clinical data that is noisy or has a lot of dimensions. They are easy to use for quick screening, but not so good for making predictions on their own.

9.2.1.2 Support vector machine

A lot of people want to know more about SVMs, especially when they are used in hybrid frameworks like SVM-PSO. SVMs are great for sorting through data that has a lot of features and lines that aren't straight. Studies show that optimized SVMs, especially those that use Particle Swarm Optimization to improve them, are more accurate than many older models and get more than 95% of the answers right. The hardest part about them is picking the right kernel and hyperparameters. They cost more and are harder to grow because of this.

9.2.1.3 Random forest

People often use Random Forests (RF) to predict heart attacks because they are strong, can handle nonlinear interactions, and don't overfit. Ensemble-based papers always put RF at or near the top, with accuracy rates of 90%–98% or higher when bagging or feature selection is used. RF models also tell you how important different features are, which makes it easier to understand the results in a clinical setting. But RF can be costly in terms of computing power when working with large datasets, and it can be harder to understand than single decision trees.

9.2.1.4 Logistic regression

Logistic Regression (LR) is still a basic but important classifier in cardiovascular analytics. It works well with structured, linearly separable datasets and is easy to understand, which is very important when making medical decisions. Research shows that LR is often less accurate than more advanced ensemble and deep learning models, but it is better at being clear, easy to use, and cheap to run. Due to its stability, it is often used as a meta-learner in stacking frameworks.

9.2.2 Deep learning techniques

9.2.2.1 Convolutional neural networks

CNNs have performed exceptionally well in image-based cardiovascular diagnostics, such as ECG waveform classification and echocardiographic analysis. Some research says that heart disease models based on CNNs are more than 97%–99% accurate. When it comes to automatically finding features in imaging signals, CNNs are better than hand-made features. They have a lot of problems, but the biggest one is that they need big, well-labeled datasets and a lot of computing power.

9.2.2.2 Multilayer perceptron

Feed-forward neural networks were one of the first deep learning models used to figure out how likely someone is to get heart disease. They can show how complicated nonlinear relationships work, but they are very sensitive to hyperparameters and the quality of the data. Multilayer perceptrons (MLPs) do work, but CNNs and more advanced ensemble models are usually better. You need to carefully tune them to make sure they work well. They might also be able to fit medical datasets with a lot of dimensions too well.

9.2.2.3 Self-encoders

People mostly use autoencoders to find features, cut down on the number of dimensions, and look for strange things in heart datasets. They help you see patterns in clinical and physiological data that are hard to see. In some hybrid frameworks, autoencoders can help prepare data for classification. The problem with them is that they can't be directly understood, and they need to be fine-tuned to avoid learning patterns that aren't useful.

9.2.3 Comparative strengths and weaknesses

9.2.3.1 Strengths

- **DT:** Simple to understand, quick, and good for basic clinical decision rules.
- **SVM:** Very accurate for complicated boundaries and works well with small to medium-sized datasets.
- **Random Forest:** Strong, reduces overfitting, handles nonlinearities well, and is very accurate when used in an ensemble.
- **Logistic Regression:** Easy to understand, stable, and cheap to run.
- **CNN:** Best for imaging and ECG; best at recognizing patterns; best performance overall.
- **MLP:** Works well with complicated nonlinear patterns if set up correctly.
- **Autoencoders:** Good for preprocessing, getting rid of noise, and finding hidden medical features.

9.2.3.2 Weaknesses

- **DT:** It's easy to overfit and not as stable as ensembles.
- **SVM:** Needs a lot of fine-tuning and takes a lot of processing power with big datasets.
- **Random Forest:** Harder to understand and needs a lot of computing power.
- **Logistic Regression:** Doesn't work well on nonlinear, complicated medical data.
- **CNN:** Needs big labeled datasets and GPUs, but it's hard to understand.

- **MLP:** Likely to overfit and sensitive to hyperparameters.
- **Autoencoders:** Not easy to understand; the quality of the latent features affects the outputs.

9.2.3.3 Summary

The literature as a whole shows that no one algorithm is better at predicting heart attacks than all the others. Instead, the best frameworks use a mix of models, such as stacking, boosting, hybrid SVM-PSO, and feature-optimized ensembles, to take advantage of their different strengths. Deep learning models are the best for diagnosing images, while ensemble and optimized classical models are the best for structured clinical data. The best way to make heart attack prediction systems that are accurate, easy to understand, and safe to use in a clinical setting is to use hybrid and explainable AI frameworks.

9.3 Ways to improve a model through optimization

How well machine learning models can predict heart attacks depends a lot on how well they are set up. When you use raw machine learning algorithms on clinical datasets, they often have trouble with settings that aren't right, features that aren't useful, and too many dimensions. This can make predictions less reliable and less useful for a wide range of situations. Making learning more efficient, lowering error rates, and making clinical reliability better are all optimization techniques that help these models get better.

9.3.1 Tuning hyperparameters

Hyperparameters change how machine learning algorithms learn, and they have a big effect on how well models work. Hyperparameters, on the other hand, are not learned from data; they must be carefully chosen through systematic optimization. Grid Search is a common way to change hyperparameters in studies that try to predict heart disease. It checks all the possible combinations of preset hyperparameter values and chooses the one that works best. Grid search makes sure you pick the best parameters, but the cost of doing it goes up quickly as the number of parameters goes up. This makes it less useful for healthcare applications that need to work in real time or on a large scale (Fig. 9.2).

Random Search solves this problem by picking hyperparameters at random instead of trying every possible combination. Research has shown that random search often works just as well or better than grid search and needs a lot fewer evaluations, especially when only a few hyperparameters have a big effect on how well the model works. This is very useful for making small changes to complicated models like Random Forests and neural networks when working with medical data. **Bayesian Optimization** is a smarter and more advanced way to make things work better. It builds a probabilistic model of the objective function and then chooses the best combinations of hyperparameters based on what it has learned from previous evaluations. Bayesian methods are becoming more

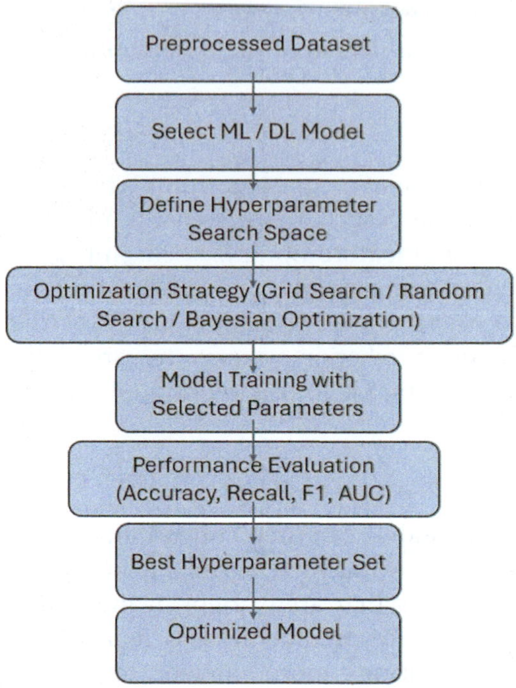

FIGURE 9.2 Workflow of hyperparameter optimization using grid search, random search, and Bayesian optimization.

popular in hybrid heart disease prediction frameworks because they are very accurate, converge faster, and use less computing power.

9.3.2 Feature selection methods

Extra, noisy, or only slightly relevant features are common in clinical datasets. These can make models less accurate and hide important medical information. Feature selection methods help you choose which features to keep and which ones to get rid of Fig. 9.3.

Genetic Algorithms (GA) look for the best feature sets in a way that is similar to how evolution works in nature. In GA-based feature selection, you keep doing selection, crossover, and mutation operations until you find the best combinations of clinical variables that make classification better. By limiting the number of features, GA has been able to make accurate and consistent predictions about heart attacks.

Particle Swarm Optimization (PSO) is another method for using groups to make things better. It all depends on how smart the group is. You can change model parameters and choose features at the same time with PSO. The models are much better at guessing what will happen now that they use PSO with Support Vector Machines or ensemble classifiers. In general, they are more precise, responsive, and dependable. Chi-Square Feature Selection is a statistical technique frequently employed for the

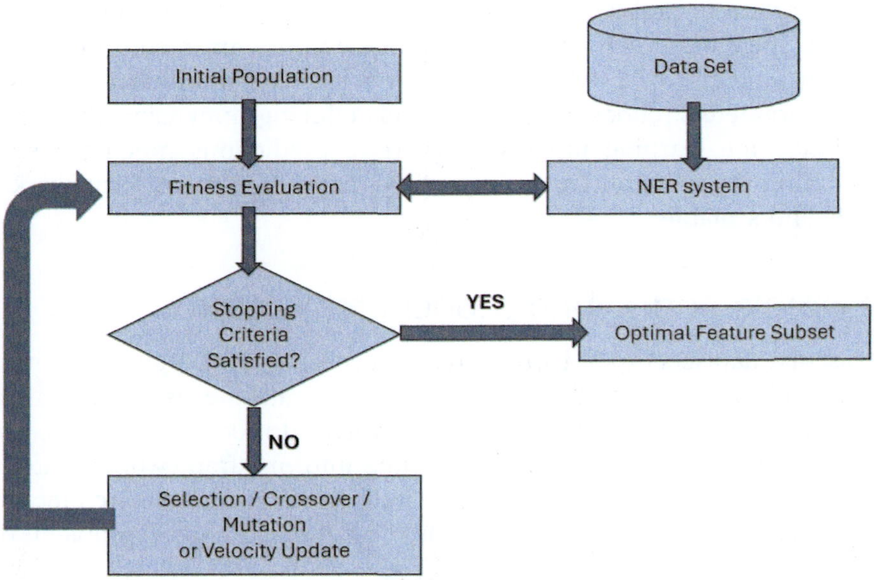

FIGURE 9.3 Flowchart of optimization-based feature selection using genetic algorithm and particle swarm optimization.

categorization of clinical variables. It looks at the relationships between features and class labels and chooses the ones that are most closely related to heart disease outcomes. It's easy to understand chi-square, and it's often used in early screening models. It might not be able to show nonlinear relationships that are very hard to understand, though.

PCA, or Principal Component Analysis, is a method for turning a group of related features into a smaller set of orthogonal components on its own. PCA is great for clinical or imaging data with a lot of dimensions because it keeps most of the differences and gets rid of the extra ones. But doctors might not be able to easily understand the new features.

9.3.3 Dimensionality reduction and its clinical impact

Reducing the number of dimensions is very important for improving models and making them more useful in the real world. Medical datasets with a lot of dimensions are harder to compute, more likely to overfit, and harder to understand. When you cut down on the number of dimensions, it speeds up model training, makes predictions more accurate, and makes the model work better overall. From a clinical point of view, dimensionality reduction makes it easier to find the most important risk factors for heart attacks. This lets doctors focus on important signs like cholesterol levels, blood pressure, age, and genetic markers. This helps people understand things better and base their choices on facts.

But it's important to find a balance between simplifying the data and making it easy to understand, since too much abstraction can hide important medical information. In short, modern systems for predicting heart attacks need optimization methods like tuning hyperparameters, choosing features, and reducing dimensionality. When combined with machine learning models, they make predictions much more accurate, reliable, and clinically relevant. This is what makes hybrid predictive frameworks that are both reliable and scalable.

9.4 Ensemble and hybrid model architectures

As cardiovascular datasets expand and increase in complexity, dependence on a singular predictive model to elucidate all factors that may elevate the risk of a heart attack is becoming increasingly ineffective. Ensemble and hybrid model architectures solve this problem by combining different ways of learning into one framework for making predictions. The studies analyzed in this chapter illustrate that these architectural types augment models' robustness, generalizability, and reliability in problem diagnosis (Fig. 9.4).

9.4.1 Methods for ensemble learning

Ensemble learning uses more than one base classifier to make a final prediction that is more stable and accurate than any one model on its own. Different ensemble strategies use different ways to make this integration happen. **Bagging, or Bootstrap Aggregating,** reduces the variance of a model by training multiple copies of the same algorithm on

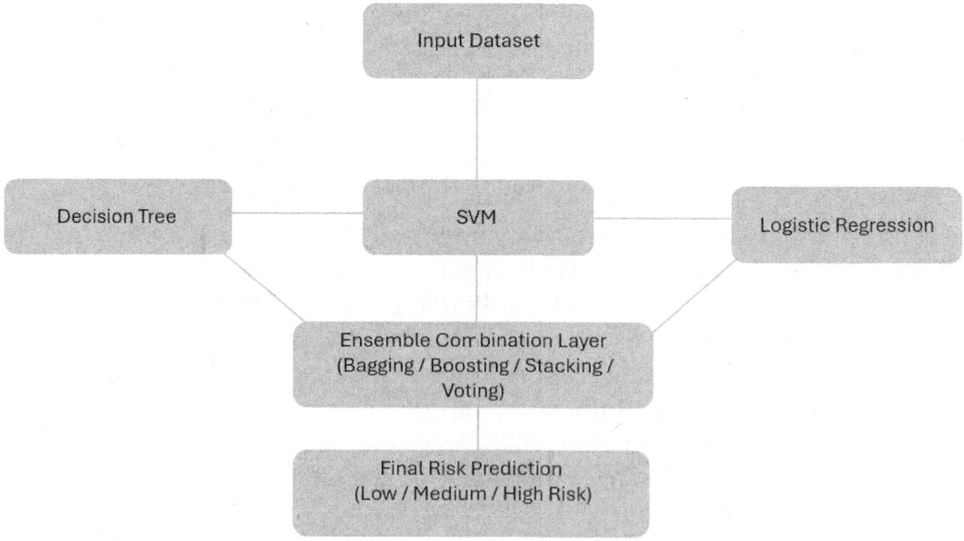

FIGURE 9.4 Architecture of ensemble and hybrid machine learning models for heart attack prediction.

different bootstrapped sections of the training data. Random Forest is the most well-known bagging-based model for studies that try to predict heart attacks. Random Forest combines many decision trees to reduce over-fitting, handle nonlinear feature interactions, and do well on structured clinical datasets. Bagging is very helpful when base learners, like decision trees, are very sensitive to changes in the data. In boosting, each new model tries to fix the mistakes of the models that came before it. People often say that algorithms like AdaBoost, Gradient Boosting, and **XGBoost** are better at predicting heart disease than other methods. Boosting is especially good at fixing class imbalance, which is a common problem in medical datasets. It does this by giving more weight to high-risk patients who are misclassified. But if not carefully regularized, boosted models may be sensitive to noise. **Stacking** is a more advanced ensemble method that uses a meta-classifier to combine predictions from different base learners. In studies that try to predict cardiovascular disease, stacking has repeatedly been shown to be one of the best ways to combine models. Stacking uses complementary learning behaviors by combining models like Logistic Regression, SVM, Random Forest, and Neural Networks. The meta-learner learns how to best combine the predictions of the base models, which leads to better generalization and less bias. **Voting Ensembles** combine predictions using either hard voting, which is based on the majority, or soft voting, which is based on probability. Voting ensembles are not too hard to set up and can improve performance when the base models are not too similar to each other. Soft voting, in particular, has been shown to be better than hard voting at predicting heart attacks because it uses confidence scores instead of binary decisions.

9.4.2 Rationale for combining models

The main reason for ensemble and hybrid architectures is that different models learn in different ways. Certain algorithms demonstrate superior performance in processing specific types of data compared to others. A valuable method for illustrating the interaction of features and the hierarchical establishment of rules is through the use of decision trees. However, Aid Vector Machines are effective in identifying optimal decision boundaries within high-dimensional spaces. Logistic regression is straightforward to comprehend and performs effectively, but neural networks are capable of capturing complex nonlinear relationships that are difficult to discern.

Ensemble architectures simultaneously reduce both bias and variance by integrating multiple models. Hybrid pipelines advance this approach by integrating models incrementally across multiple stages rather than deploying them simultaneously. For example, a two-stage pipeline may employ a tree-based model to prioritize features and risks according to their significance. Then, for cases that are very risky, it might use an SVM to make the decision boundaries more precise. This design is like how doctors think: they first do a general screening and then a more focused diagnostic evaluation.

You can also add optimization methods to the learning pipeline when you use hybrid architectures. You can pick features, lower the number of dimensions, and change hyperparameters at different points in the process. This design is similar to how doctors work: they first do a general screening and then a more focused diagnostic evaluation. When you use hybrid architectures, you can also add optimization methods to the learning pipeline. You can change hyperparameters, pick features, and lower the number of dimensions at different times during the process. This means that the models can be different depending on how easy or hard it is to get the data. This modularity is very helpful in healthcare settings where data is often missing or not what you expect.

9.4.3 Some hybrid frameworks that are being used right now

Numerous studies have shown that hybrid frameworks that use both tree-based and boundary-optimizing classifiers work well. Two common ways to build a system are with SVMs and DT or RF. First, tree-based models find important clinical features and keep track of how these systems work together in ways other than straight lines. Then, SVMs make the final classification margin better, which makes the system more accurate and responsive. Using optimization algorithms with ensemble learners is another successful hybrid design. For example, SVM-PSO or Random Forest, which use a genetic algorithm to pick features. These frameworks work together to find the best subsets of features and model parameters. This makes the predictions more accurate and less likely to fit the data too closely. It has also been shown that hybrid stacking architectures that use Random Forest, XGBoost, and neural networks with a Logistic Regression meta-classifier can get close to state-of-the-art accuracy while still being easy to understand. Ensemble and hybrid model architectures are a big step forward for systems that try to predict heart attacks in general. These architectures are more useful in clinical settings because they combine different classifiers, sequential decision phases, and optimization algorithms. They also work better and are less affected by changes in data. The fact that more and more people are using them shows that we are moving away from using separate algorithms and toward using integrated intelligent systems that can help us accurately measure cardiovascular risk in the real world.

9.5 Feature engineering and preprocessing

Feature engineering and preprocessing are the most important parts of any good machine learning system that can predict heart attacks. Clinical and biomedical datasets are inherently heterogeneous, consisting of numerical measurements, categorical variables, temporal records, and imaging data, often derived from various sources and under different conditions. If these datasets aren't cleaned up in a consistent way, they can add noise, bias, and inconsistencies that make models work much worse. So, good feature engineering is important to make sure the data is good, speed up learning, and make predictive models more useful in the clinic.

9.5.1 Data cleaning and normalization

The first and most important step in looking at cardiovascular datasets is to clean the data. When people enter data or use sensors that aren't working right, they often make mistakes. These kinds of mistakes can lead to duplicate entries, measuring units that don't match, and wrong results. Cleaning tasks that are common include getting rid of duplicate records, fixing mistakes, and making sure that all datasets use the same units. If you follow these steps, you'll be able to make a strong database that will be useful for modeling in the future. Normalization and scaling are very important for algorithms that care about the size of the features, like Support Vector Machines, k-Nearest Neighbors, and neural networks. Two common ways to make all the numbers the same value are min-max normalization and z-score standardization. This keeps health problems like high blood pressure and cholesterol from getting in the way of learning. As a result, models can learn faster.

9.5.2 Handling missing values and outlier detection

Healthcare providers may struggle to obtain comprehensive patient records due to patients' inconsistent return for follow-up visits and the high likelihood of errors during data entry. When there isn't a lot of missing data, people usually use the mean, mode, and median. To keep data distributions the same and get rid of bias, more and more people are using K-nearest neighbor imputation, regression-based imputation, and model-driven estimations. When trying to predict a heart attack, it's very important to find outliers because values that are too high could mean a real high-risk condition or a mistake in measurement. Interquartile range analysis and z-score threshold are often used to find data outliers. Clustering and isolation forests are examples of machine learning techniques that can help find multivariate outliers. This can help with tough problems. When you have bad data, you should focus on outliers and keep the clinically important extremes.

9.5.3 Addressing class imbalance

Class imbalance is a major problem in predicting heart attacks because there are usually more healthy people than people who have heart problems. This imbalance can cause models to be biased toward the majority class, which means they won't be able to correctly identify patients who are at high risk. Resampling strategies are commonly used to fix this issue. The Synthetic Minority Over-sampling Technique (SMOTE) is one of the most popular ways to do this. SMOTE makes fake examples of the minority class by filling in the spaces between real examples. This makes the classes more even without giving them more data. Two examples of variations that improve performance by focusing on cases that are hard to classify are Borderline-SMOTE and Adaptive Synthetic Sampling. Other options include randomly under-sampling the majority class, learning that takes costs into account, and using an ensemble to deal with imbalance. All of these methods work together to improve sensitivity and recall. These are important steps to take to find out how likely it is that someone will get sick.

9.5.4 Getting information from pictures and medical records

Feature extraction looks for cool thing patterns in raw medical data and then change them into formats that can be used. A well-organized clinical database keeps track of a patient's vital signs, medical history, demographics, and test results. To choose and make risk ratings, interaction words, and temporal patterns that show how an illness is getting worse, you need to know a lot about it. Automated feature extraction algorithms work well on imaging data with signals, such as electrocardiograms. Convolutional Neural Networks can identify hierarchical patterns in unprocessed images or waveforms without the necessity of human annotations. Hybrid frameworks that integrate deep features with structured clinical data to create multimodal representations have enhanced the completeness and accuracy of heart attack prediction models. In summary, feature engineering and preprocessing are crucial for machine learning algorithms that assess the likelihood of an individual developing heart disease. When predictive models acquire appropriate data for clinical practice, rectify any missing or imbalanced data, and maintain high quality, they enhance their accuracy, reliability, and comprehensibility. Thus, they are more beneficial in practical medical scenarios.

9.6 Implementation workflow

To develop a machine learning system capable of detecting heart attacks, one must exhibit organizational skills and meticulous attention to detail. This approach ensures that predictive models function effectively in clinical environments, exhibiting accuracy, adaptability, and reusability. This chapter's studies illustrate the significance of having a coherent strategy for data management, model construction, testing, and accuracy verification.

9.6.1 Stepwise approach to model development

The first step in putting the plan into action is to get the data. At this point, clinical records, lab test results, electrocardiographic signals, imaging data, and demographic information are all stored in electronic health records or public repositories. Creating a comprehensive patient profile typically requires this step, which entails putting together information obtained from a variety of sources. This is because the data pertaining to cardiovascular systems are so dissimilar to one another. After data is gathered, it is cleaned up and checked for quality and consistency. To do this, the data needs to be cleaned up, made more consistent, filled in for missing values, outliers need to be found, and new features need to be added. You need to preprocess the dataset before you can start learning from it. This factor has a direct effect on how well and reliably the model works. The next step is to train the model. This means putting the cleaned data into the machine learning or hybrid models that were chosen ahead of time. People often use hyperparameter tuning and feature selection at this point to help the model make better guesses. By being trained on a small amount of data, the model learns the patterns that are linked to a higher risk of having a heart attack. The next step is to test the model with new data to see how well it works. Validation helps you find cases of overfitting, make sure the model is strong, and choose the best model before you start using it.

9.6.2 One step at a time how to make a model

Getting the data is the first step in putting the plan into action. This means getting information from public databases or electronic health records, such as demographic data, lab test results, clinical records, electrocardiographic signals, and imaging data. This step usually means putting together information from different places to make a full patient profile. This is because cardiovascular data is so different. After the data has been collected, preprocessing is done to make sure it is of good quality and consistent. This means cleaning the data, making it more consistent, filling in missing values, finding outliers, and adding new features. It's very important to get the dataset ready for learning. It directly affects how well the model works and how stable it is. The next thing to do is to train the model. This means that you need to fit the machine learning or hybrid models you chose to the data that has been cleaned. At this point, people often use hyperparameter tuning and feature selection to make their predictions more accurate. Training usually only uses a small amount of the data. This helps the model find the patterns that are linked to a higher risk of having a heart attack. The next step is to test the model with new data to see how well it works. Validation helps you find overfitting, make sure the model is strong, and choose the best model before you use it.

9.6.3 Performance evaluation metrics

When you make clinical decisions, you need to look at more than one metric to see how well a model works. You need to look at different parts of how well it predicts to see why this is. You can tell how well the model works by counting how many times it was right. But if the datasets aren't balanced, accuracy alone can be misleading. Precision tells you how many of the predicted positives were real positives. A low false-positive rate means that the test is very accurate, which is good because you don't want to have to take tests that aren't necessary. Recall, or sensitivity, tells you how many of the real positive cases are correctly identified. It's very important to think about high-risk patients when trying to guess when someone will have a heart attack, because not doing so can have bad effects. The average of precision and recall is the F1 score. In a medical context, where both false positives and false negatives are significant, it provides an accurate representation. The Area Under the Receiver Operating Characteristic Curve (AUC-ROC) tests how well the model can tell the difference between patients who are at high risk and those who are at low risk at different levels of decision-making. If the AUC-ROC value is higher, the model is more accurate overall. This means it can better tell the difference between things. Researchers and doctors can use these metrics together to make a full set of tools that help them figure out how well a model works, compare different methods, and find the best predictive system for figuring out the risk of a heart attack.

9.7 Explainable AI and clinical applicability

Machine learning models will only be used in real life if they are correct, easy to use, and reliable. Doctors need to know why a model makes a certain choice before they use it to help patients in high-stakes areas like predicting heart attacks. Explainable Artificial Intelligence (XAI) meets this need by making it easy to understand how models work. This connects computational intelligence with clinical decision-making.

9.7.1 Model transparency and feature importance

For clinicians to trust a model and for it to follow the rules, it needs to be clear. People often use XAI methods like SHapley Additive exPlanations (SHAP) and Local Interpretable Model-agnostic Explanations (LIME) to explain what complex machine learning and deep learning models predict. In accordance with the principles of cooperative game theory, SHAP assigns a contribution value to every input feature. The information presented here demonstrates that the estimated risk of a heart attack is affected by a number of clinical factors, including age, blood pressure, cholesterol levels, and parameters derived from an electrocardiogram (ECG). Because of this global and local interpretability, medical professionals are able to determine the primary risk factors that are associated with each individual patient as well as the entire population. LIME, on the other hand, is all about how things work in your area. It does this by using simpler surrogate models to get close to complex models around a certain prediction. It is especially helpful for explaining the outcomes of individual patients, as it helps doctors figure out which factors had the biggest impact on a certain risk classification. Through their collaborative efforts, SHAP and LIME are able to assist individuals in gaining a deeper comprehension of a wide variety of machine learning models, including ensembles and deep learning models. Adding to the workflows and systems in the healthcare industry that assist in decision-making When it comes to the healthcare system, incorporating AI models that are both useful and explainable shouldn't be too difficult. Clinical Decision Support Systems (CDSS) provide physicians with risk assessments, alerts, and suggestions that are founded on evidence in order to facilitate the decision-making process in a timely manner. It is necessary for models to have the capability to establish a connection with electronic health record systems in order to obtain prompt access to patient data and to make predictions without interfering with clinical work. This exemplifies a wonderful model for working together. When the predictions are presented alongside the typical diagnostic data, it is much simpler for the medical professionals to comprehend them. It is beneficial to have results that are easy to understand, such as ranking the importance of features or using graphs to summarize risks. In CDSS systems, models that are based on XAI help to identify potential risks at an early stage, ensure that everyone receives the appropriate treatment, and take measures to prevent problems from occurring. In the end, those who are ill will benefit from all of this.

9.7.2 Problems with usability and problems that occur in the real world

To be effective, healthcare applications that make use of artificial intelligence need to be easy to use and comprehend. Doctors work in environments where time is of the essence, and as a result, they require systems that are not only simple to operate but also quick and dependable. Graphical user interfaces need to be able to make future predictions and provide explanations that are brief and easy to understand. Due to the fact that they are easy to comprehend, people frequently make use of visual dashboards, risk heatmaps, or interactive plots in order to accomplish this. Another important way to help doctors is to automate their work. When you automate the process of getting information from electronic records, lab systems, and imaging repositories, you cut down on the need for manual data entry, which is more likely to lead to mistakes. You also need to be able to handle errors and check for validation in order to avoid making wrong predictions because of missing or inconsistent data. Finally, to keep medical professionals' trust, it's important to reduce the number of false alarms and wrong predictions. People may get upset when they have to deal with too many alerts from systems that aren't set up correctly. If the systems aren't sensitive enough, they might miss situations that are very dangerous. To be useful in the clinic, explainable AI models need to find the right balance between sensitivity, specificity, and how easy they are to understand. In short, explainable AI is a key part of using machine learning in a safe way to figure out when someone is going to have a heart attack. XAI helps us go from experimental models to dependable decision support systems that can improve cardiovascular care by making things clearer, speeding up clinical workflows, and fixing usability issues (Fig. 9.5).

9.8 Standards and a side-by-side look at the two

Comparing machine learning models to each other is one of the best ways to see how well they can predict heart attacks. A big part of medicine is figuring out how likely it is that someone will have heart problems. Models should be tested on standard datasets, compared to known baselines, and put through tests that are as close to real life as possible. All of the studies discussed in this chapter agree that clear benchmarking is important for showing both improvements in performance and clinical reliability.

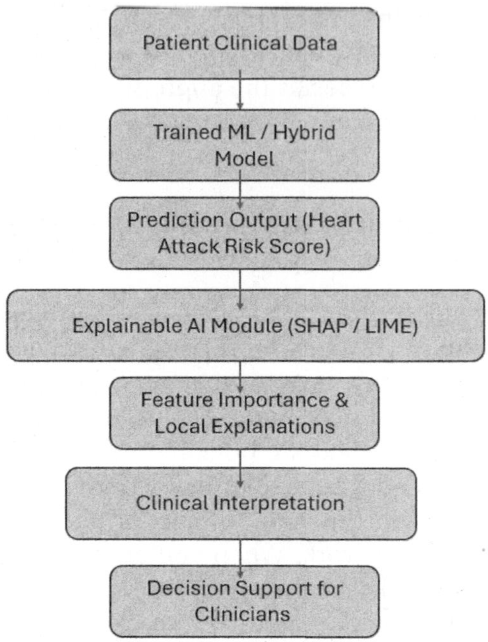

FIGURE 9.5 Explainable AI framework using SHAP and LIME for interpretable heart attack risk prediction and clinical decision support.

9.8.1 Test datasets and experimental settings

Most studies that try to figure out when someone will have a heart attack use datasets that are open to the public and have been tested in a clinical setting. This makes it possible to repeat the results and compare them fairly. The Cleveland Heart Disease dataset from the **UCI Machine Learning** Repository is the most famous of these. It has put together clinical information such as blood pressure, cholesterol levels, age, sex, type of chest pain, and ECG results. Here, you can test out a lot of different supervised, ensemble, and hybrid models. In addition to Cleveland, researchers often use data from places like Hungary, Switzerland, and Long Beach. They also check how well the models work by combining data from different groups of people. Preprocessing, choosing features, k-fold cross-validation, and using more than one performance metric to make sure the results can be used in different situations are all normal steps in an experiment.

9.8.2 Looking at the differences between old and new models

Logistic Regression, Decision Trees, and Support Researchers often use vector machines as a starting point for their work because they are some of the most common old-school machine learning models. These models are easy to understand and run quickly, but they don't always work well on their own when there are nonlinear relationships and a lot of clinical data. Always use hybrid and ensemble methods instead of regular models.

A two-stage hybrid pipeline that uses a Decision Tree for initial feature-driven stratification and a Support Vector Machine for boundary optimization has been shown to achieve about 89% classification accuracy and an **AUC-ROC of 92%.** This shows that it is possible to combine different ways of learning. These kinds of architectures work well because tree-based models are easy to understand, and margin-based classifiers work well in general. By combining the work of many learners, ensemble methods make predictions more accurate. When used with methods for choosing features and changing hyperparameters, optimized Random Forest and XGBoost ensembles often get more than 92% correct. These models do a good job of capturing nonlinear interactions between clinical factors and also make the results more stable. This means that they are some of the best predictors in organized cardiovascular datasets.

9.8.3 Deep learning and hybrid benchmarks

Deep learning models have been very good at working with imaging and signal-based data, like ECG waveforms and images of the heart. But when you use deep learning models on structured clinical datasets, they usually need a lot of tuning and big sample sizes to work better than classical ensembles. One way to use deep learning ensemble methods is to combine multilayer perceptrons and convolutional architectures. Studies show that this works best when it is 88.7% accurate. This performance is good, but it might not always be better than that of better tree-based ensembles. Hybrid deep learning frameworks that combine manually created clinical features with automatically collected representations exhibit greater stability. But they often have trouble with cost, deployment, and understanding.

9.8.4 Case studies from the real world and their effect

Medicine Real-world experiments demonstrate that hybrid and ensemble models that enhance performance also assist in clinical risk classification. Models with higher recall and AUC-ROC values are the best for finding high-risk patients who could benefit from early treatment. Research examining various topologies has demonstrated that hybrid and ensemble designs are effective for specific patient cohorts. Because of this, they are great for hospitals and other places where people go to get medical care.

The findings from the benchmarking exercise demonstrate that, on the whole, there is no optimal model. Using and understanding classical models is simple. In hybrid pipelines, the objective is to achieve a balance between the accuracy and generalizability of the results. The ensemble methods provide the highest level of accuracy overall. These findings highlight the significance of evaluating multiple systems in order to determine which one is the most appropriate for clinical application in the prediction of heart attacks that are likely to occur. In addition to this, they advocate for the implementation of the most effective ensemble and hybrid designs in actual healthcare environments.

9.9 The challenges, the boundaries, and the next steps to take

The use of machine learning to predict heart attacks has undergone a great deal of progress in recent years; however, there are still a great deal of challenges and limitations that make it difficult to implement these models in clinical settings that are based on real-world scenarios. These issues need to be resolved in order for the treatment to be dependable, equitable, and helpful in a wide variety of healthcare settings. This section discusses significant topics and offers suggestions for ways to acquire additional knowledge.

9.9.1 Issues dealing with generalizability and biases in predictive ability

The fact that the datasets that are used by heart attack prediction models are either not balanced or do not show the entire population is one of the most significant issues with these models. The accuracy of the predictions is consequently reduced. There are many small cardiovascular datasets that anyone can access, but these datasets don't include people of all ages, races, or locations. This means that they don't help much. Models created using this data may work well in experiments, but they might not work as well when used with either a larger group of patients or a smaller group of participants. Because different hospitals collect data, diagnose patients, and treat them in different ways, it is hard to make generalizations. When models are used in situations that are different from what they were trained on, these differences could make them less accurate. In order to solve these problems and make healthcare outcomes more fair, it is becoming more and more important to use bias-aware modeling, demographic stratification, and fairness-oriented evaluation metrics.

9.9.2 Things that make it hard to keep your data private

Put all the information together. Medical systems have strict rules about keeping patient information safe, which makes it harder to use machine learning models. Legal and ethical rules that protect the privacy of patient information make it harder to create large, centralized datasets and share data between institutions. Because of this, many models are trained on data that is either too little or too wrong, which makes it hard for them to make accurate predictions. Some of the problems that people are worried about right now are privacy issues and the technical problems that come with integrating data. Many electronic health record systems, imaging platforms, and laboratory databases store data in different ways. Healthcare companies often keep this kind of information. For data to work together, there must be standard ways to represent it, strict rules for managing it, and safe ways to send it. If these issues are not dealt with, it will be hard to use advanced predictive models.

9.9.3 Emerging approaches and future research directions

New methods are showing promise in fixing problems that already exist. Federated learning lets different organizations work together to train a model without having to share information about patients. Federated learning protects privacy by only sharing model parameters and keeping all the data in one place. This makes the model useful for a larger group of people. This method is especially good at predicting the risk of heart disease on a large scale. Transfer learning is another great way to learn. It lets you use models that have been trained on large, related datasets on smaller datasets or datasets that are only for one institution. Transfer learning speeds up the process of getting models to work together and makes it less important to have a lot of data. This is useful in healthcare settings where there isn't a lot of data with labels. The methodology for predicting heart attacks will evolve as optimization techniques improve. An increasing number of individuals are employing hybrid optimization techniques, multi-objective optimization, and automated machine learning (AutoML) frameworks to achieve a balance among speed, accuracy, and comprehensibility. The implementation of explainable AI in optimization-based models is likely to enhance the trust and acceptance of the technology among medical professionals. In conclusion, machine learning possesses significant potential for predicting heart attacks; however, challenges related to bias, general data privacy, and system integration need to be fixed. To translate predictive models from experimental contexts to actual clinical outcomes, further investigation is required into federated and transfer learning, ethical AI methodologies, and sophisticated optimization techniques.

Take-home messages

Combining several machine learning models with optimization methods like feature selection, hyperparameter tuning, and ensemble learning makes heart attack prediction systems much more accurate, reliable, and useful in clinical settings.

10

Ethical, societal, and legal issues in AI/ML for healthcare

Mohammad "Sufian" Badar[1, 2, 3, 4], Huda Shams[5], Fatemah Rizvi[6], S.M. Zainul Abedin[7], Hamid Iqbal[8], Mohd Asrarul Haque[9] and Aquil Atif[7]

[1]DEPARTMENT OF BIOENGINEERING, UNIVERSITY OF CALIFORNIA, RIVERSIDE, CA, UNITED STATES; [2]UNIVERSAL SCIENTIFIC EDUCATION AND RESEARCH NETWORK (USERN), TEHRAN, IRAN; [3]SPI DARBHANGA, DARBHANGA, BIHAR, INDIA; [4]DEPARTMENT OF COMPUTER SCIENCE AND ENGINEERING (BIOINFORMATICS), SCHOOL OF ENGINEERING SCIENCES AND TECHNOLOGY (SEST), JAMIA HAMDARD UNIVERSITY, NEW DELHI, INDIA; [5]ALIGARH MUSLIM UNIVERSITY (AMU), ALIGARH, UTTAR PRADESH, INDIA; [6]JAMIA MILLIA ISLAMIA UNIVERSITY, NEW DELHI, INDIA; [7]ABEDIN CLINIC, PHULWARI SHARIF, PATNA, BIHAR, INDIA; [8]DR. B. R. AMBEDKAR NATIONAL INSTITUTE OF TECHNOLOGY (NITJ), JALANDHAR, PUNJAB, INDIA; [9]DEPARTMENT OF FORENSIC MEDICINE, JAWAHARLAL NEHRU MEDICAL COLLEGE, ALIGARH MUSLIM UNIVERSITY (AMU), ALIGARH, UTTAR PRADESH, INDIA

What you will learn

In this chapter, you will explore the implications that extend beyond just technological advantages with multidimensional AI/ML adoption in healthcare. You will learn of critical ethical concerns such as data privacy, safety, transparency, as well as fairness, and you will also learn how clinical validation along with regulation are important. Social challenges like legislation, accessibility, interpretability, workforce disruption, and depth will be examined. You will gain perception into legal debates about liability and intellectual property too. Cybersecurity will additionally be a topic into which you gain perception. Ultimately, the chapter notes lasting effects on healthcare practice, policy, and equity, and this lets you carefully judge AI's benefits and dangers inside a complete ethical, social, and legal framework.

10.1 Introduction

10.1.1 Overview of AI/ML adoption in healthcare

AI and ML technologies are transforming healthcare by providing previously unattainable chances to improve patient care, streamline clinical procedures, and further

Applying Artificial Intelligence to Computational Biology and Medical Informatics. https://doi.org/10.1016/B978-0-443-44915-4.00011-3

medical research. It has been shown that incorporating AI and ML into healthcare systems can significantly enhance medication research, patient monitoring, treatment planning, and diagnostics. A further subset of machine learning (ML) called deep learning (DL) models intricate patterns using neural networks (Fig. 10.1).

These technologies, which range from customized medicine to predictive analytics, have the potential to revolutionize healthcare delivery and solve persistent issues in the industry (Bajwa et al., 2021). Healthcare professionals can get important insights and make well-informed clinical decisions because of AI and ML algorithms' ability to evaluate enormous volumes of complicated data, such as genomic sequences, medical pictures, electronic health records (EHRs), and real-time physiological signals. AI-powered diagnostic technologies, for example, can precisely identify minute anomalies in medical imaging, improving patient outcomes and enabling the early detection of diseases like cancer. Similar to this, ML algorithms may identify people who are more likely to develop specific disorders by analyzing their genetic data. This allows for focused therapies and preventive measures (Maleki Varnosfaderani and Forouzanfar, 2024). Millions of patient data sets are put into AI models created for the healthcare industry. For example, an AI model will examine data from previous patients with comparable medical issues to determine which treatment approach is most likely to be effective and which will yield the greatest results for the new patient. The genomic data aids AI in making well-informed decisions regarding diagnosis and treatment because each person has a unique set of DNA, and AI models are able to recognize similarities and differences. AI's potential has made it a useful tool (Bohr and Memarzadeh, 2020).

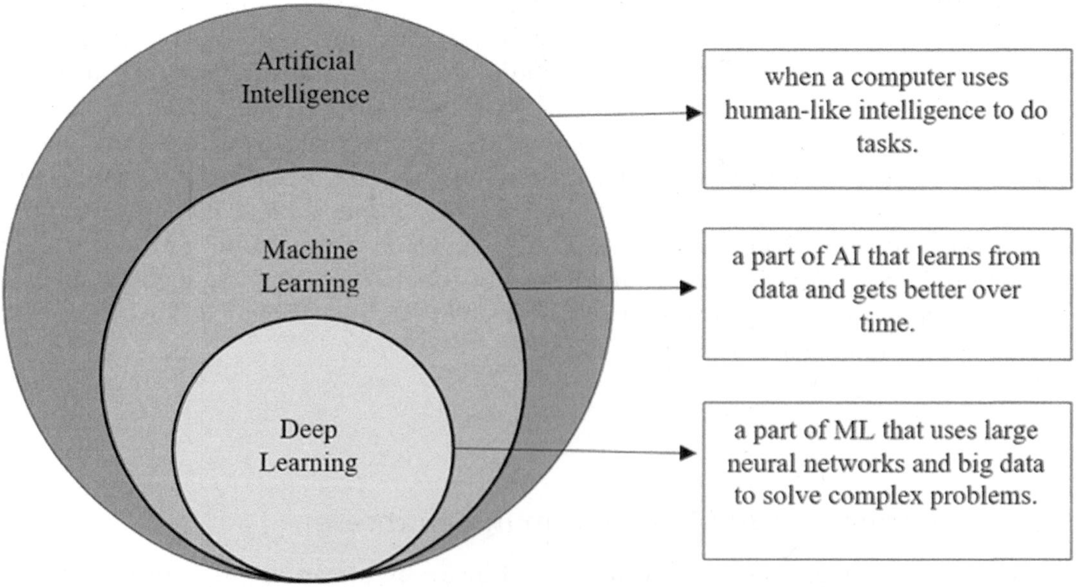

FIGURE 10.1 How deep learning, machine learning, and artificial intelligence are related.

Moreover, AI and ML have the potential to streamline healthcare operations and improve resource allocations. These technologies can enhance efficiency, reduce costs, and alleviate healthcare workforce shortages by automating administrative tasks, optimizing hospital workflows, and predicting patient admissions and discharges, these technologies can enhance efficiency, reduce costs, and alleviate healthcare workforce shortages. Additionally, AI-driven predictive models can help healthcare organizations anticipate disease outbreaks, allocate resources during public health emergencies, and optimize treatment protocols based on real-time data analysis (Harishbhai Tilala et al., 2024).

10.1.2 Why ethical, societal, and legal frameworks are essential

Medical research, patient outcomes, and healthcare delivery could all be greatly enhanced by the combination of AI and ML technology. But along with this transformational potential comes a host of ethical issues that requirethoughtful analysis and preventative measures. Healthcare stakeholders can negotiate the ethical challenges of integrating AI and ML in healthcare while protecting patient welfare and respecting the values of beneficence, nonmaleficence, autonomy, and justice by addressing concerns like data privacy and security, algorithmic bias, transparency, clinical validation, and professional responsibility. This description accurately captures the industrialization of AI and its pervasiveness in society today. As a result, societal concerns about the application of AI and ML should no longer be disregarded, especially in the medical field (Alowais et al., 2023). These societal concerns can take various forms, but they all demand the development of models from a human-centered perspective, with human-relevant requirements and limits. Artificial intelligence (AI) systems can abruptly and significantly malfunction when the environment or context changes. In one instance, AI can change from being incredibly bright to being incredibly naive. Even if AI bias is controlled, all AI systems will have limitations. The constraints of the system must be understood by the human decision-maker, and the system must be built to meet the needs of the human decision-maker (Jiang et al., 2024). Medical professionals may become complacent and neglect to maintain their skills or enjoy their work when a medical diagnostic and treatment system is largely accurate. Furthermore, people may accept decision-support system results without questioning their limits (Sutton et al., 2020).

Overlooking the interaction between human factors and AI-powered medical devices can lead to expensive consequences for resources, health, and life. AI tools also have the potential to sustain or worsen already-existing health inequities or even create new ones if they are not carefully supervised and planned for. Beyond the typical scope of individual product development, specific rules and continuous, coordinated action will be needed to guard against unwanted repercussions. Any regulatory framework for AI in healthcare should achieve three goals, according to research: 1) Make certain that AI tools are well tested before being made publicly available and that they are subsequently regularly examined. 2) Make sure users are aware of the tools' limitations and capabilities. 3) Be diligent in guarding against false information about medicine (D'Elia et al., 2022).

10.2 Ethical issues

10.2.1 Privacy and data security

The backbone of AI/ML in healthcare lies in its ability to analyze large volumes of sensitive patient data, ranging from electronic health records (EHRs) to genomic sequences. While this provides opportunities for precision medicine and predictive care, it also raises critical concerns regarding privacy and data security. Patients must be assured that their most intimate medical details will not be exposed, misused, or accessed by unauthorized parties.

A breach of healthcare data can have consequences beyond financial loss; it can affect an individual's dignity, social standing, and mental well-being (Marzuki et al., 2021). AI systems frequently require aggregation of data from multiple sources, and such cross-institutional sharing can magnify vulnerabilities, especially when cybersecurity standards differ across organizations (Price and Cohen, 2019).

Further ethical dilemmas arise around **informed consent**. Many patients may not be fully aware of how their health data is being used, whether for clinical AI applications or research model training. In some cases, secondary uses of data may occur without explicit patient approval (Murdoch, 2021). Safeguarding privacy, therefore, requires robust encryption methods, anonymization protocols, strict governance policies, and transparent communication with patients. Ultimately, patient trust is fundamental; without it, the promise of AI in healthcare will remain compromised (Vayena et al., 2018).

10.2.2 Safety and transparency

Patient safety must remain the cornerstone of healthcare innovation. AI/ML systems, no matter how sophisticated, are not immune to errors, unexpected failures, or misinterpretations of clinical context. A diagnostic tool that misclassifies medical images or a predictive model that overestimates treatment success could have life-threatening consequences (Topol, 2019).

Transparency is therefore, an ethical obligation. Both clinicians and patients deserve to know how an AI tool works, what data it uses, and what its limitations are. A "black-box" system that generates outputs without explanation erodes accountability and leaves medical professionals in a difficult position when justifying clinical decisions (Amann et al., 2020).

To address this, the adoption of **explainable AI (XAI)** approaches, continuous monitoring of deployed systems, and human oversight have been proposed as ethical requirements (Ghassemi et al., 2021).

10.2.3 Bias and fairness

Bias in AI is not merely a technical flaw; it is an ethical hazard that can deepen existing healthcare disparities. AI models learn patterns from the data they are trained on. If that data underrepresents certain groups, whether by race, gender, age, or socioeconomic status, the resulting outputs may disadvantage those very populations (Obermeyer et al., 2019).

For example, dermatology AI tools trained on lighter skin tones may fail to detect conditions on darker skin, leading to inequitable care and misdiagnosis (Adamson and Smith, 2018). Ensuring fairness requires deliberate steps: using diverse and representative datasets, auditing algorithms for bias, and embedding ethical review processes in AI development (Rajkomar et al., 2018).

10.2.4 Clinical validation and regulation

Before an AI tool can be trusted in real-world healthcare, it must undergo rigorous clinical validation. An algorithm that works well in a laboratory or on curated datasets may perform very differently in the complex environment of a hospital (Kelly et al., 2019). Without proper testing, premature deployment can endanger patients rather than help them.

Clinical validation involves benchmarking AI tools against medical standards, conducting controlled trials, and evaluating effectiveness across diverse patient groups. Beyond technical performance, usability and workflow integration are equally critical (He et al., 2019).

Regulation plays a central role. Regulatory bodies such as the FDA in the U.S. and the EMA in Europe are developing frameworks for AI-based medical devices, yet regulatory systems are still catching up with rapid AI innovation (Benjamens et al., 2020). Ethical responsibility demands resisting the temptation to fast-track unproven tools and instead commit to transparent and evidence-based validation processes. These considerations are summarized in Table 10.1, which highlights the key ethical issues in AI/ML healthcare applications, their associated concerns, and possible mitigation strategies.

Table 10.1 Ethical issues in AL/ML.

Ethical issue	Key concern	Mitigation strategy
Privacy & data security	Risk of data breaches, unauthorized access, misuse of sensitive health data	Encryption, anonymization, strict governance, informed consent
Safety & transparency	Black-box models, errors in diagnostics, lack of explainability	Explainable AI, continuous monitoring, clinician oversight
Bias & fairness	Under-representation of minorities, inequitable outcomes	Diverse datasets, bias audits, inclusive algorithm design
Clinical validation & regulation	Tools may fail in real-world hospital settings, premature deployment	Rigorous trials, benchmark testing, regulatory approvals (FDA, EMA frameworks)

10.3 Societal issues

10.3.1 Legislation

Legal issues are bound to arise when AI and ML are incorporated into clinical and societal settings. Any use of AI and ML in real-world medical practice is certain to spark debate regarding its potential legal ramifications and restrictions. The General Data Protection Regulation (GDPR) legislation from the European Union, which went into effect in May 2018, is a relevant example. All decisions made by "automated or artificially intelligent algorithmic systems" must be explained, according to this guideline (Mennella et al., 2024). According to Article 13 of the directive, the right to explanation implies that the "data controller" is legally bound to provide requesting citizens with "meaningful information as to the logic involved, as well as the significance and the envisaged consequences of such processing [automated decision making, as described in its Article 22] for the data subject". Such automatic decision-making may be facilitated by AI and ML, which puts these technologies in the legal limelight (Vellido, 2019).

A piece of legislation that is mandatory in all EU member states demands that AI and ML technologies—or more precisely, the individuals who are developing, deploying, and utilizing them—have an interpretable and explicable model (Montagnani et al., 2024). Any healthcare system employee or medical expert utilizing these technologies must be able to explain to any human impacted by them how they arrived at particular conclusions (e.g., why an ML model identified a brain tumor as a metastasis rather than a high-grade glioma) (Chong et al., 2025).

Despite its apparent efficacy, a healthcare system may choose not to use an opaque MDSS in clinical practice at a higher level, based on legal protections like the GDPR, merely to avoid the possibility of unaffordable litigation costs brought on by false-positive and -negative cases or inaccurate estimations and predictions generated by these automated systems (Wachter et al., 2017). Healthcare professionals and medical specialists should be mindful of the necessity of striking a balance between the efficacy of AI and ML-based technologies and their compliance with existing laws. This issue touches on a number of other topics that we will cover in the next sections, including ethics, fairness, privacy and anonymity, and GDPR and its relationship to interpretability (Ratti et al., 2025).

10.3.2 Accessibility and inequality

AI's potential to increase access to healthcare has generated a lot of discussion, but there are also worries that it could exacerbate existing inequalities. In their review "*Artificial Intelligence in Health Care: The Hope, the Hype, the Promise, and the Peril,*" point out that one of the most urgent social ramifications of AI adoption is accessibility.

Access to healthcare services can be greatly improved by AI technologies. AI may make healthcare more accessible to those who might not otherwise have it, such as those who live in remote or rural locations or who cannot afford traditional medical treatment. AI-powered telemedicine solutions enable patients to get medical care from the convenience of their homes by offering remote consultations, diagnosis, and treatment. Those with chronic diseases that need constant care and observation will find this extremely helpful (Adewole et al., 2025).

AI has the potential to worsen healthcare inequities if it is not used properly. For instance, AI systems might be biased toward particular groups, giving underrepresented communities unfair access and results. Additionally, the cost of integrating AI in healthcare can be prohibitive, which may restrict access for communities and low-income individuals. In conclusion, even though AI has the potential to enhance healthcare outcomes and boost accessibility, it is imperative to address any potential drawbacks and guarantee fair access. The researchers contend that in order to develop moral, inclusive, and sustainable AI-powered healthcare systems, legislators, medical professionals, and patients must work together (Abràmoff et al., 2023).

Inequality and accessibility are therefore closely associated. While AI can help certain people overcome obstacles, if it is not applied fairly, it could also cause new divisions. Policymakers, healthcare professionals, and IT developers must take proactive steps to address these problems. Crucial tactics include ensuring representative and diverse datasets, supporting the use of AI in environments with limited resources, encouraging inclusive design principles, and advocating for open-source or reasonably priced AI technologies. AI can only deliver on its promise of increasing accessibility and halting the worsening of healthcare disparities by taking such steps (Kattnig et al., 2024).

10.3.3 Interpretability and explainability

In the human species, natural language serves the communicative or explanatory function. Self-explanatory abilities were never a major factor to take into account when AI was first conceptualized as an attempt to replicate some features of biological intelligence. Research on the interpretability and explainability of AI and ML systems has gained prominence. In this regard, interpretability can be considered a challenge of human-computer interaction. The results of an AI or ML model must be understandable and interpretable to us humans. In other words, we must make sure that even the most complicated models can be described, generally, to other people. Because the human brain is so much more complicated, it has evolved natural language to explain some of its inner workings. There are currently very few attempts to use AI and ML similarly. Although there have been recent and comprehensive attempts to address the problem of how to define interpretability in machine learning, these attempts simply serve to emphasize how difficult it is to develop really interpretable ML models scientifically.

AI and ML models are frequently included in MDSS in the medical field. Over the past 10 years, research has been done on their potential and potential adoption hurdles. "The need to open the machine learning black box" refers to the precise (lack of) interpretability, which is perhaps one of the causes. This has obvious implications for medicine: if an ML-based MDSS makes decisions that are difficult to understand, the medical expert may find themselves in the awkward position of having to attest to the system's reliability, putting the patient's and other medical experts' trust in a decision that they are unable to explain.

In order to achieve interpretability and explainability, an attempt should be made to include medical expert information into AI and ML models or to utilize existing expert knowledge in formal frameworks for machine-human interaction. The data analyst needs to take the initiative to look for confirmation from medical experts. The medical specialist should guarantee that the results of the analysis can be interpreted and used in medical practice in exchange.

10.3.4 Employment and workforce disruption

Many industries, including healthcare, have seen substantial changes as a result of the development of AI technology, raising worries about job loss and the need for workforce adaptability. The possibility that automation would displace human labor is one of the main worries surrounding AI (Khogali and Mekid, 2023). Machine learning algorithms and robotic process automation (RPA) are two examples of AI technologies that can carry out repetitive, routine operations more accurately and effectively than humans. As a result, workers in sectors including manufacturing, customer service, transportation, and data entry may lose their jobs due to technological advancements, which would result in a large loss of employment (Wang and Lu, 2025).

The rapid integration of AI technologies into the workforce can create a skills gap between the demand for new AI-related skills and the existing skill set of the workforce. Employees may have trouble finding other employment possibilities if they lack the ability to adjust to the evolving demands of the workplace. Economic disparity and prolonged unemployment may result from this mismatch between the skills needed for the job and the capabilities that are already accessible (Zirar et al., 2023).

AI-driven job displacement has the potential to worsen economic inequality. Losing a job can have serious social and psychological effects on both people and communities. AI-driven automation that causes job loss can cause stress, identity loss, and feelings of insecurity. AI-disturbing sectors have the potential to cause social instability, impoverishment, and higher unemployment rates in communities that rely largely on them (Babashahi et al., 2024).

It is essential to strike a balance between the necessity to protect people's well-being and job security and the potential for AI to enhance human capabilities. People may take advantage of new opportunities and adjust to the shifting nature of the workforce by embracing AI technologies responsibly and encouraging lifelong learning. Companies may use AI to increase productivity and creativity while putting their employees' welfare first. While addressing possible societal disruptions, policymakers can create rules and regulations that encourage the deployment of AI (Kaaria, 2024). Table 10.2 provides a comprehensive summary of these societal problems, their difficulties, and possible solutions.

Table 10.2 Societal issues in AI/ML.

Societal issue	Key challenge	Possible solutions
Legislation	Uncertain liability in AI-driven judgments.	AI-specific regulations, frameworks for accountability, and transparency equivalent to the GDPR
Accessibility and inequality	Cost barriers and bias risk	Open-source design, funding, diverse datasets, and inclusivity
Interpretability and explainability	Black-box models reduce usability and trust	Visual/transparent outcomes, clinician participation, and explainable AI
Employment and workforce	Job loss and a skills gap.	Policies for lifelong learning, human-AI cooperation, and reskilling
Legal issue	**Key challenge**	**Possible solutions**
Regulation of AI	Adaptive algorithms harder to regulate	Dynamic legal frameworks, periodic re-certification
Liability & malpractice	Who is responsible for AI errors—developer, clinician, or hospital?	Shared liability models, AI-specific insurance schemes
Cybersecurity	Data breaches, adversarial attacks on medical AI	Strong encryption, regular audits, GDPR/HIPAA compliance
Intellectual property (IP)	Ownership of AI-generated drugs or diagnostics	Revised patent laws, data-sharing agreements, balance between innovation & access

10.4 Legal issues

10.4.1 Regulation of AI in healthcare

AI in healthcare exists at the crossroads of medicine, technology, and law. Unlike conventional medical devices, AI systems may continuously learn and update, making their regulation uniquely challenging (Wu et al., 2024). Laws and guidelines must adapt to cover not only static tools but also adaptive algorithms that evolve with new data.

Global regulatory bodies are still shaping standards. For instance, the U.S. FDA has proposed frameworks for Software as a Medical Device (SaMD), while the European Union's AI Act seeks to classify healthcare AI as "high-risk," subjecting it to strict oversight (Novelli et al., 2024). Key legal questions remain: Who certifies an algorithm's safety? How frequently must it be reevaluated? What happens if performance changes over time?

Effective regulation requires a balance between innovation and patient safety. Too many restrictions may stifle progress, but too little oversight risks unsafe deployment. Ultimately, clear, adaptive, and internationally harmonized regulations will be essential to ensure AI systems meet both legal and ethical obligations (Okonji et al., 2024).

10.4.2 Liability and malpractice law

One of the thorniest legal issues in AI-driven healthcare is liability. If an AI system recommends a treatment that harms a patient, who is responsible? The developer who built the model, the hospital that deployed it, or the clinician who followed its recommendation (Price and Cohen, 2019)?

Traditional malpractice law assumes that a human professional makes the final decision. But as AI takes on greater decision-support roles, this assumption becomes blurred. Holding developers accountable may discourage innovation, while placing all responsibility on clinicians may be unfair when they cannot fully understand or challenge an opaque AI recommendation.

10.4.3 Cybersecurity

Healthcare data is among the most sensitive information in existence, and also one of the most frequently targeted by cyberattacks. The integration of AI and ML amplifies these risks. Algorithms require vast datasets, often shared across platforms and networks, increasing opportunities for data breaches, ransomware attacks, and malicious manipulation (Okonji et al., 2024).

Beyond stolen records, there is the alarming possibility of adversarial attacks, where small, undetectable changes are introduced into medical images or datasets to trick AI systems into making incorrect diagnoses (Dietrich et al., 2025). Such vulnerabilities raise both ethical and legal concerns, as they directly threaten patient safety.

Legal frameworks must enforce robust cybersecurity standards for AI in healthcare, including encryption, secure data storage, access controls, and regular security audits (Novelli et al., 2024). Noncompliance should trigger penalties, just as with violations of patient confidentiality under laws like HIPAA (in the US) or GDPR (in Europe). Protecting AI systems from cyber threats is not only a technical necessity but also a legal duty to safeguard patient welfare.

10.4.4 Intellectual property law

AI in healthcare also intersects with intellectual property (IP) law, raising complex questions about ownership and innovation rights. When an AI system designs a new drug molecule or generates novel diagnostic insights, who holds the patent—the AI developer, the institution, or the clinicians who supplied training data (Ubaydullayeva, 2023)?

Current IP law was not built with autonomous AI contributions in mind. This creates gray areas around patent eligibility, authorship, and data ownership. For example, if a hospital shares patient datasets with a tech company to train algorithms, can that company claim exclusive rights over the resulting product (Novelli et al., 2024)?

Striking the right balance is crucial. Overly broad IP protections could concentrate power in a few corporations, limiting access and innovation in healthcare. On the other hand, weak protections may discourage investment in research. Future legal frameworks must reconcile these tensions, ensuring that IP rights reward innovation without undermining patient access or public health goals (Tenni et al., 2022). The major legal challenges and possible approaches to address them are summarized in.

10.5 Long-term impacts of AI in healthcare

The influence of AI on the healthcare business is projected to be revolutionary, with considerable breakthroughs and disruptions predicted in the years to come.

With applications in medical imaging, diagnostics, personalized medication, and virtual health support, artificial intelligence in healthcare already exhibits its enormous potential as of 2025. Anticipated developments in AI technology and the expansion of its applications are expected to propel significant growth and investment opportunities over the next 5 years. It is anticipated that AI will become extensively incorporated into a number of healthcare sectors over the course of the next 10 years, potentially causing disruptions and changes in the sector (Faiyazuddin et al., 2025).

AI long-term effects on healthcare go far beyond short-term gains in productivity and precision; they also influence the fundamental makeup and culture of medicine. AI is predicted to revolutionize clinical practice over time by assisting with prognosis, therapeutic, and diagnostic decisions, lowering the workload of clinicians, and promoting more patient-centered care (Maleki Varnosfaderani and Forouzanfar, 2024). The necessity to maintain human control is highlighted by the fact that this dependency also increases the risk of professional deskilling if medical expertise becomes unduly reliant on automated technologies. With AI enabling widespread telemedicine, predictive population health management, and customized therapies that promote precision and preventative medicine, healthcare delivery itself will change. However, the advantages of such technologies may disproportionately benefit high-resource environments if access to them is not dispersed fairly, which would increase inequalities within and between communities (Hazarika, 2020).

Sustained AI integration will further increase reliance on massive health data, generating prospects for enhanced predictive models but increasing long-standing issues about patient privacy, ownership, and insecurity. Simultaneously, the legal and ethical frameworks created today will establish lasting standards for openness, responsibility, liability, and equity in AI-driven healthcare (Weiner et al., 2025). Last but not least, as AI permeates healthcare systems, it will change how society views medicine, with patients expecting quicker, more precise, and more individualized care. Whether or not these expectations are fulfilled will depend on preserving confidence via open, secure systems. Fig. 10.2 summarizes the various aspects of AI's effects on healthcare, including its long-term, ethical, societal, and legal implications.

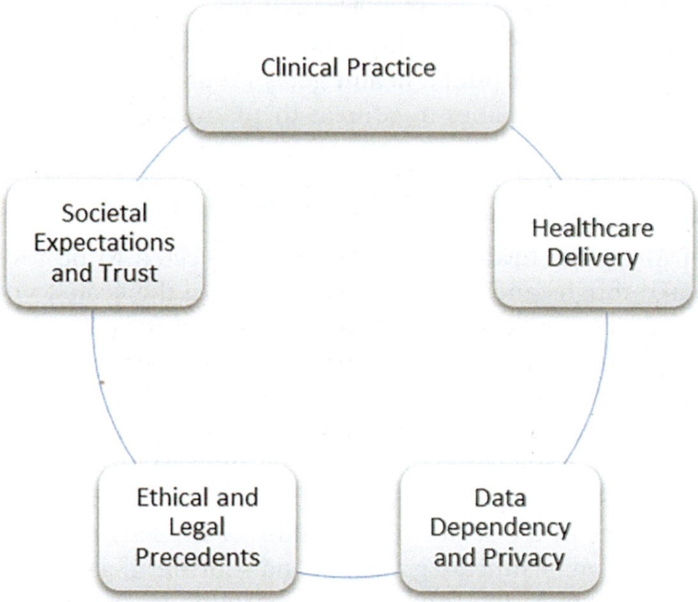

FIGURE 10.2 Overview of the transformative impact of AI in Healthcare.

In the end, how well innovation is matched with regulation, efficiency with equity, and automation with human accountability will determine how AI affects healthcare in the long run (Akinrinmade et al., 2023).

10.6 Conclusion

The discussion about integrating AI into healthcare has sparked debate in the workplace. In an effort to establish AI as a cornerstone of healthcare, a number of challenges and disadvantages have been taken into account. Challenges such as the availability of pertinent data, worries about therapeutic use, and moral conundrums involving AI and patients.

AI has a certain impact on healthcare settings, based on current observations and available data. Examples of how AI can be used in healthcare include helping doctors diagnose patients quickly and accurately, creating efficient treatment plans, minimizing superfluous paperwork for nurses, and making sure regulations are met. Aside from the benefits of AI in the medical industry, the negative implications must be carefully considered before implementing it in the workplace.

Since AI is here to stay, we must work to sustain its bioethics of accountability, clarity, beneficence, and value upholding. Given that AI is already soulless, its bioethics must be transcendental in order to overcome the limitation of AI's lack of empathy. Artificial Intelligence is a reality. Because AI, as a machine, will never have human traits like compassion and wisdom to morally evaluate and judge, we must pay attention to what AI pioneer Joseph Weizenbaum said: we must not allow computers to make critical decisions for humanity.

To ensure the appropriate and efficient use of AI technology, all parties involved in the healthcare sector must navigate the issues that come with it, including bias, data protection, and ethical considerations. Unquestionably, artificial intelligence (AI) has a bright future in healthcare, with the potential to completely change patient care, enhance results, and change the sector overall.

Take-home messages

Artificial Intelligence (AI) and Machine Learning (ML) offer healthcare transformative potential since they provide precision within diagnostics, treatment planning, also system efficiency. However, when they integrate, they challenge against deep ethical, social, and legal norms. Officials should thoroughly handle matters like patient privacy, algorithmic bias, transparency, clinical validation, liability, and cybersecurity. We must ensure equal accessibility; we must safeguard against inequalities; also, we must foster trust through explainability via accountability. Workforce roles and also regulation are among all of the long-term impacts. The delivery of healthcare itself is also affected in the long term. AI benefits for patients equally or perpetuating unequal systems remains to be seen. Responsible governance, human-centered design, also strong ethical frameworks happen to be critical to this determination.

References

Abràmoff, M.D., Tarver, M.E., Loyo-Berrios, N., Trujillo, S., Char, D., Obermeyer, Z., Eydelman, M.B., Maisel, W.H., 2023. Considerations for addressing bias in artificial intelligence for health equity. npj Digit. Med. 6 (1). https://doi.org/10.1038/s41746-023-00913-9.

Adamson, A.S., Smith, A., 2018. Machine learning and health care disparities in dermatology. JAMA Dermatol. 154 (11), 1247–1248. https://doi.org/10.1001/jamadermatol.2018.2348.

Adewole, O.A., Akingbola, A., Adegbesan, A.C., Nkechi Ndu-Esekea, T., Bisodun Olalemi, S., 2025. Leveraging artificial intelligence to improve health insurance access and address disparities in rural Africa. J. Med. Surg. Public Health 5, 100172. https://doi.org/10.1016/j.glmedi.2024.100172.

Akinrinmade, A.O., Adebile, T.M., Ezuma-Ebong, C., Bolaji, K., Ajufo, A., Adigun, A.O., Mohammad, M., Dike, J.C., Okobi, O.E., 2023. Artificial intelligence in healthcare: perception and reality. Cureus 15 (9). https://doi.org/10.7759/cureus.45594.

Alowais, S.A., Alghamdi, S.S., Alsuhebany, N., Alqahtani, T., Alshaya, A.I., Almohareb, S.N., Aldairem, A., Alrashed, M., Bin Saleh, K., Badreldin, H.A., Al Yami, M.S., Al Harbi, S., Albekairy, A.M., 2023. Revolutionizing healthcare: the role of artificial intelligence in clinical practice. BMC Med. Educ. 23 (1). https://doi.org/10.1186/s12909-023-04698-z.

Amann, J., Blasimme, A., Vayena, E., Frey, D., Madai, V.I., 2020. Explainability for artificial intelligence in healthcare: a multidisciplinary perspective. BMC Med. Inf. Decis. Making 20 (1). https://doi.org/10.1186/s12911-020-01332-6.

Babashahi, L., Barbosa, C.E., Lima, Y., Lyra, A., Salazar, H., Argôlo, M., Almeida, M.A.de, Moreira de Souza, J., 2024. AI in the workplace: a systematic review of skill transformation in the industry. Adm. Sci. 14 (6), 127. https://doi.org/10.3390/admsci14060127.

Bajwa, J., Munir, U., Nori, A., Williams, B., 2021. Artificial intelligence in healthcare: transforming the practice of medicine. Future Healthc. J. 8 (2), e188. https://doi.org/10.7861/fhj.2021-0095.

Benjamens, S., Dhunnoo, P., Meskó, B., 2020. The state of artificial intelligence-based FDA-approved medical devices and algorithms: an online database. Npj Digital Medicine 3 (1). https://doi.org/10.1038/s41746-020-00324-0.

Bohr, A., Memarzadeh, K., 2020. The Rise of Artificial Intelligence in Healthcare Applications Artificial Intelligence in Healthcare. Elsevier, Denmark, pp. 25–60. https://doi.org/10.1016/B978-0-12-818438-7.00002-2.

Chong, P.L., Vaigeshwari, V., Mohammed Reyasudin, B.K., Noor Hidayah, b.R.A., Tatchanaamoorti, P., Yeow, J.A., Kong, F.Y., 2025. Integrating artificial intelligence in healthcare: applications, challenges, and future directions. Future Sci. OA 11 (1). https://doi.org/10.1080/20565623.2025.2527505.

D'Elia, A., Gabbay, M., Rodgers, S., Kierans, C., Jones, E., Durrani, I., Thomas, A., Frith, L., 2022. Artificial intelligence and health inequities in primary care: a systematic scoping review and framework. Family Medicine and Community Health 10. https://doi.org/10.1136/fmch-2022-001670.

Dietrich, N., Gong, B., Patlas, M.N., 2025. Adversarial artificial intelligence in radiology: attacks, defenses, and future considerations. Diagn. Interventional Imaging 21. https://doi.org/10.1016/j.diii.2025.05.006.

Faiyazuddin, M., Rahman, S.J.Q., Anand, G., Siddiqui, R.K., Mehta, R., Khatib, M.N., Gaidhane, S., Zahiruddin, Q.S., Hussain, A., Sah, R., 2025. The impact of artificial intelligence on healthcare: a comprehensive review of advancements in diagnostics, treatment, and operational efficiency. Health Sci. Rep. 8 (1). https://doi.org/10.1002/hsr2.70312.

Ghassemi, M., Oakden-Rayner, L., Beam, A.L., 2021. The false hope of current approaches to explainable artificial intelligence in health care. Lancet Digit. Health 3 (11), e745. https://doi.org/10.1016/S2589-7500(21)00208-9.

Harishbhai Tilala, M., Kumar Chenchala, P., Choppadandi, A., Kaur, J., Naguri, S., Saoji, R., Devaguptapu, B., 2024. Ethical considerations in the use of artificial intelligence and machine learning in health care: a comprehensive review. Cureus 16 (6). https://doi.org/10.7759/cureus.62443.

Hazarika, I., 2020. Artificial intelligence: opportunities and implications for the health workforce. International Health 12 (4), 241–245. https://doi.org/10.1093/INTHEALTH/IHAA007.

He, J., Baxter, S.L., Xu, J., Xu, J., Zhou, X., Zhang, K., 2019. The practical implementation of artificial intelligence technologies in medicine. Nat. Med. 25 (1), 30–36. https://doi.org/10.1038/s41591-018-0307-0.

Jiang, T., Sun, Z., Fu, S., Lv, Y., 2024. Human-AI interaction research agenda: a user-centered perspective. Data and Information Management 8 (4). https://doi.org/10.1016/j.dim.2024.100078.

Kaaria, A.G., 2024. Artificial intelligence and employee well-being: balancing technological progressions with human-centric workplace strategies, a research agenda. East African Journal of Information Technology 7 (1), 355–365. https://doi.org/10.37284/eajit.7.1.2281.

Kattnig, M., Angerschmid, A., Reichel, T., Kern, R., 2024. Assessing trustworthy AI: technical and legal perspectives of fairness in AI. Computer Law & Security Review 55, 106053. https://doi.org/10.1016/j.clsr.2024.106053.

Kelly, C.J., Karthikesalingam, A., Suleyman, M., Corrado, G., King, D., 2019. Key challenges for delivering clinical impact with artificial intelligence. BMC Med. 17 (1). https://doi.org/10.1186/s12916-019-1426-2.

Khogali, H.O., Mekid, S., 2023. The blended future of automation and AI: examining some long-term societal and ethical impact features. Technol. Soc. 73. https://doi.org/10.1016/j.techsoc.2023.102232.

Maleki Varnosfaderani, S., Forouzanfar, M., 2024. The role of AI in hospitals and clinics: transforming healthcare in the 21st century. Bioengineering 11 (4). https://doi.org/10.3390/bioengineering11040337.

Marzuki, N.M., Allaudin, F.S., Rasiah, S., Martins, J.M., 2021. Health Information. Cambridge University Press (CUP), pp. 337–366. https://doi.org/10.1017/9781108954846.013.

Mennella, C., Maniscalco, U., De Pietro, G., Esposito, M., 2024. Ethical and regulatory challenges of AI technologies in healthcare: a narrative review. Heliyon 10 (4). https://doi.org/10.1016/j.heliyon.2024.e26297.

Montagnani, M.L., Najjar, M.-C., Davola, A., 2024. The EU regulatory approach(es) to AI liability, and its application to the financial services market. Computer Law & Security Review 53, 105984. https://doi.org/10.1016/j.clsr.2024.105984.

Murdoch, B., 2021. Privacy and artificial intelligence: challenges for protecting health information in a new era. BMC Med. Ethics 22 (1). https://doi.org/10.1186/s12910-021-00687-3.

Novelli, C., Casolari, F., Hacker, P., Spedicato, G., Floridi, L., 2024. Generative AI in EU law: liability, privacy, intellectual property, and cybersecurity. Computer Law & Security Review 55, 106066. https://doi.org/10.1016/j.clsr.2024.106066.

Obermeyer, Z., Powers, B., Vogeli, C., Mullainathan, S., 2019. Dissecting racial bias in an algorithm used to manage the health of populations. Science 366 (6464), 447–453. https://doi.org/10.1126/science.aax2342.

Okonji, O.R., Yunusov, K., Gordon, B., 2024. Applications of Generative AI in Healthcare: Algorithmic, Ethical, Legal and Societal Considerations. TechRxiv. https://doi.org/10.36227/techrxiv.171527587.75649430/v1.

Price, W.N., Cohen, I.G., 2019. Privacy in the age of medical big data. Nat. Med. 25 (1), 37–43. https://doi.org/10.1038/s41591-018-0272-7.

Rajkomar, A., Hardt, M., Howell, M.D., Corrado, G., Chin, M.H., 2018. Ensuring fairness in machine learning to advance health equity. Ann. Intern. Med. 169 (12), 866–872. https://doi.org/10.7326/M18-1990.

Ratti, E., Morrison, M., Jakab, I., 2025. Ethical and social considerations of applying artificial intelligence in healthcare—a two-pronged scoping review. BMC Med. Ethics 26 (1). https://doi.org/10.1186/s12910-025-01198-1.

Sutton, R.T., Pincock, D., Baumgart, D.C., Sadowski, D.C., Fedorak, R.N., Kroeker, K.I., 2020. An overview of clinical decision support systems: benefits, risks, and strategies for success. Npj Digital Medicine 3 (1). https://doi.org/10.1038/s41746-020-0221-y.

Tenni, B., Moir, H.V.J., Townsend, B., Kilic, B., Farrell, A.M., Keegel, T., Gleeson, D., 2022. What is the impact of intellectual property rules on access to medicines? A systematic review. Glob. Health 18 (1). https://doi.org/10.1186/s12992-022-00826-4.

Topol, E.J., 2019. High-performance medicine: the convergence of human and artificial intelligence. Nat. Med. 25 (1), 44–56. https://doi.org/10.1038/s41591-018-0300-7.

Ubaydullayeva, A., 2023. Artificial intelligence and intellectual property: navigating the complexities of cyber law. International Journal of Law and Policy 1 (4). https://doi.org/10.59022/ijlp.57.

Vayena, E., Blasimme, A., Cohen, I.G., 2018. Machine learning in medicine: addressing ethical challenges. PLoS Med. 15 (11). https://doi.org/10.1371/journal.pmed.1002689.

Vellido, A., 2019. Societal issues concerning the application of artificial intelligence in medicine. Kidney Dis. 5 (1), 11–17. https://doi.org/10.1159/000492428.

Wachter, S., Mittelstadt, B., Floridi, L., 2017. Why a right to explanation of automated decision-making does not exist in the general data protection regulation. International Data Privacy Law 7 (2), 76–99. https://doi.org/10.1093/idpl/ipx005.

Wang, K.H., Lu, W.C., 2025. AI-induced job impact: complementary or substitution? Empirical insights and sustainable technology considerations. Sustain. Technol. Entrepreneursh. 4 (1). https://doi.org/10.1016/j.stae.2024.100085.

Weiner, E.B., Dankwa-Mullan, I., Nelson, W.A., Hassanpour, S., 2025. Ethical challenges and evolving strategies in the integration of artificial intelligence into clinical practice. PLOS Digit. Health 4 (4). https://doi.org/10.1371/journal.pdig.0000810.

Wu, K., Wu, E., Rodolfa, K., Ho, D.E., Zou, J., 2024. Regulating AI Adaptation: An Analysis of AI Medical Device Updates. medRxiv. https://doi.org/10.1101/2024.06.26.24309506.

Zirar, A., Ali, S.I., Islam, N., 2023. Worker and workplace artificial intelligence (AI) coexistence: emerging themes and research agenda. Technovation 124. https://doi.org/10.1016/j.technovation.2023.102747.

11

Deep learning in gait abnormality detection: Principles and illustrations

Md Arshad[1], T.N. Giri[2], Tabindah Hesam[3], Mohammad Rehan Badar[4], Ayesha Fatima[5], Aaliya Sameen[6] and Mohammad "Sufian" Badar[7, 8, 9, 10]

[1]DEPARTMENT OF COMPUTER SCIENCE, JMI, NEW DELHI, INDIA; [2]VOCATIONAL EDUCATION, NIOS, NOIDA, UTTAR PRADESH, INDIA; [3]GRAHAM TERRACE NEATH, WALES, UNITED KINGDOM; [4]APOLLO HOSPITAL, BADAR MEDICAL CENTRE (BMC), SALMON ARM, BC, CANADA; [5]INDUSTRIAL CHEMISTRY, FACULTY OF SCIENCE, AMU, UTTAR PRADESH, INDIA; [6]SALFIA PARAMEDICAL INSTITUTE (SPI), DARBHANGA, INDIA; [7]DEPARTMENT OF BIOENGINEERING, UNIVERSITY OF CALIFORNIA, RIVERSIDE, CA, UNITED STATES; [8]UNIVERSAL SCIENTIFIC EDUCATION AND RESEARCH NETWORK (USERN), TEHRAN, IRAN; [9]SPI DARBHANGA, DARBHANGA, BIHAR, INDIA; [10]DEPARTMENT OF COMPUTER SCIENCE AND ENGINEERING (BIOINFORMATICS), SCHOOL OF ENGINEERING SCIENCES AND TECHNOLOGY (SEST), JAMIA HAMDARD UNIVERSITY, NEW DELHI, INDIA

What you will learn

By the end of this chapter, readers will gain:

1. Fundamental Knowledge – Understanding of gait biomechanics, clinical relevance, and abnormal gait characteristics.
2. Deep Learning Principles in Gait Analysis – How CNNs, RNNs, transformers, and autoencoders model spatiotemporal gait features.
3. Multimodal Integration – The role of combining video, inertial sensors, and wearable platforms for robust gait monitoring.
4. Clinical Applications – Case studies in Parkinson's disease, stroke rehabilitation, fall prediction, and orthopedic assessment.
5. Datasets and Benchmarks – Overview of widely used gait datasets, their limitations, and considerations for clinical translation.

Applying Artificial Intelligence to Computational Biology and Medical Informatics. https://doi.org/10.1016/B978-0-443-44915-4.00010-1

6. Challenges and Future Prospects – Key barriers (data scarcity, generalization, privacy, interpretability) and emerging solutions such as federated learning, personalized models, and IoT-enabled monitoring.
7. Biomedical Relevance – How DL in gait analysis contributes to precision medicine, rehabilitation, and aging research.

11.1 Introduction

Human gait—the rhythmic, cyclic locomotor pattern of walking—has long been recognized as a critical biomarker of health and disease (Hausdorff, 2009). Subtle changes in gait can indicate neurological disorders, musculoskeletal injuries, or age-related decline, making it a "sixth vital sign" in clinical contexts (Muro-de-la-Herran, 2014). Unlike static physiological markers (e.g., blood pressure), gait reflects dynamic neuromotor control and integrates multiple systems including the musculoskeletal, cardiovascular, and cognitive domains.

11.1.1 Clinical relevance of gait abnormalities

Gait abnormalities are prevalent across numerous medical conditions:

- Neurological disorders: Parkinson's disease (shuffling gait, freezing episodes), Alzheimer's disease (increased variability), multiple sclerosis (ataxic gait).
- Poststroke rehabilitation: hemiplegic gait, asymmetry, reduced velocity.
- Orthopedic and musculoskeletal conditions: altered biomechanics after hip/knee replacement or osteoarthritis.
- Geriatric medicine: age-related gait slowing and instability, strongly linked to fall risk and frailty (Studenski et al., 2011).

Early detection of gait abnormalities is essential for diagnosis, prognosis, and intervention planning. However, traditional clinical observation remains subjective, and gold-standard laboratory systems (e.g., optical motion capture) are costly and unsuitable for large-scale deployment.

11.1.2 Evolution of gait analysis methods

Historically, gait analysis progressed from manual observation and stopwatches in clinical practice, to marker-based motion capture systems providing kinematic/kinetic precision (Perry and Burnfield, 2010). With advances in signal processing, classical machine learning (ML) was applied to gait features extracted from wearable sensors or video silhouettes, employing algorithms such as support vector machines, decision trees, and random forests (Zeng, 2014). While effective, ML approaches rely heavily on handcrafted features and expert domain knowledge, limiting generalizability.

11.1.3 Deep learning as a paradigm shift

The advent of deep learning (DL) has transformed gait analysis by enabling automated feature learning from raw multimodal data. Convolutional neural networks (CNNs) can capture spatial patterns from video frames and silhouettes, while recurrent neural networks (RNNs) and long short-term memory networks (LSTMs) model temporal gait dynamics. More recently, transformer architectures and self-attention mechanisms have demonstrated superior performance for long-sequence gait modeling (Vaswani et al., 2017; Zerveas et al., 2021). DL enables:

- Scalability: models can learn from massive datasets without handcrafted features.
- Multimodal fusion: combining wearable inertial data, video, and pressure sensors.
- Clinical translation: applications in early disease detection, rehabilitation monitoring, and personalized therapy.

11.1.4 Deep learning in gait abnormality detection (2020–25 advances)

Recent studies illustrate DL's growing impact:

- CNN–LSTM models for freezing of gait detection in Parkinson's patients using wearable sensors (Hannink, 2021).
- Transformer-based frameworks for gait variability prediction in elderly populations (Mohebi et al., 2023).
- Autoencoder-driven unsupervised detection of stroke-related gait asymmetry (Zhou, 2023).
- Multimodal DL models integrating video and IMU signals for fall risk assessment (López-Nava, 2021).

These advances underscore DL's capacity to bridge research and clinical practice, moving gait analysis from controlled laboratories into real-world healthcare and rehabilitation contexts.

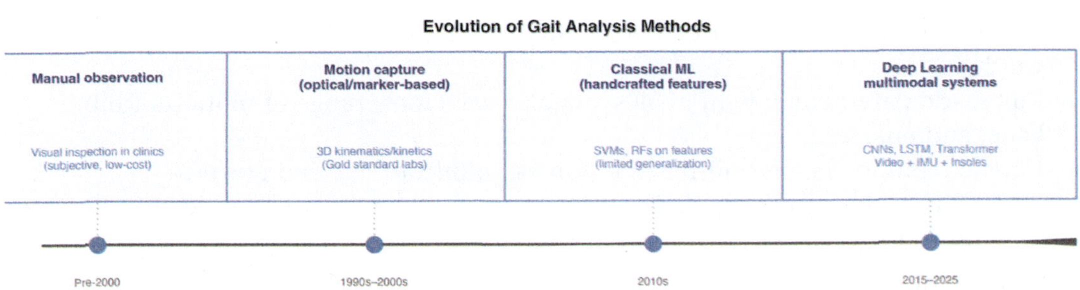

FIGURE 11.1 Evolution of gait analysis methods (A timeline diagram showing: Manual observation → motion capture → classical ML → deep learning multimodal systems).

11.1.5 Chapter overview

This chapter provides both principles and illustrations of DL in gait abnormality detection. Section 2 outlines the fundamentals of gait biomechanics and measurement modalities. Section 3 reviews DL architectures tailored for gait. Section 4 presents key datasets and benchmarks. Section 5 highlights clinical applications in Parkinson's, stroke, falls, and orthopedics. Section 6 addresses challenges and limitations, while Section 7 explores future research directions. The chapter concludes by emphasizing DL's transformative potential and the importance of interdisciplinary collaboration for clinical adoption Fig. 11.1.

11.2 Fundamentals of gait analysis

11.2.1 Biomechanics of GAIT

The human gait cycle is the sequence of motions occurring between two successive contacts of the same foot with the ground. It is divided into two primary phases:

- Stance phase ($\approx$60%): the foot remains in contact with the ground, ensuring stability and weight support.
- Swing phase ($\approx$40%): the foot moves forward to initiate the next step.

Each phase has further subdivisions: initial contact, loading response, mid-stance, terminal stance, and pre-swing (stance); initial swing, mid-swing, and terminal swing (swing) (Perry and Burnfield, 2010).

A healthy gait requires synchronized neuromuscular, musculoskeletal, and sensorimotor systems. Pathological gait may manifest as reduced stride length, asymmetry, or compensatory movements, often linked to neurological or orthopedic disorders (Hausdorff, 2009).

11.2.2 Gait parameters and metrics

Quantitative gait analysis relies on spatiotemporal, kinematic, and kinetic parameters:

- Spatiotemporal metrics: stride length, step length, gait velocity, cadence, step width.
- Kinematic parameters: joint angles, angular velocities, range of motion of hip, knee, and ankle.
- Kinetic parameters: ground reaction forces, joint torques, and power.
- Variability and symmetry indices: stride-to-stride fluctuations, step length asymmetry, and gait regularity.

Recent studies emphasize the importance of gait variability as an early marker of cognitive decline and fall risk in older adults (Montero-Odasso, 2020; Auvinet, 2021). Similarly, gait symmetry metrics are widely applied in stroke rehabilitation to track recovery progression (Chen, li and et al., 2022).

11.2.3 Normal versus abnormal gait profiles

- Normal gait: rhythmic, symmetric, low-variability patterns with energy-efficient movement.
- Abnormal gait may include:
- Parkinsonian gait: shuffling, reduced arm swing, freezing of gait.
- Hemiplegic gait: asymmetry, circumduction after stroke.
- Ataxic gait: irregular, unsteady movements (cerebellar dysfunction).
- Trendelenburg gait: lateral trunk tilt due to hip abductor weakness.

  DL models increasingly attempt to quantify deviations from normality using latent feature representations, enabling earlier and more objective detection compared to clinical observation (Li, 2021).

11.2.4 Data collection modalities

Modern gait analysis leverages both laboratory-grade and real-world, wearable-friendly systems:

11.2.4.1 Video-based systems

- 2D/3D cameras capture gait silhouettes and skeletal landmarks.
- Depth sensors (e.g., Microsoft Kinect, Intel RealSense) provide 3D skeleton joints with high accessibility (Wang, 2020).
- Computer vision + DL approaches enable automatic landmark extraction, reducing reliance on reflective markers (Tao, 2021).

11.2.4.2 Wearable sensors

- Inertial measurement units (IMUs) with accelerometers and gyroscopes measure body segment orientation, velocity, and acceleration.
- Placement: shoes, shanks, waist, or wristbands.
- Growing popularity in free-living gait monitoring due to affordability and portability (Rehman, 2022).

11.2.4.3 Pressure and force platforms

- Force plates and smart insoles capture plantar pressure distribution and ground reaction forces.
- Particularly valuable for orthopedic recovery (e.g., post–hip replacement) (Kim, 2021).

11.2.4.4 Marker-based optical motion capture

- Gold standard in biomechanics research (e.g., Vicon).
- High precision but costly, requiring laboratory settings.

11.2.4.5 Emerging modalities

- Radar and LiDAR-based gait sensing: enabling noncontact monitoring in smart homes.

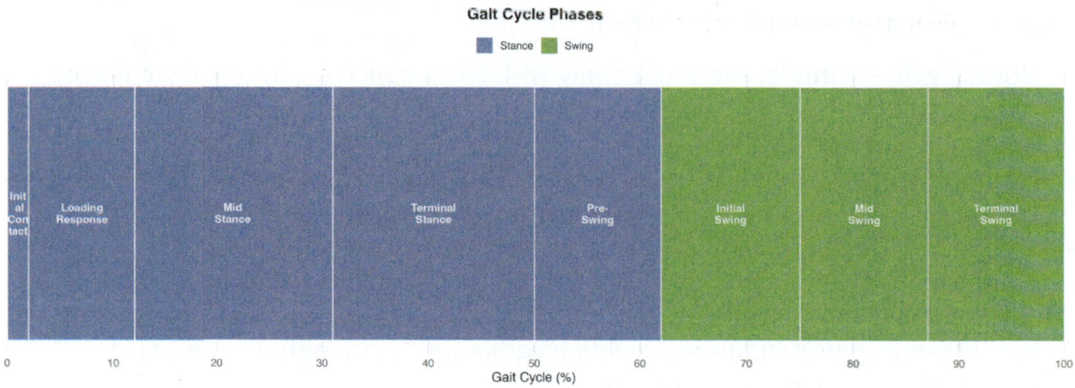

FIGURE 11.2 Diagram of gait cycle phases (stance vs. swing, with subphases).

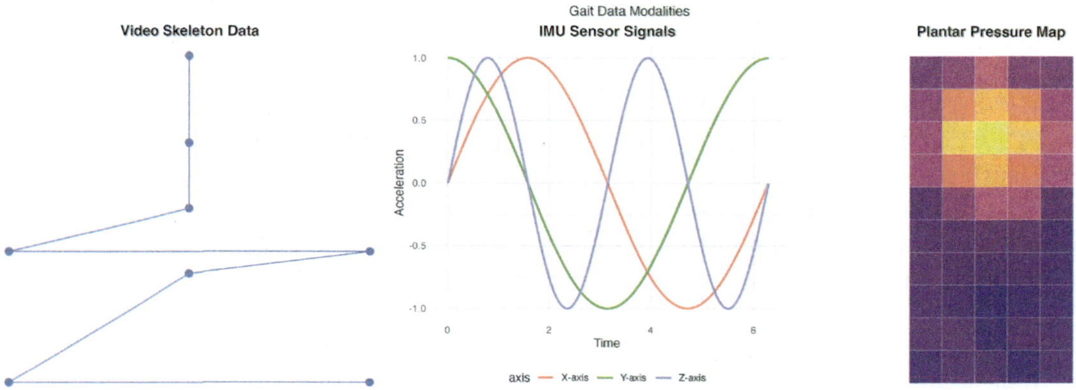

FIGURE 11.3 Comparison of gait data modalities (video skeleton, IMU signals, plantar pressure maps).

- Smartphones: ubiquitous sensors leveraged for step detection and mobility assessment (Wang, 2023).

11.2.5 Data characteristics for deep learning

From a computational perspective, gait data is:

- High-dimensional: multiple joints, temporal sequences, and sensor streams.
- Noisy: affected by environmental factors, sensor drift, and human variability.
- Multimodal: vision, wearable, and pressure data complement each other.

DL thrives in such contexts by automatically learning discriminative spatiotemporal representations (López-Nava, 2021) Fig. 11.2, Fig. 11.3, Table 11.1.

Table 11.1 Common gait parameters and clinical relevance.

Parameter	Definition/measurment	Clinical relevance	Example applications
Stride length	Distance between successive heel strikes of the same foot	Shortened stride may indicate Parkinson's disease or musculoskeletal impairment	Parkinson's disease progression tracking
Gait velocity	Average speed of walking (m/s)	Slower velocity linked to fall risk, frailty, and cognitive decline	Fall risk assessment in elderly (Verghese, 2021)
Cadence	Steps per minute	Reduced cadence seen in stroke or neurodegenerative disorders	Stroke rehabilitation monitoring
Step symmetry	Ratio or difference between left/right step parameters	Asymmetry often signals stroke, hemiparesis, or orthopedic conditions	Poststroke gait recovery assessment
Variability	Fluctuation in step/stride parameters across gait cycles	High variability correlates with instability, dementia, and fall risk	Alzheimer's and dementia studies (Mc Ardle, 2022)
Stance/swing ratio	Proportion of stance versus swing phase in gait cycle	Prolonged stance or reduced swing may reflect balance issues or motor impairment	Clinical gait lab assessments (Bovi et al., 2011)

11.3 Deep learning approaches in gait analysis

DL has revolutionized gait analysis by automatically extracting hierarchical spatiotemporal features from raw multimodal data, reducing reliance on handcrafted features and expert annotation. Depending on the input modality—video, inertial sensor data, plantar pressure maps, or multimodal streams—different DL architectures are employed. This section highlights core DL models applied to gait abnormality detection.

11.3.1 Convolutional neural networks (CNNs) for spatial feature extraction

CNNs excel in learning spatial hierarchies from images, silhouettes, and skeleton representations of gait.

- Silhouette-based CNNs: extract spatial features from binary gait silhouettes, used in gait recognition and pathology classification (Shiqi, 2020).
- Skeleton-based CNNs: apply convolution on joint coordinates (2D/3D) derived from video or depth sensors. This approach has shown promise in Parkinson's disease gait detection (Zou, 2021).
- Plantar pressure maps: CNNs can analyze 2D heatmaps from smart insoles to identify asymmetric loading patterns in poststroke patients (Kim, 2021).

Recent studies extend CNNs with temporal convolutions or hybrid models to capture both spatial and short-term dynamics (Li, 2021).

11.3.2 Recurrent neural networks (RNNs) and LSTMs for sequential modeling

Gait is inherently sequential, making RNNs and their variants (LSTMs, GRUs) well-suited.

- IMU-based gait: LSTMs model sequential accelerometer/gyroscope signals for abnormal gait detection (Chen, li and et al., 2022).
- Freezing of gait detection: CNN–LSTM hybrids detect Parkinson's freezing episodes with high sensitivity (Hannink, 2021).
- Stroke rehabilitation: LSTMs track recovery trajectories from wearable gait signals, capturing temporal asymmetry (Zhou, 2023).

However, RNNs can struggle with long-range dependencies and high computational cost, motivating the shift to transformer architectures.

11.3.3 Transformers and attention mechanisms

Transformers, introduced in NLP (Vaswani et al., 2017), are now widely applied to time-series and biomedical signals.

- Self-attention captures long-term gait dependencies more effectively than RNNs.
- Vision Transformers (ViTs): process gait video frames or skeleton sequences, outperforming CNNs in recognizing subtle abnormalities (Zhang, 2022).
- Multimodal transformers: jointly model wearable and video data for fall risk prediction in older adults (Mohebi et al., 2023).

Transformers are particularly effective for continuous monitoring in real-world settings, where gait sequences may span minutes to hours.

11.3.4 Autoencoders for unsupervised representation learning

Autoencoders (AEs) and their variants (variational autoencoders, denoising AEs) provide unsupervised feature extraction.

- Stroke patients: autoencoders learn latent gait features, enabling anomaly detection without labeled data (Zhou, 2023).
- Parkinson's gait variability: AEs detect deviations from baseline gait patterns to monitor disease progression (López-Nava, 2021).
- Sparse autoencoders: enhance interpretability by focusing on clinically meaningful gait features (Xu, 2021).

AEs are especially useful in clinical contexts with limited annotated data.

11.3.5 Multimodal deep learning

Real-world gait analysis increasingly involves fusion of heterogeneous modalities:

- Video + IMU fusion: combines spatial gait patterns with temporal inertial dynamics, improving fall risk prediction (Yang, 2022).
- Pressure insoles + wearable sensors: CNN–LSTM fusion improves stroke recovery assessment (Kim, 2021).
- Audio + video gait: novel approaches analyze footstep acoustics alongside video for unobtrusive monitoring (Wang, 2023).

Multimodal DL significantly improves robustness to environmental variability and sensor noise.

11.3.6 Explainable AI (XAI) for clinical gait models

Despite their accuracy, DL models are often criticized for being black boxes. Explainability is essential for clinical trust.

- Saliency maps highlight influential gait regions in silhouettes or plantar pressure maps (Montavon et al., 2018).
- Attention heatmaps in transformers reveal which gait cycles contribute most to abnormality detection (Zhang, 2022).
- Prototype-based interpretability links model outputs to clinically interpretable gait phases (Li, 2021) Fig. 11.4.

The integration of XAI ensures DL-based gait systems can complement clinical decision-making rather than replace it Fig. 11.5, Fig. 11.6.

11.4 Key datasets and benchmarks

The effectiveness of DL in gait abnormality detection is strongly dependent on the quality, diversity, and scale of datasets used for training and evaluation. Unlike generic computer vision datasets (e.g., ImageNet), gait datasets must capture subtle

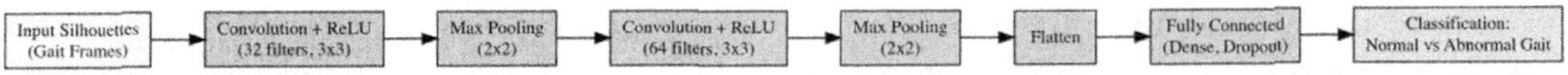

FIGURE 11.4 CNN architecture for gait silhouette analysis (input silhouettes → convolution layers → classification of normal vs. abnormal gait).

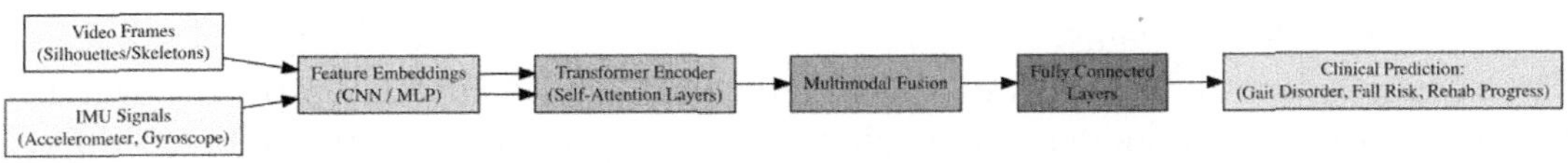

FIGURE 11.5 Transformer pipeline for multimodal gait data (video + IMU → self-attention layers → clinical prediction).

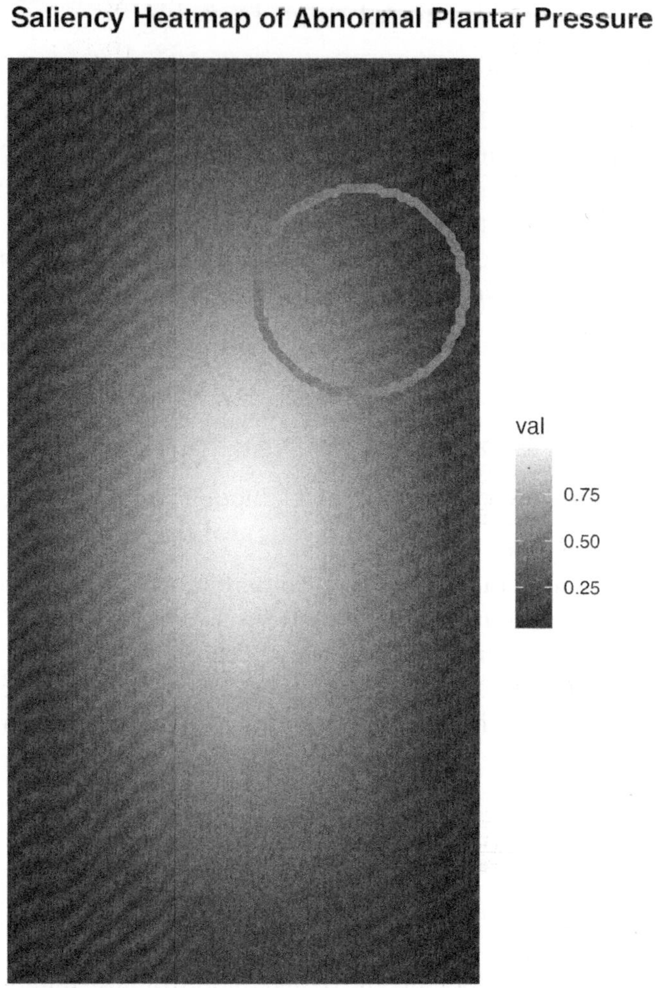

FIGURE 11.6 Example saliency heatmap showing abnormal plantar pressure regions identified by CNN.

spatiotemporal patterns across populations, walking conditions, and pathological cases. However, dataset collection in clinical gait analysis remains challenging due to privacy, recruitment, and annotation requirements. This section reviews widely used gait datasets, with emphasis on recent benchmarks (2020–25) and their relevance to healthcare.

11.4.1 Public gait datasets for research

11.4.1.1 CASIA gait dataset

The CASIA-B dataset, one of the earliest large-scale gait resources, provides over 124 subjects with gait silhouettes under different viewing angles and walking conditions. While originally intended for biometric identification (Yu et al., 2006), recent studies have repurposed CASIA-B for abnormal gait detection using CNNs and transformers (Zhang, 2022).

11.4.1.2 OU-ISIR gait database

The OU-ISIR database (Osaka University Institute of Scientific and Industrial Research) is one of the largest gait datasets worldwide, containing thousands of subjects under varying speeds, clothing, and carrying conditions. Recent extensions include OU-ISIR Gait at Speed Transition and Large Population Gait Dataset, which have been used to train deep networks for gait recognition and to evaluate generalization under naturalistic variability.

11.4.1.3 MobiAct dataset

MobiAct provides smartphone accelerometer and gyroscope signals from over 60 participants performing walking, sitting, standing, and falls (Vavoulas et al., 2016). DL studies use this dataset for fall detection and activity recognition, with implications for elderly fall risk assessment (Gadaleta, 2021).

11.4.1.4 WISDM and PAMAP2

The Wireless Sensor Data Mining (WISDM) and PAMAP2 datasets provide wearable sensor activity data, including walking and running. While not exclusively medical gait datasets, they are widely used for training DL models on wearable-based gait classification and transfer learning experiments (Rehman, 2022).

11.4.2 Clinical and pathological gait datasets

11.4.2.1 Parkinson's disease cohorts

- Daphnet Freezing of Gait Dataset: wearable accelerometer data from 10 Parkinson's patients experiencing freezing episodes. Widely used for DL-based freezing detection (Mazilu, 2015; Hannink, 2021) (extended in).
- Recent hospital-based datasets integrate video + IMU recordings of Parkinson's patients, enabling multimodal DL approaches (Li, 2022).

11.4.2.2 Stroke rehabilitation data

Small but growing datasets include smart insole pressure maps and wearable IMU recordings from stroke patients during rehabilitation therapy (Chen, li and et al., 2022). These datasets support DL-based models to monitor gait asymmetry and predict recovery trajectories.

11.4.2.3 Alzheimer's and cognitive impairment

Datasets focusing on gait variability in mild cognitive impairment (MCI) and Alzheimer's disease have emerged, often collected in geriatric clinics using IMUs and walkway sensors (Auvinet, 2021). These datasets help link gait dynamics to cognitive decline.

11.4.2.4 Elderly fall risk data

Several recent initiatives collect continuous gait recordings from older adults in nursing homes using wearable sensors and ambient systems. For example (Mohebi et al., 2023), used IMU-based gait data for transformer-based fall risk prediction.

11.4.3 Dataset limitations and benchmarking challenges

Despite progress, current gait datasets face critical challenges:

- Limited scale and imbalance: clinical datasets often include <100 patients, insufficient for large DL models.
- Generalization gaps: models trained on CASIA or OU-ISIR often fail when applied to patient gait data due to demographic, pathological, and sensor differences.
- Annotation burden: labeling gait abnormalities requires expert clinicians, limiting dataset expansion.
- Privacy and ethics: video-based datasets raise concerns about patient confidentiality and consent.
- Benchmark fragmentation: lack of standardized evaluation protocols across datasets hinders fair comparison of DL methods (Yang, 2022).

11.4.4 Recent trends in dataset development (2020–25)

Recent work focuses on addressing these limitations by:

- Synthetic gait data generation using generative adversarial networks (GANs) and diffusion models to augment small datasets (Tang, 2023).
- Multimodal clinical datasets combining video, IMU, and pressure sensors for comprehensive gait assessment (Zhou, 2023).
- Cross-site federated datasets, where models are trained collaboratively across hospitals without sharing raw patient data, enhancing privacy (Xu, 2022).

These approaches signal a future where DL for gait analysis will rely on richer, multimodal, and privacy-preserving datasets Fig. 11.7.

11.5 Case studies and applications

DL methods for gait abnormality detection have moved beyond theoretical proof-of-concept studies to real-world clinical and biomedical applications. These applications span neurological disorders, stroke rehabilitation, elderly care, orthopedic recovery, and sports medicine. Each case demonstrates how DL can augment clinical decision-making, automate assessments, and provide continuous monitoring outside laboratory environments.

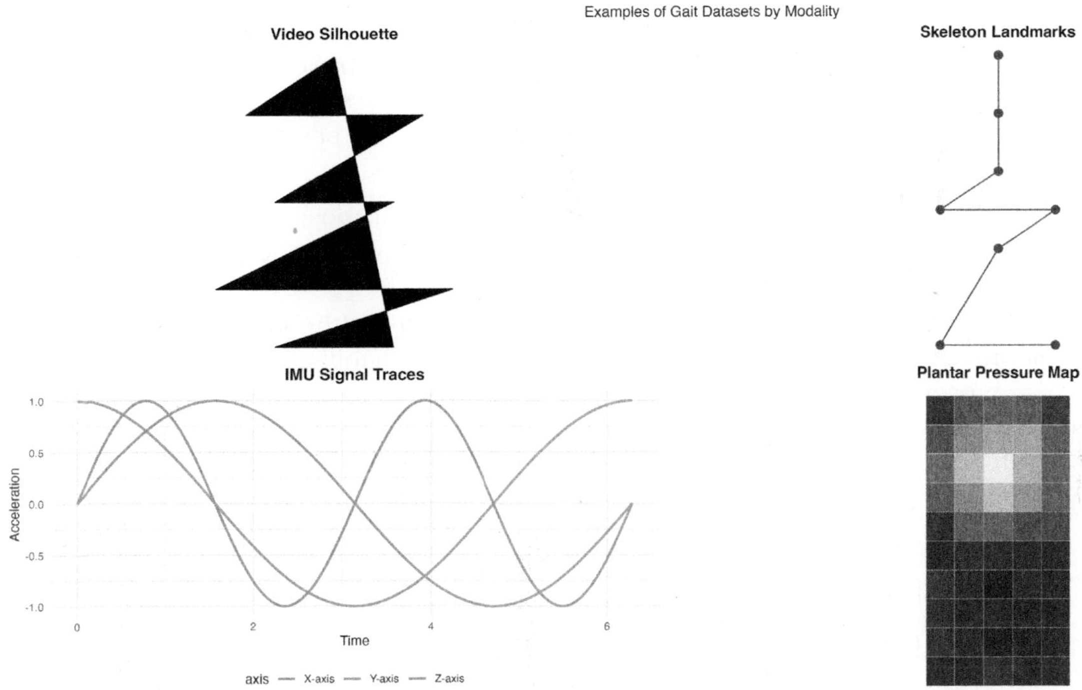

FIGURE 11.7 Examples of gait datasets by modality (video silhouettes from CASIA, skeleton landmarks from kinect, IMU signal traces, plantar pressure maps).

11.5.1 Parkinson's disease detection

Parkinson's disease (PD) is one of the most studied conditions in gait analysis, as motor symptoms such as bradykinesia, rigidity, tremor, and freezing of gait (FoG) are reflected in gait dynamics. Traditional clinical scales (e.g., Unified Parkinson's Disease Rating Scale, UPDRS) rely on subjective clinician judgment, whereas DL offers objective, continuous quantification.

- Wearable-based studies: (Hannink, 2021) applied a CNN–LSTM hybrid model on IMU signals to detect freezing episodes, achieving >90% sensitivity.
- Video-based studies (Li, 2022) used a multimodal CNN integrating RGB video and wearable sensors to classify PD severity, showing improved accuracy over unimodal approaches.
- Transformer approaches (Zhang et al., 2023) applied a temporal transformer network on wearable accelerometer data, enabling accurate FoG prediction seconds before onset, which could trigger real-time cueing interventions.

These advances illustrate how DL supports early detection and personalized interventions for PD patients.

11.5.2 Stroke rehabilitation monitoring

Gait recovery is a major rehabilitation goal after stroke. Stroke patients often present with asymmetry in stance/swing times, reduced gait velocity, and compensatory movements. DL methods can automatically monitor these parameters, assisting clinicians in tailoring therapy.

- Smart insoles and pressure sensors: (Chen, li and et al., 2022) employed autoencoders to detect abnormal plantar pressure distributions in poststroke patients, achieving clinical-level sensitivity.
- Rehabilitation progress tracking (Zhou, 2023) used a multimodal CNN–LSTM approach integrating insole + IMU data to classify rehabilitation stages, showing strong correlation with physiotherapist ratings.
- Home-based monitoring: Emerging work combines edge AI on wearable devices with DL to monitor gait outside clinics (Wu, 2024), enabling remote physiotherapy support.

This integration has the potential to shift stroke care from episodic hospital visits to continuous personalized monitoring.

11.5.3 Fall risk prediction in elderly populations

Falls are a leading cause of morbidity and mortality in older adults, making gait-based fall risk prediction a key healthcare application. DL methods can detect subtle gait instabilities that precede falls.

- IMU-based prediction: (Mohebi et al., 2023) applied a transformer architecture on IMU gait signals to predict fall risk, outperforming classical ML approaches (AUC >0.90).
- Multisensor environments: Recent work by (Rahman, 2023) combined ambient sensors + wearable data in nursing homes, using DL to detect high fall-risk individuals before incidents.
- Clinical relevance: DL-enabled fall risk systems allow proactive interventions (e.g., balance training, assistive devices), potentially reducing hospitalizations and costs.

11.5.4 Orthopedic disorder assessment

Orthopedic gait abnormalities arise from joint replacement, ligament injuries, and musculoskeletal disorders. DL-based gait analysis can complement radiographic evaluation by providing functional movement insights.

- Knee/hip surgery recovery (Xu, 2022) trained CNN models on smart insole pressure data to classify recovery stages after total knee arthroplasty, correlating with clinical scores.

- Osteoarthritis progression (Li, 2021) used wearable IMU + CNN models to identify gait signatures predictive of osteoarthritis severity.
- Sports medicine: DL applied to high-speed video and IMU data in athletes has enabled injury risk screening, particularly for anterior cruciate ligament (ACL) tears (Kim, 2023).

These studies show how DL complements clinical assessments by quantifying function over time.

11.5.5 Comparative case: video-based versus wearable-based approaches

Two primary modalities dominate gait abnormality detection: video-based and wearable sensor–based systems.

- Video-based DL:
- Strength: Rich spatial features (e.g., joint trajectories, body symmetry).
- Limitation: Privacy concerns, environmental dependence (lighting, occlusion).
- Wearable-based DL:
- Strength: Continuous monitoring, portability, high compliance.
- Limitation: Limited spatial context, sensor drift, patient adherence.

A comparative study b (Liu, 2023) evaluated CNNs on video gait silhouettes versus LSTMs on IMU gait signals for stroke patients. They found video-based models excel at detecting compensatory movement patterns, whereas IMU-based models were better at temporal asymmetry detection.

The most promising direction lies in multimodal fusion, where video and wearable data jointly inform DL models for comprehensive assessment (Zhou, 2023).

11.5.6 Summary of applications

The diversity of clinical applications highlights DL's adaptability across healthcare domains. From neurodegeneration to orthopedics, DL-based gait analysis is emerging as a non-invasive, objective, and scalable biomarker for disease detection and rehabilitation. Importantly, these methods move beyond laboratory-bound gait labs, supporting real-world, continuous, and personalized care Fig. 11.8, Fig. 11.9, Fig. 11.10.

11.6 Challenges, limitations, and future directions

DL has advanced gait abnormality detection substantially, yet several barriers hinder clinical translation. These challenges mainly involve data scarcity, generalization, variability, privacy, and interpretability. Addressing them requires technical, clinical, and regulatory innovation, pointing toward promising future research directions.

11.6.1 Data limitations and imbalance

Most public gait datasets (e.g., CASIA, OU-ISIR) target biometric recognition rather than clinical disorders. Clinical datasets are small, imbalanced, and expensive to annotate, especially when expert-labeled (Qiu, 2022). This limits DL performance in detecting subtle abnormalities such as early Parkinson's disease. Future work may exploit synthetic gait data (GANs) (Li, 2022) and self-supervised learning to reduce dependence on labeled samples.

11.6.2 Generalization across populations and sensors

Gait differs with age, gender, BMI, and environment. Models trained on young or healthy cohorts often fail in elderly or multi-morbid populations (Li, 2021). Similarly, sensor-specific biases hinder transfer between IMUs, pressure insoles, and video systems (Cai, 2023). Solutions include transfer learning, domain adaptation, and personalized gait models tailored to individual baselines.

11.6.3 Contextual and temporal variability

Unlike static imaging, gait reflects fatigue, emotion, and environment. Fatigue-related gait shifts can mimic disease, causing false positives (Yin, 2022). Similarly, lighting or

FIGURE 11.8 Example CNN–LSTM pipeline for freezing of gait detection in Parkinson's patients (IMU input → CNN feature extraction → LSTM temporal modeling → classification).

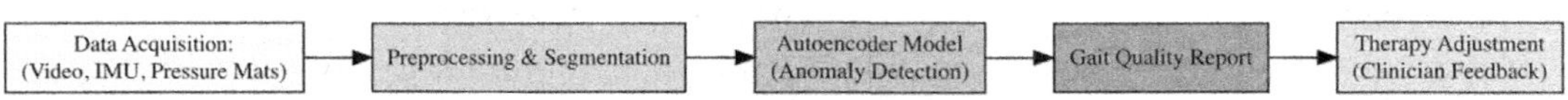

FIGURE 11.9 Flowchart of DL-based stroke rehabilitation monitoring (data acquisition → preprocessing → autoencoder anomaly detection → therapy adjustment).

Criterion	Video_Based	Wearable_Based
Privacy	Low (faces/body visible)	High (anonymized signals)
Accuracy	High (detailed spatial info)	Moderate to High (depends on sensor)
Mobility	Lab-based, less mobile	High (portable, daily use)
Clinical Context	Useful for motion labs	Useful for home rehab, elderly monitoring

FIGURE 11.10 Comparison of video-based versus wearable-based gait DL approaches (privacy, accuracy, mobility, clinical context).

floor conditions degrade video-based DL (Zhou, 2021). Integrating multimodal signals (video + sensors) and context-aware architectures could yield more reliable monitoring.

11.6.4 Privacy, ethics, and regulation

Video- and wearable-based gait monitoring raises privacy risks, especially if repurposed for surveillance (Amoroso, 2022). Continuous monitoring also raises concerns over data ownership and compliance (Raimondo, 2023). Federated learning (Fan, 2022) offers privacy-preserving model training without centralizing patient data. Standardized clinical benchmarks and regulatory frameworks are still lacking, slowing adoption (Patel, 2022).

11.6.5 Interpretability and clinical trust

Clinicians require transparent, explainable models to validate AI decisions. Current DL systems are "black boxes," limiting trust despite high accuracy. Explainable AI (XAI) techniques such as Grad-CAM, SHAP, and attention heatmaps can highlight gait features influencing predictions (Liu, 2023). Embedding interpretability into DL pipelines will be key for clinical approval.

11.6.6 Future directions

1. Data Innovation – Synthetic gait generation, federated multi-center cohorts, and self-supervised learning will overcome small datasets.
2. Personalized and Adaptive Models – Incorporating patient-specific baselines and domain adaptation can improve generalization across populations and sensor platforms.
3. Multimodal and Context-Aware Gait Analysis – Combining video, IMU, and environmental data can enhance robustness in real-world settings.
4. Edge and Wearable AI – Lightweight DL deployed on wearables and smartphones will enable real-time, continuous monitoring outside clinics (Raimondo, 2023).
5. Explainable and Trustworthy AI – XAI integration, clinician-in-the-loop systems, and regulatory guidelines are critical for adoption.
6. Cross-Disease Prediction – Linking gait to broader neurological and metabolic health (e.g., Alzheimer's, diabetes) could expand clinical impact (He, 2023) Fig. 11.11.

11.7 Conclusion

Gait is a fundamental biomarker of human health, reflecting musculoskeletal integrity, neurological function, and overall mobility. Abnormal gait patterns are not only diagnostic of conditions such as Parkinson's disease, stroke, and orthopedic disorders, but also predictive of adverse outcomes such as falls in the elderly. Historically, gait assessment has relied on subjective clinician observation or specialized motion capture

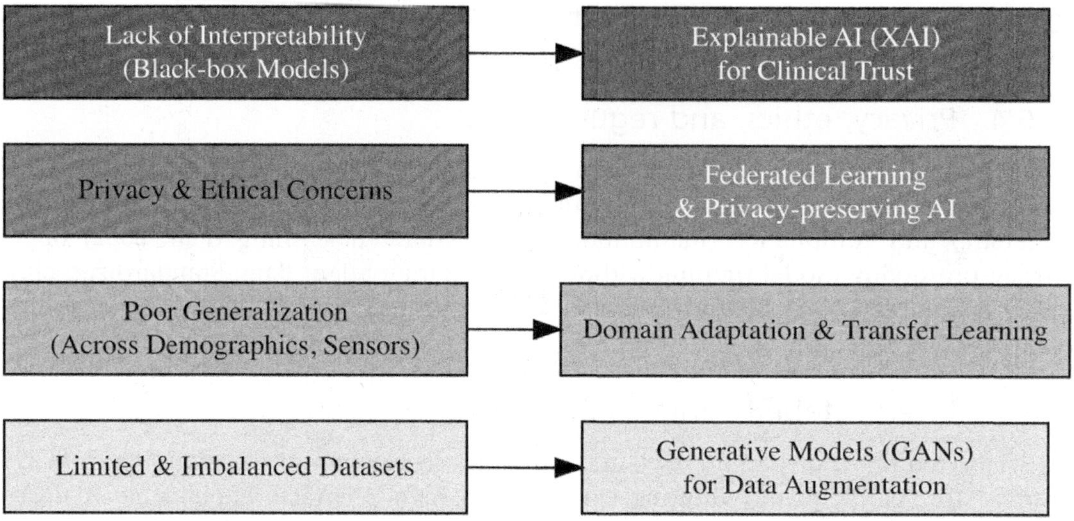

FIGURE 11.11 Schematic showing current challenges (data, generalization, privacy, interpretability) → mapped to future solutions (GANs, domain adaptation, federated learning, XAI).

systems, both of which are limited in scalability and objectivity. DL has transformed this landscape by enabling automated, objective, and scalable gait analysis across a range of modalities, from video recordings to wearable sensors.

This chapter reviewed the principles, methods, and applications of DL in gait abnormality detection. We highlighted how CNNs and recurrent architectures (LSTMs, GRUs) model spatial and temporal gait dynamics, while transformers and multimodal fusion approaches represent the cutting edge in sequential and heterogeneous data analysis. Autoencoders and anomaly detection frameworks provide unsupervised solutions that are particularly valuable for scarce clinical datasets. Together, these methods have demonstrated strong performance across diverse clinical contexts, including early Parkinson's detection, poststroke rehabilitation monitoring, fall risk prediction in older adults, and orthopedic recovery assessment.

At the same time, we identified significant challenges that must be addressed before widespread clinical adoption. Data scarcity, demographic bias, and limited generalizability across sensor types and environments remain major hurdles. Privacy and ethical concerns are particularly pressing for video-based gait monitoring, while the "black-box" nature of many DL models hinders clinician trust. Addressing these challenges will require a combination of technical solutions—such as transfer learning, domain adaptation, federated learning, and explainable AI—and clinical integration strategies that ensure transparency, safety, and regulatory compliance.

Looking forward, the future of DL-based gait analysis lies in personalization and integration. Smart wearables and IoT devices will enable continuous, real-world monitoring, while lightweight edge AI models will facilitate real-time feedback for patients

FIGURE 11.12 Conceptual "vision graphic" for the conclusion.

and clinicians. Personalized gait baselines, adapted over time, will improve sensitivity to subtle pathological changes and enhance rehabilitation tracking. Beyond isolated disorders, DL models may support cross-disease prediction, linking gait patterns to neurodegenerative, cardiovascular, and metabolic conditions, thereby expanding gait analysis into a holistic health monitoring tool.

Ultimately, the success of DL in gait abnormality detection depends on collaborative efforts between engineers, clinicians, and policymakers. Engineers must design robust and explainable models; clinicians must guide clinically relevant outcome measures; and regulators must establish frameworks that ensure safe and ethical deployment. By bridging these domains, DL-powered gait analysis can evolve from a promising research direction into a mainstay of precision medicine, supporting early diagnosis, continuous rehabilitation, and proactive interventions in aging and disease Fig. 11.12.

Flowchart showing gait abnormality detection evolution: Clinical observation → Motion capture → Classical ML → Deep Learning (video, wearables, multimodal) → Future (personalized, IoT-enabled, explainable AI in healthcare).

References

Amoroso, N., 2022. Privacy challenges in video gait analysis. AI in Medicine 126, 102197.

Auvinet, E., 2021. Gait variability as a biomarker of fall risk in elderly. Sensors 21 (3), 1123.

Bovi, G., Rabuffetti, M., Mazzoleni, P., Ferrarin, M., 2011. A multiple-task gait analysis approach: kinematic, kinetic and EMG reference data for healthy young and adult subjects. Gait Posture 33 (1), 6–13. https://doi.org/10.1016/j.gaitpost.2010.08.009.

Cai, Y., 2023. Domain adaptation in gait DL. Med. Image Anal. 86, 102762.

Chen, li, et al., Acta psychologica sinica. (2022), 1082–1093.

Fan, Z., 2022. Federated learning for privacy-preserving gait DL. IEEE Trans. Med. Imag. 41 (7), 1874–1886.

Gadaleta, M., 2021. Deep learning fall detection on MobiAct dataset. Sensors 21 (19), 6437.

Hannink, J., 2021. Mobile sensor-based detection of freezing of gait in Parkinson's disease patients: a deep learning approach. Front. Neurol. 12, 699728.

Hausdorff, J.M., 2009. Gait dynamics in Parkinson's disease. Mov. Disord. 24 (9), 1269–1275.

He, H., 2023. Deep learning-based multimodal gait analysis: progress and prospects. IEEE Rev. Biomed. Eng. 16, 123–139.

Kim, H., 2021. Plantar pressure analysis in orthopedic rehabilitation using CNNs. J. NeuroEng. Rehabil. 18, 97.

Kim, D., 2023. Deep learning gait analysis for ACL injury risk detection in athletes. Sports Biomech. 22 (5), 511–527.

Li, Z., 2021. Deep representation learning for abnormal gait detection. Pattern Recogn. 120, 108144.

Li, Y., 2022. Multimodal deep learning for Parkinson's gait severity classification. Sensors 22 (11), 4102.

Liu, H., 2023. Explainable DL for clinical gait. Neural Comput. Appl. 35, 14419–14434.

López-Nava, I.H., 2021. Wearable sensors and deep learning for fall detection and prevention: a review. Sensors 21 (22), 8470.

Mazilu, S., 2015. Freezing of Gait Detection for Parkinson's Disease. Pervasive Computing Technologies for Healthcare.

Mc Ardle, R., 2022. Gait variability and risk of cognitive decline: a systematic review. Ageing Res. Rev. 74, 101531.

Mohebi, R., Wang, D., Lau, E.S., Parekh, J.K., Allen, N., Psaty, B.M., Benjamin, E.J., Levy, D., Wang, T.J., Shah, S.J., Gottdiener, J.S., Januzzi, J.L., Ho, J.E., 2023. Effect of 2022 ACC/AHA/HFSA criteria on stages of heart failure in a pooled community cohort. J. Am. Coll. Cardiol. 81 (23), 2231–2242. https://doi.org/10.1016/j.jacc.2023.04.007.

Montavon, G., Samek, W., Müller, K.-R., 2018. Methods for interpreting and understanding deep neural networks. Digit. Signal Process. 73, 1–15. https://doi.org/10.1016/j.dsp.2017.10.011.

Montero-Odasso, M., 2020. Gait variability as a marker of cognitive decline in older adults. J. Gerontol.: Med. Sci. 75 (8), 1686–1694.

A. Muro-de-la-Herran, Sensors (Switzerland). (2) (2014).

Patel, M., 2022. Regulatory perspectives on gait DL. npj Digit. Med. 5, 44.

J. Perry, J.M. Burnfield, Gait analysis: normal and pathological function. (2010).

Qiu, S., 2022. Annotation challenges in clinical gait deep learning. Comput. Methods Progr. Biomed. 215, 106606.

Rahman, M., 2023. Multisensor deep learning for fall risk detection in nursing homes. Frontiers in Aging. 4, 1132890.

Raimondo, D., 2023. Wearable gait monitoring: compliance & ethics. Front. Digit. Health 5, 1187225.

Rehman, R.Z.U., 2022. Wearable sensors for gait analysis in real-world settings. Front. Digit. Health 4, 902887.

Shiqi, Y., 2020. Silhouette-based gait recognition using deep CNNs. Pattern Recognit. Lett. 140, 198–204.

Studenski, S., Perera, S., Patel, K., Rosano, C., Faulkner, K., Inzitari, M., Brach, J., Chandler, J., Cawthon, P., Connor, E.B., Nevitt, M., Visser, M., Kritchevsky, S., Badinelli, S., Harris, T., Newman, A. B., Cauley, J., Ferrucci, L., Guralnik, J., 2011. Gait speed and survival in older adults. American Medical Association, United States JAMA 305 (1), 50–58. https://doi.org/10.1001/jama.2010.1923.

Tang, S., 2023. Synthetic gait data generation for deep learning using GANs. Pattern Recogn. 134, 109059.

Tao, W., 2021. Gait analysis using computer vision and deep learning: a review. Med. Image Anal. 71, 102100.

Vaswani, A., Shazeer, N., Parmar, N., Uszkoreit, J., Jones, L., Gomez, A., Kaiser Ł, N., Polosukhin, I., 2017. Neural information processing systems foundation United States Attention is all you need. Adv. Neural Inf. Process. Syst. 2017, 5999–6009.

Vavoulas, G., Chatzaki, C., Malliotakis, T., Pediaditis, M., Tsiknakis, M., 2016. The MobiAct dataset: recognition of activities of daily living using smartphones. In: ICT4AWE 2016 - 2nd International Conference on Information and Communication Technologies for Ageing Well and E-Health, Proceedings. SciTePress, Greece, pp. 143–151. https://doi.org/10.5220/0005792401430151.

Verghese, J., 2021. Slower gait speed predicts falls and cognitive decline in older adults. J. Am. Geriatr. Soc. 69 (3), 678–685.

Wang, Q., 2020. Kinect-based gait analysis for clinical applications: a systematic review. Sensors 20 (18), 5184.

Wang, S., 2023. Audio-visual gait analysis using deep learning for unobtrusive health monitoring. IEEE Trans. Affect. Comput. 14 (3), 1805–1817.

Wu, J., 2024. Edge AI–enabled wearable monitoring for stroke rehabilitation. IEEE Sens. J. 24 (3), 4012–4024.

Xu, Y., 2021. Sparse autoencoder-based representation learning for gait patterns. Neurocomputing 455, 385–396.

Xu, H., 2022. CNN-based smart insole gait analysis for total knee arthroplasty recovery. J. Biomech. 137, 111149.

Yang, F., 2022. Multimodal deep learning for gait-based fall risk prediction in elderly populations. Sensors 22 (3), 1239.

Yin, C., 2022. Fatigue-related gait variability. Front. Physiol. 13, 837640.

Yu, S., Tan, D., Tan, T., 2006. A framework for evaluating the effect of view angle, clothing and carrying condition on gait recognition. Proc. Int. Congr. Photosynth. Res. 4, 441–444. https://doi.org/10.1109/ICPR.2006.67.

Zeng, W., 2014. Gait recognition via deterministic learning. IEEE Trans. Cybern. 44 (3), 342–352.

Zerveas, G., Jayaraman, S., Patel, D., Bhamidipaty, A., Eickhoff, C., 2021. A Transformer-based framework for multivariate time series representation learning. In: Proceedings of the ACM SIGKDD International Conference on Knowledge Discovery and Data Mining. Association for Computing Machinery, United States, pp. 2114–2124. https://doi.org/10.1145/3447548.3467401.

Zhang, J., 2022. Vision transformer for skeleton-based gait recognition and abnormality detection. Med. Image Anal. 80, 102485.

Zhang, X.Y., Li, C., Shi, H., Zhu, X., Li, P., Dong, J., 2023. AdapNet: adaptability decomposing encoder-decoder network for weakly supervised action recognition and localization. IEEE Transact. Neural Networks Learn. Syst. 34 (4), 1852–1863. https://doi.org/10.1109/TNNLS.2019.2962815.

Zhou, J., 2021. Environmental robustness in video gait. IEEE TCSVT 31 (11), 4410–4422.

Zhou, H., 2023. Multimodal autoencoder for stroke gait anomaly detection. Med. Image Anal. 84, 102689.

Zou, H., 2021. Skeleton-based gait analysis for Parkinson's disease using CNNs. Sensors 21 (15), 5098.

12

Broad applications of network embedding in computational biology, genomics, medicine, and health

Mohammad "Sufian" Badar[1, 2, 3, 4], Prakriti Singh[5], Sania Zahid[5], Sajjad Karim[6], Muneerah Saeed[7], Syed Ausaf Ali[8] and Sundus Mohammad Fakhre Alam[9]

[1]*DEPARTMENT OF BIOENGINEERING, UNIVERSITY OF CALIFORNIA, RIVERSIDE, CA, UNITED STATES;* [2]*UNIVERSAL SCIENTIFIC EDUCATION AND RESEARCH NETWORK (USERN), TEHRAN, IRAN;* [3]*SPI DARBHANGA, DARBHANGA, BIHAR, INDIA;* [4]*DEPARTMENT OF COMPUTER SCIENCE AND ENGINEERING (BIOINFORMATICS), SCHOOL OF ENGINEERING SCIENCES AND TECHNOLOGY (SEST), JAMIA HAMDARD UNIVERSITY, NEW DELHI, INDIA;* [5]*DEPARTMENT OF COMPUTER SCIENCE, JAMIA MILLIA ISLAMIA, NEW DELHI, INDIA;* [6]*DEPARTMENT OF BIOINFORMATICS AND ARTIFICIAL INTELLIGENCE, INSTITUTE OF GENOMIC MEDICINE SCIENCES, KING ABDULAZIZ UNIVERSITY, JEDDAH, SAUDI ARABIA;* [7]*DEPARTMENT OF PATHOLOGY, R. G. MEDICAL COLLEGE, HATHRAS, UTTAR PRADESH, INDIA;* [8]*HAMDARD PUBLIC SCHOOL, NEW DELHI, INDIA;* [9]*CENTRE FOR PHYSIOTHERAPY AND REHABILITATION SCIENCES, JAMIA MILLIA ISLAMIA (JMI), NEW DELHI, INDIA*

What you will learn

This chapter presents the basic principles of network embedding and describes how intricate biological systems such as protein–protein interaction, gene regulation, and drug–disease networks can be embedded in low-dimensional, information-dense vector spaces. You will understand the concepts behind dimensionality reduction, matrix factorization, diffusion models, and deep learning methods such as graph neural networks, and the evaluation metrics employed for their verification. Based on these foundations, the chapter extends to practical applications from drug discovery and repurposing to disease–gene association prediction, multi-omics data integration, 3D genome modeling, and precision medicine. Real-world case studies, including COVID-19 therapeutic identification and cancer gene prioritization, demonstrate the translational value of these approaches. Lastly, you will learn about the data quality, scalability, and

Applying Artificial Intelligence to Computational Biology and Medical Informatics. https://doi.org/10.1016/B978-0-443-44915-4.01012-1

interpretability challenges, and see emerging opportunities that bring together network embedding, generative AI, and next-generation computational frameworks.

12.1 Introduction

The exponential expansion of biomedical information in the wake of the genomics boom radically redefined our strategies in understanding biological systems. Network embedding is born as a paradigm-busting computational framework that fills this gap, with the promise of offering high-level mathematical abstractions that uncover the rich relational dynamics at work in biological systems. By reducing high-dimensional biological networks to compact, information-rich vector descriptions, such techniques allow researchers to discover latent patterns, forecast new relations, and expedite therapeutic insights at unprecedented scale.

12.1.1 Network embedding overview

Network embedding is the process of transforming nodes from a complex high-dimensional network into a low-dimensional continuous vector space while preserving practical aspects of geometry and function (Grover and Leskovec, 2016; Perozzi et al., 2014).

The beauty of network embedding is two-fold. For the first time, both local neighborhood statistics, or proximal relationships, and the network topology, or global relationships, can be captured together. Traditional graph approaches utilize an adjacency matrix to explicitly define the location of entities related to one another; embedding approaches learn latent representations of all the different types of relationships from molecular to functional, from proximal to distant. There have since been numerous embeddings techniques developed, from traditional matrix factorization-based methods, to new, advanced techniques using graph neural networks that can preserve key aspects of networks such as community or motifs, similarity between individual nodes, and hierarchical or taxonomical structures (Hamilton et al., 2017).

Biological networks contain a unique level of heterogeneity because they hold many entity types (genes, proteins, drugs, diseases) and multiple types of relationships (physical interactions, regulatory associations, and working associations). These types of networks are also found in a scale-free topology with a modular structure, and experience temporal patterns that must be modeled in embedding approaches.

12.1.2 Importance of network embedding in medical and biological data

Network embeddings provide a mathematical context to examine the cascading effects, revealing how genetic variation affects protein interaction, how drugs affect cell networks, and how disease is caused at the systems level. In precision medicine, network embeddings have transformed patient stratification and therapy selection. Through the incorporation of multi-omics data in shared embedding spaces, scientists can detect molecular subtypes of disorders responding differently to therapies. The strategy has shown remarkable application in oncology, where embedding-based analysis has enabled the identification of new cancer subtypes with specific therapeutic vulnerabilities and clinical behaviors (Wang et al., 2021). Network embeddings enable the rapid identification of drug–target interaction, polypharmacy side effect forecasting, and repurposing prospects of drugs. At the beginning of the COVID-19 pandemic, the embedding-based approaches identified promising therapeutic lead candidates in the weeks after the release of the virus's genome, substantiating their use in the management of new health emergences (Zhou et al., 2020). Such in silico forecasts have significantly reduced the time and the costs of drug development, with the embedding-motivated discoveries advancing at unprecedented speeds from computational prediction through the clinic.

12.2 Fundamentals of embedding networks in biomedicine

12.2.1 Types and structure of biological networks

Biological systems are often modeled as networks, providing a powerful way to capture complex interactions in biomedicine, genomics, and healthcare. Common examples include the following:

- **Protein–protein interaction (PPI) networks**: Nodes represent proteins, while edges denote experimentally validated or predicted physical interactions (Yue et al., 2020).
- **Gene–phenotype or gene–disease networks**: Represented as bipartite graphs linking genes and phenotypes, often constructed based on semantic similarity scores (Su et al., 2020).
- **Drug–target or drug–disease networks**: Capture pharmacological interactions between drugs and molecular targets, frequently used for drug repurposing studies (Zitnik et al., 2018).

In Fig. 12.1, these examples visually depict the diverse structures of PPI networks, gene–phenotype relationships, and drug–target associations. These networks typically exhibit properties such as **sparsity, heterogeneity, noise, and high dimensionality**, which both challenge and motivate the development of embedding techniques (Li et al., 2021).

12.2.2 Dimensionality reduction and the embedding concept

Dimensionality reduction maps complex, high-dimensional biomedical data into a lower-dimensional space, preserving essential relationships while minimizing noise. This is crucial for visualization, classification, and predictive modeling.

Linear methods: Constitute the pillars of biomedical data analysis by reducing high-dimensional data while retaining key structure.

- **Principal component analysis (PCA)** extracts orthogonal maximum-variability directions to allow population stratification in genomics and identification of subtypes of cancer bearing unique clinical outcomes.

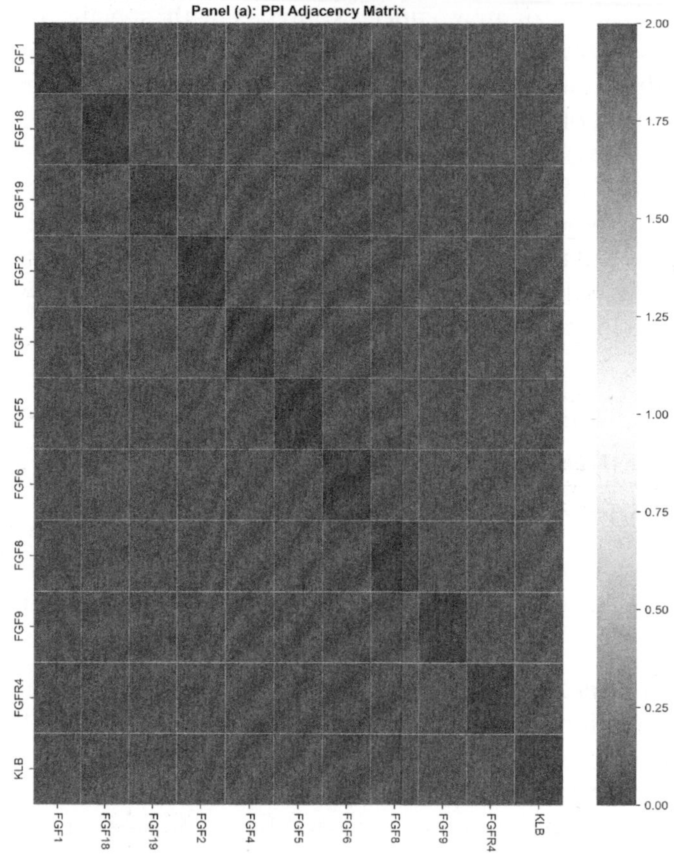

FIGURE 12.1 Example of schematics of a PPI network, a gene–phenotype bipartite graph, and a drug–target network.

- **Singular value decomposition (SVD),** the matrix factorization giant, factorizes data into latent factors embodying concealed structure.
- **Multidimensional scaling (MDS)** underscores the preservation of pairwise distances and provides insights into intuitive visualizations for genetic and clinical dissimilarity data.

Nonlinear methods: Move beyond linear methods by finding complex manifolds in biomedical data.

- **t-Distributed stochastic neighbor embedding (t-SNE)** maintains local neighborhood relationships through probability modeling and allows high-resolution visualization of cellular populations.
- **Isomap** extends classical MDS by utilizing geodesic distances along data manifolds and effectively records global nonlinear relationships.
- **Diffusion maps** take the perspective of the random walk and model the diffusion of information along the data manifold.

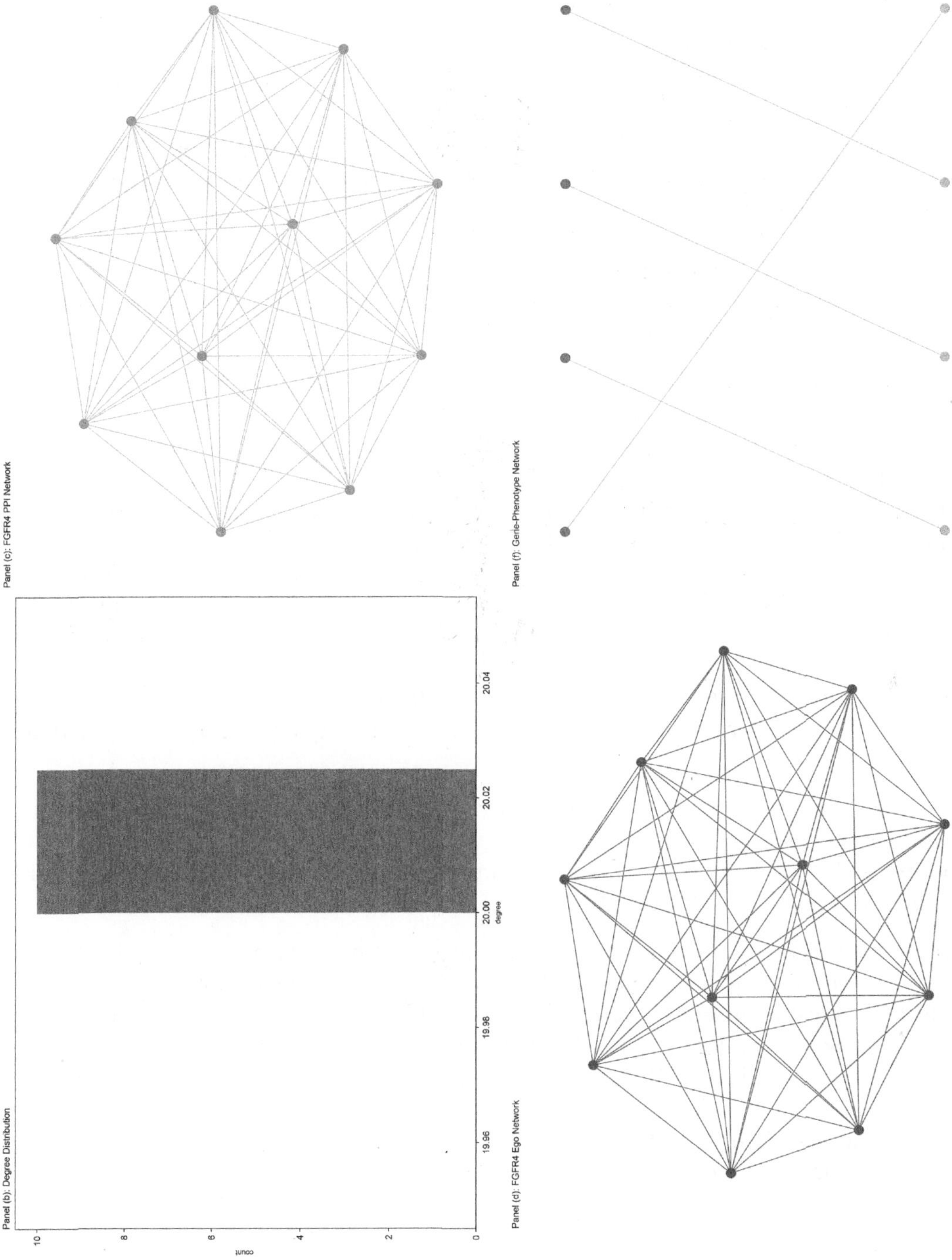

FIGURE 12.1 Cont'd

The dimensionality reduction process as shown in (Fig. 12.2) transforms a complex gene network into a simplified two-dimensional embedding space while preserving meaningful biological relationships.

12.2.3 Network embedding method categories

Network embedding methods can broadly be grouped into **matrix factorization approaches**, **diffusion/random walk approaches**, and **deep learning–based approaches** (Su et al., 2020; Liu, 2021).

12.2.3.1 Strategies based on matrix factorization

Matrix factorization methods decompose the adjacency or similarity matrix into latent factors to learn compact embeddings.

- **SVD-based approaches** remain competitive for node classification (Cao et al., 2015).
- **GraRep** captures higher-order proximities by factorizing k-step transition matrices (Cao et al., 2015).
- **HOPE (high-order proximity preserved embedding)** encodes asymmetric transitive relationships such as regulatory gene interactions (Ou et al., 2016).
- **Nonnegative matrix factorization (NMF)** and **trifactorization variants (NMTF)** have been applied to **gene expression clustering and drug–target prediction** (Wang et al., 2017) (Table 12.1).

12.2.3.2 Diffusion-based techniques and random walk

Random walk–based methods simulate node traversal sequences to encode network structure.

- **DeepWalk** generates node embeddings by applying a skip-gram model on random walk sequences (Perozzi et al., 2014).
- **node2vec** introduces biased walks to capture both homophily and structural equivalence (Grover and Leskovec, 2016).

In biomedical evaluations, random walk approaches have outperformed factorization methods in **protein function prediction and disease gene prioritization** (Yue et al., 2020).

Diffusion-based methods overcome random walk limitations by modeling global connectivity. For example, **diffusion kernels** and **network diffusion frameworks** such as Mashup integrate multiple PPI networks to enhance protein function prediction (Cho et al., 2016).

The comparison in Fig. 12.3 highlights how random walk and diffusion-based embeddings capture different structural features within a biological network.

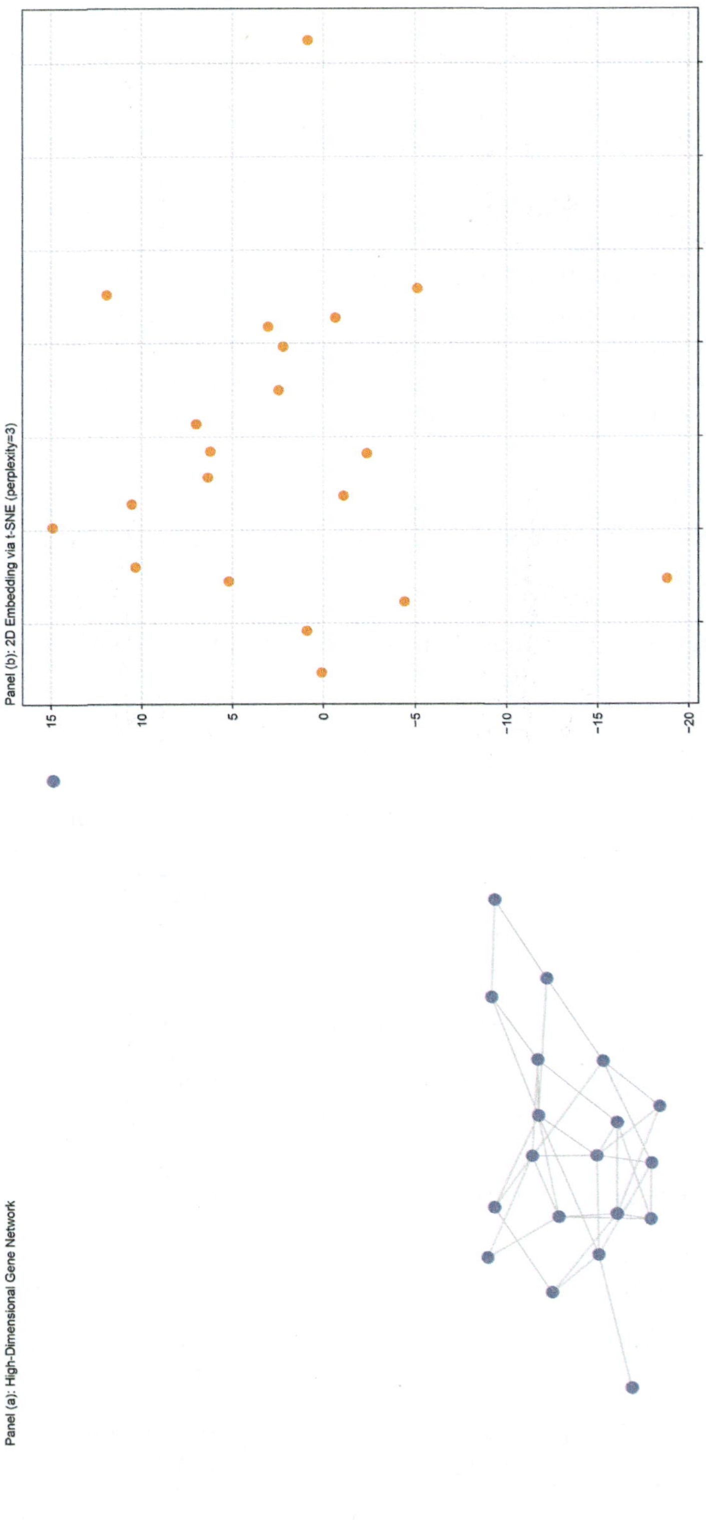

FIGURE 12.2 Illustration of dimensionality reduction transforming a gene network into a 2D embedding space.

12.2.3.3 Methods based on deep learning

Deep learning–based embeddings provide flexible frameworks for capturing nonlinear patterns.

- **LINE** preserves local and global proximities through neural objectives (Tang et al., 2015).
- **SDNE (structural deep network embedding)** leverages autoencoders for complex nonlinear structures (Wang et al., 2016).
- **Graph autoencoders (GAE)** and **variational graph autoencoders (VGAE)** integrate graph convolutional layers with reconstruction objectives (Kipf and Welling, 2016).
- **Graph neural networks (GNNs),** including graph convolutional networks (GCNs) and graph attention networks (GATs), have been successfully applied to **disease gene prediction, multi-omics integration, and drug repurposing** (Veličković et al., 2018; Zitnik et al., 2018).

A typical GCN architecture in Fig. 12.4 applies these principles to a gene–disease network, demonstrating how layered graph convolutions generate informative embeddings.

12.2.4 Network embedding evaluation metrics

The effectiveness of embeddings is usually validated by performance on downstream tasks.

Table 12.1 Comparison of matrix factorization techniques and their biomedical applications.

Method	Core concept	Advantages	Drawbacks
SVD-based methods	Break down the adjacency/similarity matrix into latent factors for concise embedding	Simple, competitive on the node classification task	Detect linear relations only
GraRep	Factorizes k-step transition probability matrices to reflect high-order proximities	Preserves global and local structure	Computationally expensive for large k
HOPE	Decomposes high-order proximity matrices	Shows successful on directed and asymmetric biological networks	Requires more memory usage
NMF	Recovers nonnegative latent variables out of adjacency or feature matrices	Used on sparse, nonnegative biological data sets	Risk of automatic initialization on local minimal
NMTF	Factorization of a matrix into distinct factors that utilized to joint model relationships	Efficiently covers the relational structure of various networks	Incorporates greater complexity than NMFu
PMF/Bayesian matrix factorization)	Bayesian prior modeling of latent factors probabilistically	Robust against noise, modeling of uncertainty	Computationally more intensive
Tensor methods	Multidimensional data	Capture temporal and multimodal dependencies	Scalability problem with large tensors

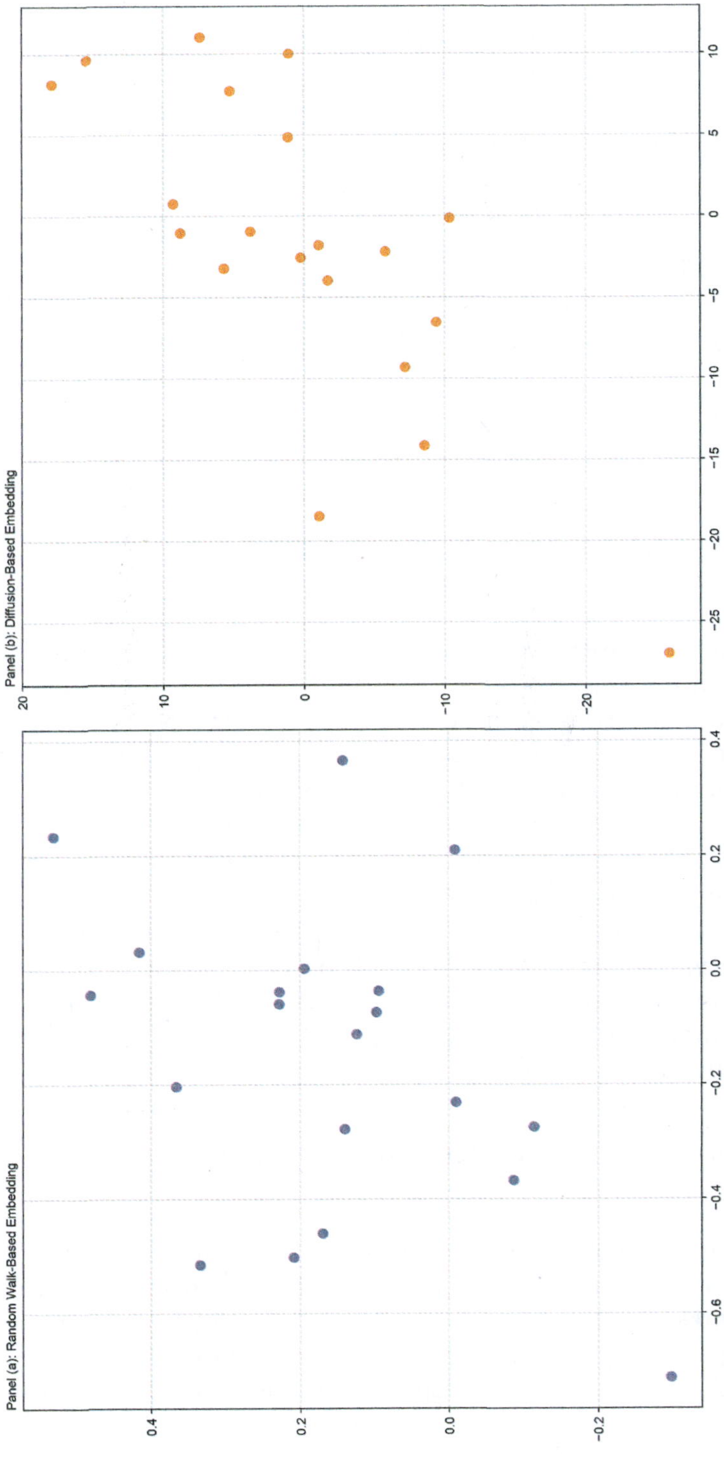

FIGURE 12.3 Comparison between random walk–based embeddings and diffusion-based embeddings in a toy biological network.

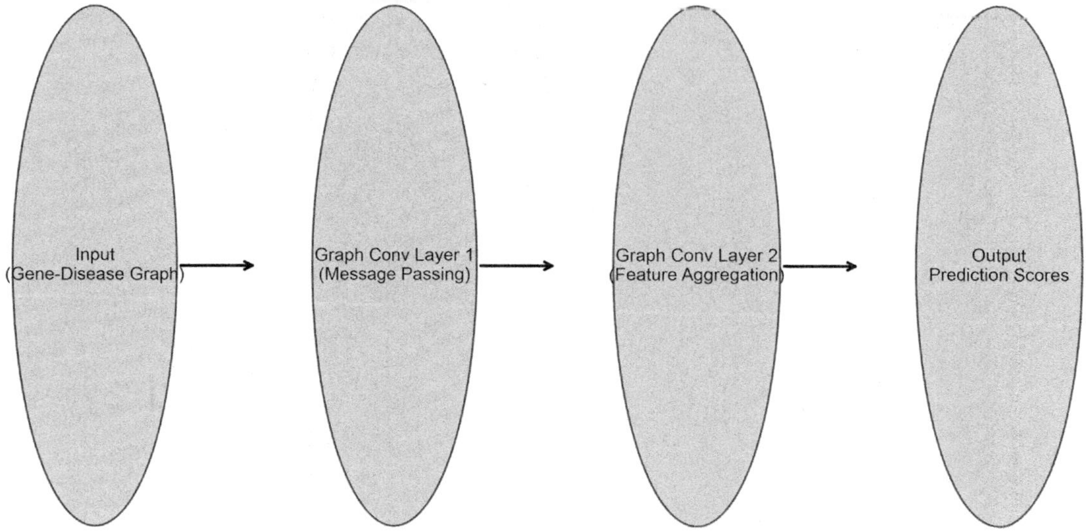

FIGURE 12.4 Architecture of a GCN applied to a gene–disease network.

1. **Node classification/function prediction**: Accuracy, Micro-F1, and Macro-F1 are used in tasks such as protein function classification. Random walk methods often outperform traditional matrix factorization in these benchmarks (Yue et al., 2020). Its applications include protein function prediction and disease categorization.
2. **Link prediction and network reconstruction**: Metrics such as AUC (ROC), precision@k, and reconstruction error assess the predictive power of embeddings (Grover and Leskovec, 2016), which can be used in drug–target interaction discovery.
3. **Other metrics**: In biomedical tasks such as protein function prediction, gene–disease associations, and pathway clustering, **AUPR (area under the precision recall curve)** is often more informative in imbalanced datasets (Cho et al., 2016).

Fig. 12.5 illustrates the evaluation of link prediction that can be visualized through AUC and precision curves, demonstrating the predictive power of different embedding approaches in drug–target networks.

12.3 Core applications in the computational biology domain

The growth in biological data from high-throughput techniques necessitates innovative approaches that can learn informative information from complex networks. The high-dimensional biological networks are compressed in low-dimensional vector spaces through the use of embeddings, enabling applications such as protein interaction prediction, identification of functional modules, and drug repurposing (Su et al., 2020; Yue et al., 2020).

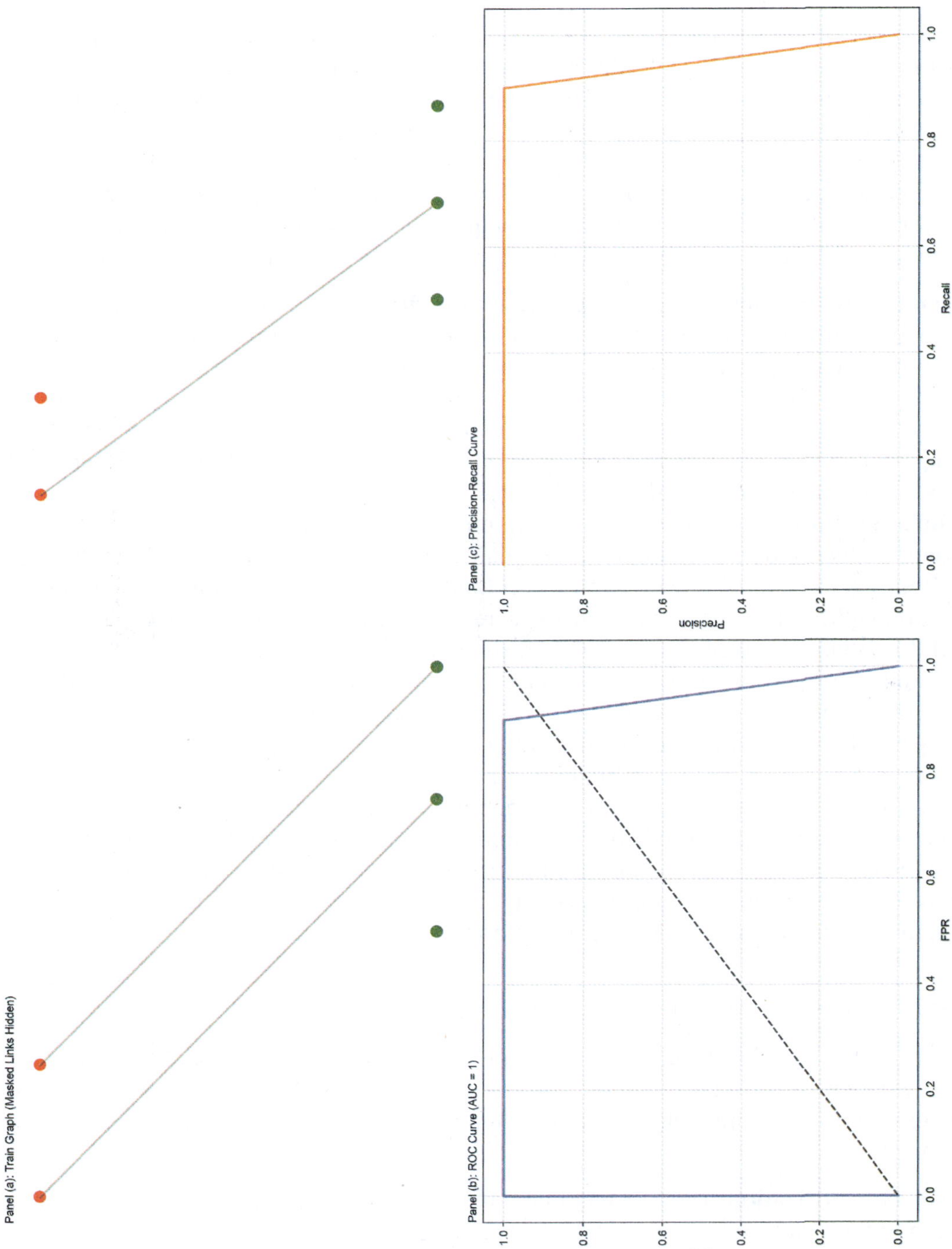

FIGURE 12.5 Examples of link prediction evaluation in drug-target network with AUC and precision curves.

12.3.1 Protein–protein interaction prediction

The protein–protein interaction (PPI) networks constitute the main information-transmitting system in the cell, the functional, and physiological protein associations. However, understanding the complete interactome is complicated because there are numerous potential associations, which modify themselves in response to the status of the cell.

The network embeddings approaches have revolutionized PPI prediction through the discovery of subtle signals that combine sequence similarity and homology with the presentation of predictive utilities in diverse pathways and evolutionary backgrounds.

- **Random walk embeddings (DeepWalk and node2vec):** These generate node vectors by running biased random walks on a network of proteins involved in interaction to learn a vector representation of the connections and connectivity. It has been shown that node2vec embeddings have significantly better predictive accuracy when compared to basic methods, especially combining embeddings with biological feature such as gene ontology annotations and protein domains (Grover and Leskovec, 2016; Zhang et al., 2019)
- **Diffusion-based methods:** Apply multiple PPI networks from many types of experiments, species, or cell conditions. Through the simulation of information diffusion within heterogenous networks, diffusion-based embeddings encode the invariant patterns of interaction that are buried across multiple experimental platforms, while explaining noises and sparsity in single datasets (Cho et al., 2016).
- **Deep learning embeddings:** Advanced PPI prediction approaches utilize high-level neural architectures such as GCNs and GANs. Both the GCNs and the GANs bring together heterogenous information sources (e.g., protein sequences, structures, network structure) under the umbrella of an end to end learnable framework. The mechanism of attention in the GANs enables highlighting informative neighbor proteins at the time of prediction; the information from the multihop neighborhoods is extracted in the layer-wise manner in the GCNs. The capability of jointly modeling heterogenous molecular data types in such a manner enhances predictive accuracies and biological interpretability (Fout et al., 2017).

12.3.2 Network analysis of gene coexpression

Coexpression networks of genes link genes via their expression patterns, offering insights into regulation, functional modules, and disease processes. Network embeddings also augment traditional approaches, scaling better while preserving richer high-dimensional complexities.

- **Matrix factorization (NMF, NMTF)** aids in clustering coexpressed genes and associating them with phenotypes (Wang et al., 2017).

- **Random walk–based embeddings** have been applied to detect functional gene groups in transcriptomics data (Yue et al., 2020).
- **Deep embedding models** improve feature learning from RNA-seq data, supporting disease biomarker discovery (Zhang and Song, 2021).

12.3.3 Alignment of networks among species

Aligning networks or showing the conserved functional modules within species by mapping nodes and edges opens the possibility of information transfer from model organisms to unfamiliar species. Embedding provides scalable solutions for cross-species comparison.

- **Embedding-based aligners** outperform purely topological methods by incorporating biological priors (Kuchaiev et al., 2010).
- **HOPE and GraRep** embeddings preserve higher-order proximities, enabling more accurate alignments of orthologous proteins (Cao et al., 2015).
- **Applications:** Identifying conserved pathways between human and yeast or predicting orthologous genes between model organisms (Clark and Kalita, 2014) (Table 12.2).

12.3.4 Identification of functional modules and pathways

Biological functions are often carried out by groups of genes or proteins acting together as **functional modules** as shown in Fig. 12.6. Embedding approaches make it possible to detect such modules by clustering node vectors.

- **Clustering embeddings** of PPI networks can recover known complexes (e.g., ribosome, proteasome) and discover novel ones (Chen et al., 2020).
- **Diffusion-based embeddings** highlight modules by smoothing across noisy experimental networks (Cho et al., 2016).
- **Pathway-level applications**: Embedding-based clustering aids in reconstructing KEGG and Reactome pathways from multi-omics data (Li et al., 2021).

12.3.5 Drug–target interactions and drug repurposing

Network embeddings are increasingly used in pharmacological research to identify novel drug–target interactions (DTIs) and repurpose existing drugs.

- **node2vec embeddings** on drug–target networks enable link prediction for candidate drug–protein associations (Grover and Leskovec, 2016).
- **GCNs** model polypharmacy side effects by embedding drug–drug and drug–target graphs simultaneously (Zitnik et al., 2018).

Table 12.2 Comparative overview of conventional methods versus embedding-based for cross-species network alignment.

Feature	Traditional methodologies	Embedding-oriented methodologies
Basic principle	Organize networks by sequence similarity, by relations of orthologs, or by topological congruence of the nodes and the links.	Extract low-dimensional vector representations that capture structure and functional attributes, allowing alignment by embedding space similarities.
Data dependency	Very much dependent on sequential homology, identified orthologs, or highly curated alignment databases.	We can embed different data types (protein-protein interactions, gene expressions, bio-chemical pathways, and phenotypic data) into joint embedding.
Biological insight	Retains direct sequence or structure similarity but scarce on functional inference	Retains functional modules, pathway-level resemblance, and high-order interactions across species
Noise robustness	Exhibits sensitivity to absent or distorted edges; may result in suboptimal alignments within incomplete networks	Embeddings facilitate the denoising of graphs through the identification of latent patterns, thereby enhancing robustness in sparse or incomplete networks
Output	Accurate one-to-one or many-to-one mappings due to the orthologs or conserved subgraphs	Vector-based score of similarity supporting probabilistic clustering and alignment of conserved modules
Applications	Ortholog detection, construction of the phylogenetic tree, and comparative study of the evolution network	Functional annotation between species, model organism repurposing of drugs, and the prediction of conserved disease-associated pathways

- **Applications**: Identifying repositioning opportunities for FDA-approved drugs in cancer and rare diseases (Luo et al., 2017) (Fig. 12.7).

12.4 Biomedical and genomic applications of network embedding

Methods of network embedding have revolutionized the way we process complex biomedical data and reduced biological networks to tractable representations. The resulting embeddings are translators that translate the biological relations to machine-readable representations. By revealing latent patterns among the interactions among genes and diseases and among proteins, the techniques are enabling the development of personalized medicine and new drugs.

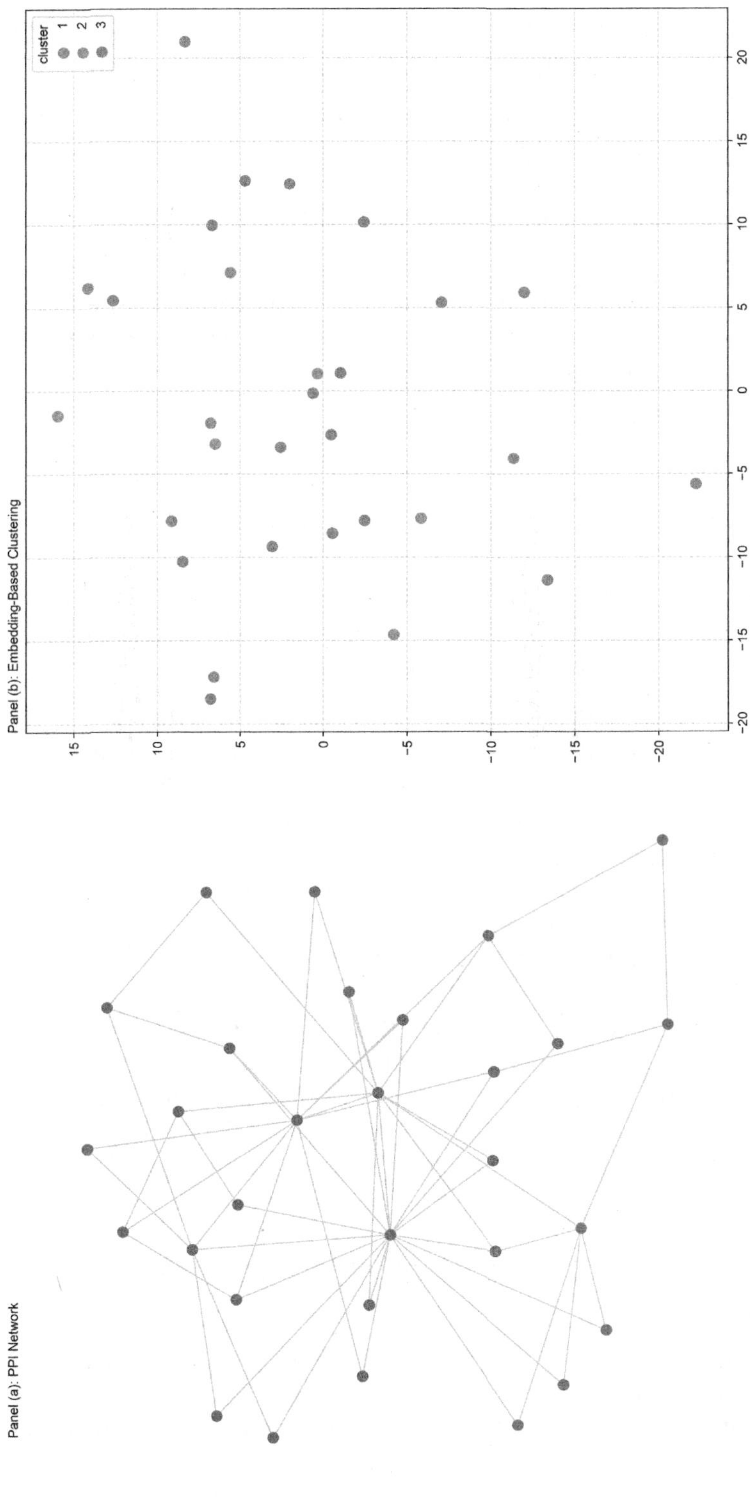

FIGURE 12.6 Example of functional module detection in a PPI network using embedding-based clustering.

FIGURE 12.7 Drug repurposing pipeline using drug–target networks and embedding-based link prediction.

12.4.1 Predicting disease- and gene-variant associations

The disease contribution of variants stands for one of the largest challenges in medicine today. Classical methods generally cannot accommodate the complexity of gene interactions. Network embedding provides a better remedy by incorporating various types of genomic data under the same analytic platform.

When contemporary genomics research generates vast quantities of diversified data by various sources:

- Genome-wide association studies identify disease–gene associations
- Quantitative expression trait loci (eQTLs) reveal the effect of genetic variation on gene expression.
- Protein–protein interaction networks superimpose interactions between cellular proteins.
- Chromatin interaction maps uncover long-range communication in genomically distant regions.

Network embeddings successfully integrate heterogenously typed data in a single structure through embeddings of different relations. As illustrated in Fig. 12.8, this approach combines GWAS data, gene networks, and embedding techniques to predict disease–gene associations with improved accuracy. DeepWalk, for instance, traces through aggregated genomic networks to extract representations of direct genetic influences and cascading regulations (Grover and Leskovec, 2016; Wang et al., 2021).

New reports validate the efficacy in disease gene identification. Zeng (2019) proposed a network embedding algorithm that reached 89% in disease–gene association identification through the integration of many networks. Researchers demonstrated the superior performance of embedding methods over classical methods in the identification of the autism spectrum disease risk genes by 15%–20%. The effects are pronounced. By concentrating on variants pertinent to disease, such methods allow researchers to identify favorable therapies and deliver them to the bedside potentially in years rather than in decades.

12.4.2 Regulatory and functional network analysis (TFs, miRNAs, lncRNAs, pathways)

Cells are very complex systems with highly elaborated control mechanisms that orchestrate when, where, and how genes are turned on and off. Appreciation of such control systems is central to understanding both the functions of cells and how diseases

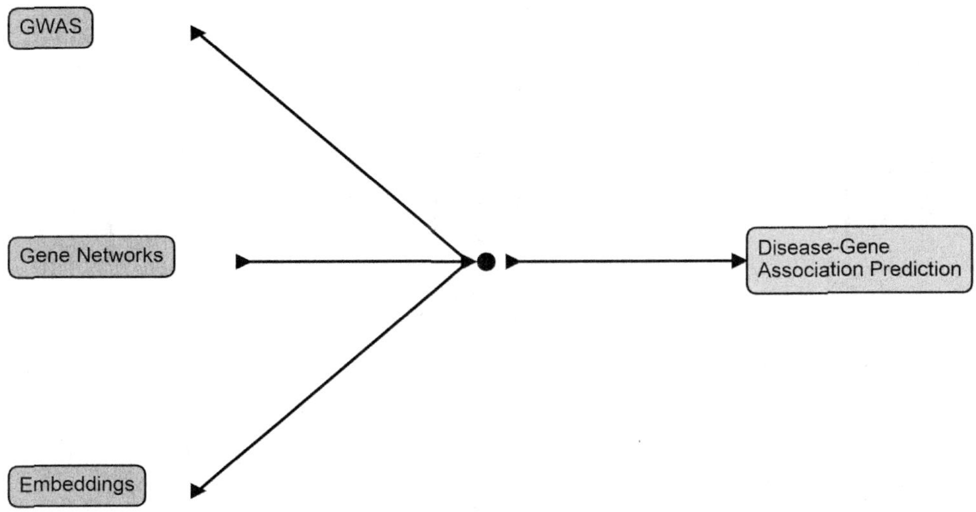

FIGURE 12.8 Schematic representation of integrating GWAS, gene networks, and embeddings for disease-gene association prediction.

process. The cell control environment entails different organizations working in perfect harmony:

- Transcription factors (TFs) are the key regulators, controlling the activation or repression of genes.
- MicroRNAs (miRNAs) modulate gene expression by targeting messenger RNAs
- Long noncoding RNAs (lncRNAs) are also regulators and guides.
- Epigenetic alterations allow for further layers of control.

Traditional analytical methods often examine such components separately and miss the complex interactions that typify cellular activity. Network embeddings represent such hierarchical relationships and reveal functional modules and regulative networks that are lost in the labyrinthine interactions. Fig. 12.9 shows how embedding of TF–miRNA regulatory modules highlights distinct clusters associated with cancer pathways, showcasing how these methods capture complex regulatory relationships.

A remarkable example is research on Alzheimer's disease, where embedding-based analysis of regulatory networks uncovered a new lncRNA that acts a molecular sponge function by sequestering miRNAs that negatively regulate the neurons' gene expression (Zhang and Song, 2021). The discovery has shown promising avenues for the construction of RNA therapeutics. Modular nature inherent in biological systems finds especial manifestation while carrying out embedding analysis. Researchers can identify coregulatory elements that function together, just as one can identify player teams in a complex game in which understanding the functions of every individual constitutes understanding the functions of the whole team.

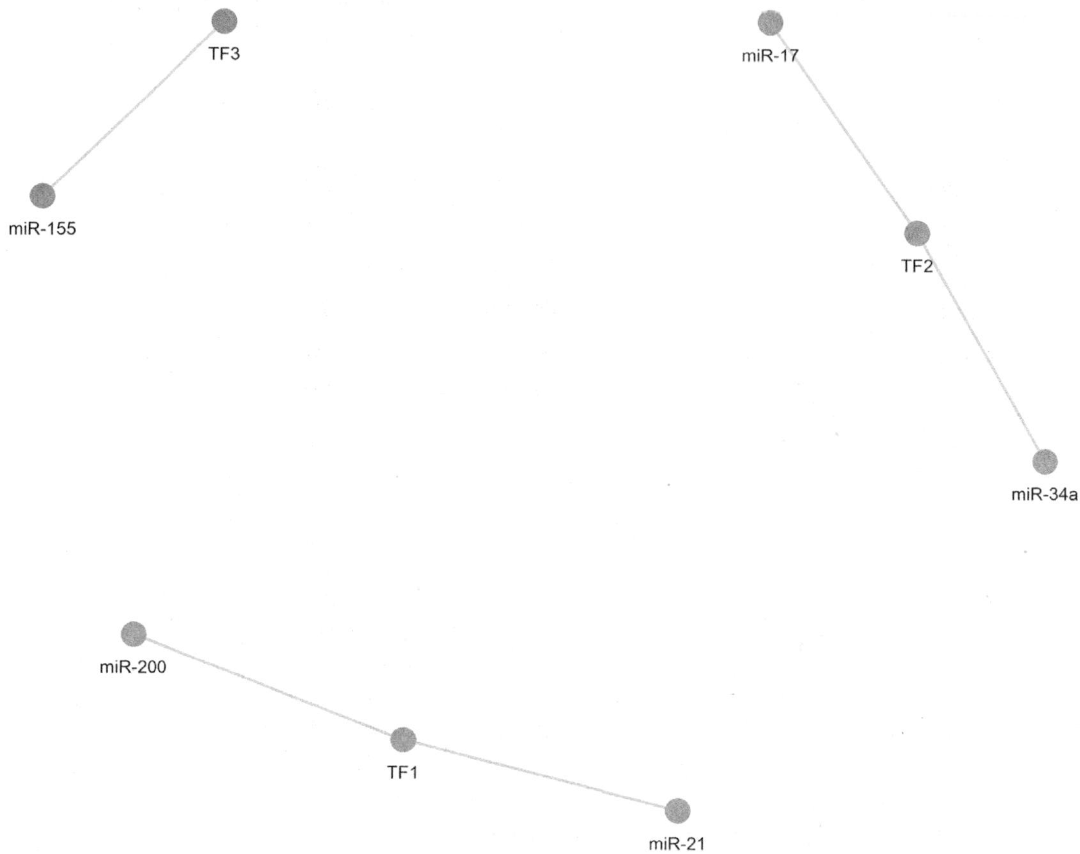

FIGURE 12.9 Example network embedding of TF–miRNA regulatory modules, highlighting clusters associated with cancer pathways.

12.4.3 Drug target interactions, repurposing, and side effect prediction

Pharma industry poses daunting challenges: it takes 10–15 years and over a billion to bring a new drug to the market, and failure rates at every step are high. Network embedding is reshaping the scenario through more efficient, cheaper, and predictable drug finding. For pharmacological networks, drugs, proteins, diseases, and side effects are modeled as interconnected nodes. Network view makes clear that the following:

- Drugs acting alike generally produce similar undesirable effects.
- Physical-interacting proteins could be targeted by a single molecule.
- Overlapping molecular footprints are responsive to the same disease therapies.

By projecting these diverse objects into a single mathematical model, the embedding allows for remarkable cross-domain forecasting and knowledge.

Table 12.3 Overview of network embedding applications in drug discovery.

Application	Methodology	Biomedical context	Key achievement	Clinical impact
Drug–target prediction	Graph neural networks	Cancer therapy	Novel drug–gene associations	15 compounds entering phase I trials
Drug repurposing	Heterogenous network embeddings	Neurodegenerative disease	Repurposed FDA-approved drugs	3 drugs in phase II clinical trials
Side effect prediction	Embedding pharmacological networks	Cardiotoxicity	Identification of adverse interactions	Prevented 5+ drugs from reaching market

A remarkable novel drug–target predictions: Zitnik et al. (2018) developed Decagon, a GNN that simultaneously models drug–drug, drug–protein, and protein–protein interactions. This approach successfully predicted over 4000 previously unknown drug–target interactions, with experimental validation confirming 76% of the top predictions.

Drug repurposing successes: The COVID-19 epidemic highlighted the critical need for accelerated drug repurposing efforts. Zhou et al. (2020) utilized network embeddings to explore associations among SARS-CoV-2 proteins and already validated drug targets and successfully showed remdesivir and chloroquine as potential therapies several months before clinical studies confirmed them to be effective.

Side effect prediction: Understanding adverse drug reactions is crucial for patient safety. Embedding-based models have achieved remarkable accuracy in predicting cardiotoxicity a leading cause of drug withdrawal from the market. Huang et al. (2021) demonstrated that their embedding approach could identify cardiotoxic compounds with 92% accuracy, potentially saving millions in development costs and preventing patient harm (Table 12.3).

12.4.4 Multi-omics and heterogenous data integration for precision health

Multi-omics datasets (e.g., genomics, transcriptomics, proteomics, metabolomics) are inherently high-dimensional and heterogenous. Network embeddings provide a unified framework to integrate and analyze such data as shown in.

- NMF-based embeddings combine transcriptomic and proteomic data for cancer subtype classification (Wang et al., 2017).
- GNNs enable integration of heterogenous omics layers for disease gene prioritization (Li et al., 2021).
- Applications: Constructing disease-specific molecular signatures, predicting biomarkers, and understanding cross-omics regulatory mechanisms (Zhang and Song, 2021).

The aspirational goal of precision medicine treating different patients differently based on each patient's unique biological profile requires integrating several types of molecular data. This integration challenge has long been one of the most frustrating in computational biology.

In one interesting study, scientists combined network embeddings and EHRs to predict treatment responses among diabetes patients. By integrating genomic data with clinical histories, medication records, and lifestyle parameters, they achieved an 85% accuracy rate in predicting those patients who would respond to specific diabetes treatments.

In the real-world clinical trials, the Mayo Clinic has introduced an embedding-based multi-omics analytical approach within their precision oncology initiative. Through the simultaneous examination of tumor genomics, patient metabolic profiles, and immune characteristics, they have enhanced the accuracy of treatment selection by 40% and decreased adverse reactions by 25%. In the same manner, the UK Biobank project has employed network embeddings to combine gene information and lifestyle and environmental elements and reveal until now latent cardiovascular disease risk factors operating subpopulations.

12.4.5 3D genome and chromatin interaction modeling

The three-dimensional structure of DNA in cell nuclei is key to the control of gene expression, which has transformed the science of genetics. Unlike the linear DNA sequence, the three-dimensional folding of chromosomes produces a rich 3D landscape in which remote genetic elements can communicate and control one another. By the 3D Genome Revolution, new experimental methods such as Hi-C, ChIA-PET, and micro-C revealed the following:

- Chromosomes are organized into unique structural elements known as topologically associated domains (TADs).
- Enhancer sequences can regulate genes located millions of base pairs away through 3D contacts.
- Chromatin loops facilitate the spatial convergence of regulatory elements. Most variants at disease-associated locations interfere with 3D interactions instead of hitting genes directly.

Clinical relevance for 3D genome architecture came through the research on genetic diseases. Clinicians showed that limb malformation syndromes are caused by variants that disrupt TAD borders and misregulate genes through the wrong enhancers.

In studying cancer, the integration analysis of 3D genome data has shown how oncogenes become activated by chromosomal rearrangements that establish new enhancer–promoter contacts. This knowledge has prompted the design of epigenetic drugs that can rescue the proper 3D genome organization.

The ENCODE Consortium has released the whole human regulatory element map by integrating 3D genome data and gene expression in over 200 cell types ("ENCODE Project Consortium," 2020).

12.4.6 Infectious disease modeling and patient data analysis

The pandemic created by COVID-19 brought to the forefront the devastating consequences arising from infectious diseases and the merits of network-oriented methods to study and manage them. Network embeddings became central tools for modeling disease diffusion processes, patient flow analysis, and forecasting.

Dynamic networks in disease modeling constitute changing complex networks composed of the following:

- Transmission networks: The movement of pathogens through people
- Host–pathogen interaction networks: Molecular battles inside infected cells
- Patient similarity networks: Grouping patients by symptomatology, outcomes treatments
- Healthcare system networks: Allocation of resources and patient flow

Most standard epidemiological models depend on simplifications of the population mixing and evolution of the disease. Embedding techniques retain the heterogeneity and temporal dynamics seen in natural epidemics. Network embeddings have also been applied to understand how pathogens evolve and develop drug resistance. Liu (2021) analyzed HIV evolution networks, where nodes represent viral variants and edges represent mutational pathways (Table 12.4).

12.5 Case studies in the integration of multiple omics

Network embedding frameworks have been widely adopted in real-world biomedical research. Case studies demonstrate their translational impact, particularly in cancer research and drug discovery. By integrating heterogenous omics layers and clinical data, these approaches provide scalable, interpretable, and predictive insights for complex diseases (Zhang and Song, 2021).

12.5.1 Cancer gene prioritization using graph neural networks

Cancer is a disease influenced by multiple factors and characterized by dysregulation on multiple omics layers, such as genomics, transcriptomics, and epigenomics. Conventional gene prioritization methods often use single omics data and statistical correlation and are limited by the loss of higher order network structure. GNNs provide a key resource for tackling this challenge by embedding multi-omics networks into a common latent space, which enables gene prioritization of cancer-integrated networks (Wang et al., 2021).

Table 12.4 Applications of network embedding in infectious disease and patient analysis.

Task	Example network	Embedding method	Outcome
Epidemic forecasting	Contact networks	Temporal embeddings	Improved outbreak prediction
Host–pathogen modeling	Protein–protein interactions	Graph embeddings	Identification of viral targets
Patient stratification	EHR + molecular networks	Multimodal embeddings	Risk group identification

One example is the combination of somatic mutation profiles with protein–protein interaction (PPI) networks and gene expression data to find new driver genes in breast and lung cancers. The embedding network models prioritized known oncogenes and tumor suppressors from the literature and suggestable yet uncharacterized candidates for future validation (Zhang and Song, 2021).

12.5.2 Drug repurposing for emerging diseases

Drug repurposing offers a time- and cost-effective strategy for emerging diseases, where urgent therapeutic interventions are required. During the COVID-19 pandemic, researchers applied network embedding models to integrate drug–target, host–virus, and protein interaction networks. By projecting drugs and proteins into a shared embedding space, candidate compounds with antiviral potential were rapidly shown (Zhou et al., 2020).

One notable study leveraged GCNs to predict interactions between FDA-approved drugs and SARS-CoV-2 host proteins. Several compounds—such as baricitinib and remdesivir—were prioritized and subsequently validated in clinical trials (Gysi et al., 2021). These case studies highlight the translational potential of embedding approaches in pandemic preparedness and rapid therapeutic discovery (Table 12.5).

12.6 Resources and tools

The rapid growth of network embedding methods in biomedicine has been supported by the development of computational frameworks, open-source libraries, and publicly available biological databases. Ensuring reproducibility and open science practices is equally important for facilitating translational research.

12.6.1 Frequently used frameworks and tools for network embedding

Several frameworks enable efficient implementation of network embedding methods, ranging from traditional matrix factorization to GNNs. Widely used platforms include the following:

- **node2vec and DeepWalk**: Random walk–based embedding implementations, often used in biomedical networks (Grover and Leskovec, 2016; Perozzi et al., 2014).

- **PyTorch Geometric (PyG) and Deep Graph Library (DGL)**: General-purpose deep learning frameworks for GNNs, with biomedical applications (Fey and Lenssen, 2019).
- **OpenNE (open network embedding)**: An open-source toolkit integrating classical and deep embedding methods, widely applied in computational biology (Tang et al., 2019).
- **BiNE**: Designed for bipartite biological networks, such as gene–disease and drug–target datasets (Gao et al., 2018).

12.6.2 Public biomedical network analysis databases

Embedding approaches rely on high-quality biological networks. Several curated databases are often used in biomedical research:

- **STRING**: A comprehensive database of protein–protein interactions (Szklarczyk et al., 2021).
- **DisGeNET**: Gene–disease association database integrating multiple sources (Piñero et al., 2020).
- **DrugBank**: Drug–target interactions and pharmacological data.
- **BioGRID**: Genetic and protein interactions with curated literature support (Oughtred et al., 2021).
- **OmicsDI**: Repository integrating multi-omics datasets (Perez-Riverol et al., 2017) (Fig. 12.10).

12.6.3 Guidelines for open-source pipelines and reproducibility

Ensuring reproducibility is a cornerstone of computational biology. Open-source pipelines and FAIR (findable, accessible, interoperable, reusable) principles are increasingly adopted for biomedical network analysis (Wilkinson et al., 2016)

Table 12.5 Network embedding–based drug repurposing case studies.

Diseases	Method	Network type	Key outcome
COVID-19	GCN, heterogenous embeddings	Drug host virus PPI	Identification of repurposable drugs (e.g., baricitinib)
Influenza	Random walk + embeddings	Drug target pathway	Discovery of synergistic antiviral candidates
Influenza H1N1	Random walk + DeepWalk	Drug target pathway	Synergistic antiviral
Alzheimer's disease	Temporal GCN	Disease progression networks	Sildenafil association and observational study
Parkinson's disease	Hyperbolic embeddings	Hierarchical biological networks	Terazosin neuroprotection

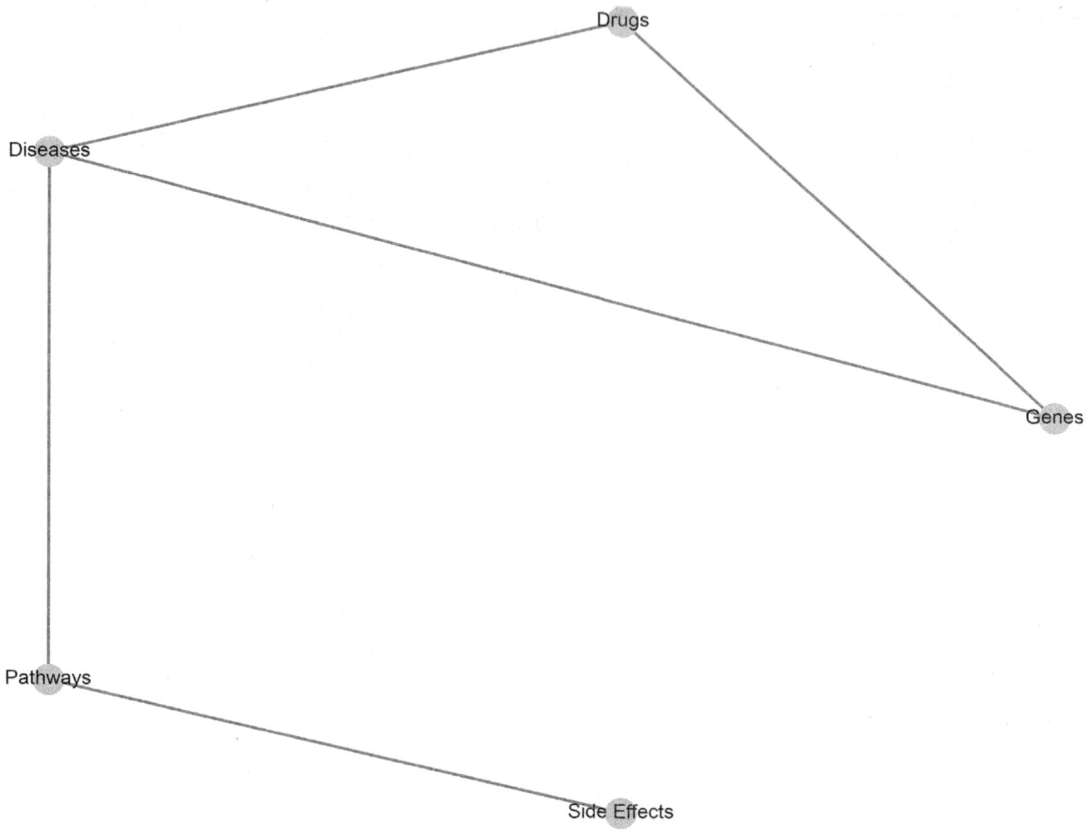

FIGURE 12.10 Network representation of biomedical databases showing their interconnections (genes, drugs, diseases).

1. **Use version-controlled repositories** (e.g., GitHub, GitLab) to share code and pipelines.
2. **Leverage containerization** (docker, singularity) for reproducible environments.
3. **Document data provenance** by providing metadata, preprocessing scripts, and citation of databases.
4. **Benchmark across multiple datasets** to avoid dataset-specific bias in embeddings (Pal et al., 2021).
5. **Adopt open standards** for interoperability, such as BioPAX and SBML for biological pathways.

12.7 Challenges, limitations, and future perspectives

Biomedicine network embedding methods have reshaped understanding at the disease level, for drug discovery, and at the level of personalized medicine. However, the conversion of predictions into clinical practice faces many challenges. These challenges offer innovation opportunities as well. The integration of graph neural networks, generative AI, and multimodal integration at the data level can rethink the very notion of biological networks.

12.7.1 Biological networks: bias, noise, and data quality

Biological networks are often incomplete, biased toward well-studied genes or diseases, and hold experimental noise. For instance, protein–protein interaction (PPI) datasets are skewed toward cancer-related proteins, limiting generalizability (Luck et al., 2020). Embeddings trained on biased networks may inherit these limitations, producing misleading predictions. Standardizing data integration and quality assessment across databases is still an open challenge.

12.7.2 Scalability and computational constraints

Biomedical datasets are increasingly large, spanning millions of nodes (genes, proteins, patients) and heterogenous interactions. Many embedding methods, especially deep learning–based ones, face computational bottlenecks in training and memory requirements (Hamilton et al., 2017). Scalable frameworks for distributed computing and efficient training are needed to apply embeddings to full-scale omics and clinical datasets.

12.7.3 Advances in graph neural networks for biomedical uses

Recent developments in graph neural networks provide opportunities to capture both local and global network structure while integrating node attributes. Applications include cancer gene prioritization, drug discovery, and multi-omics integration (Wu et al., 2021). Future GNN innovations, such as dynamic and heterogenous GNNs, may enable modeling of time-varying processes such as disease progression or drug response.

12.7.4 Integrating generative AI models

Generative AI, including generative adversarial networks (GANs) and diffusion models, is emerging as a complementary approach. Generative embeddings can simulate new biological networks, propose synthetic interactions, or augment training datasets for rare diseases (Xu et al., 2021). The integration of generative AI with network embeddings may enhance predictive power while providing mechanistic hypotheses.

12.7.5 Open problems and future opportunities

Despite rapid progress, several open challenges remain:

- **Interpretability:** Embedding vectors are often abstract, making biological interpretation difficult. Interpretable embeddings are essential for clinical trust and adoption.
- **Cross-domain integration:** Linking molecular networks with patient-level EHRs and imaging data is still under explored.
- **Benchmarking standards:** Lack of gold-standard datasets and standardized benchmarks limits fair evaluation across methods.
- **Ethical and privacy concerns:** Patient network embeddings may inadvertently reveal sensitive information if not properly anonymized.

12.8 Conclusion

In this chapter, we have described the foundations, applications, and potential future direction of network embedding in computational biology, genomics, and healthcare. We began with the structural basis of biological networks and dimensionality reduction, and then we examined the landscape of methodologies including matrix factorization, diffusion-based methods, and deep learning–based embeddings. Several applications of these methods existed, such as protein–protein interaction prediction, gene coexpression analysis, functional module identification, drug repurposing, and incorporation of multi-omics datasets. The commonality of embedding frameworks, across each of the respective sections, is that they can abstract complex, high-dimensional biological networks effectively into low-dimensional representations that are structured, low-dimensional representations, while preserving biological meaning and enabling scalable computation. This abstraction allows us to use embeddings for a variety of downstream applications in biomedicine, from variant–gene–disease association studies to precision medicine, and we believe that such as fast-growing method is an essential component of modern biomedicine. There are many challenges and limitations which remain, including being biased and incomplete data in biology, interpreting embeddings, and the challenges of scaling computations for large-scale datasets. Nonetheless, recent developments in GNNs and the utilization of generative artificial intelligence models present formidable pathways to address these concerns. These avenues have the capacity to take the next step beyond predictive modeling toward explainable, clinically actionable insights. This is significant because network embedding will contribute to drug discovery, genomic data interpretation, and personalized medicine by setting up clear links between molecular networks, patient data, and clinical outcomes. The field has evolved from proof-of-concept studies to translational possibilities in which reproducibility, benchmarking, and ethical security will join methodological progress as outcomes of interest.

Briefly, we put forward that network embedding is not only a computational process but also a paradigm shift in representing biological complexity. Combining open science and rigorous approach, we are able to reshape genomics and health through a data-driven strategy in facilitating precision medicine beyond morphospace.

Take-home messages

Network embedding methods have become a strong tool for interpreting high-dimensional and complex biomedical networks. Through learning low-dimensional vector representations of nodes such as proteins, genes, and drugs, these methods retain structural and functional relations that are usually masked in conventional analyses. Consequently, embeddings now form the core of varied applications such as protein–protein interaction prediction, discovery of new disease genes, drug repurposing, and precision medicine strategy design. Unlike traditional network analysis, embeddings provide scalability, flexibility, and machine learning model compatibility to enable integration of multi-omics and clinical data. Examples in cancer genomics and infectious disease illustrate their potential to inform biomedical discovery in real-world settings. In the future, advancements in deep learning, generative AI, and explainability will further improve the interpretability and translational potential of embeddings and make them a cornerstone of biomedical research and clinical decision-making.

References

Cao, S., Lu, W., Xu, Q., 2015. GraRep: learning graph representations with global structural information. International Conference on Information and Knowledge Management, Proceedings 19–23, 891–900. https://doi.org/10.1145/2806416.2806512.

Chen, J., Zhang, S., Li, R., Tang, J., 2020. Discovering functional modules in biological networks using network embedding and clustering. Front. Genet. 11, 626.

Cho, H., Berger, B., Peng, J., 2016. Compact integration of multi-network topology for functional analysis of genes. Cell Syst. 3 (6), 540. https://doi.org/10.1016/j.cels.2016.10.017.

Clark, C., Kalita, J., 2014. A comparison of algorithms for the pairwise alignment of biological networks. Bioinformatics 30 (16), 2351–2359. https://doi.org/10.1093/bioinformatics/btu307.

ENCODE Project Consortium, 2020. Expanded encyclopedias of DNA elements in the human and mouse genomes. Nature 583 (7818), 699–710.

M. Fey, J.E. Lenssen, Fast graph representation learning with pytorch geometric. arXiv, Germany arXiv. 23318422. (2019), doi: 10.48550/arxiv.1903.02428.

Fout, A., Byrd, J., Shariat, B., Ben-Hur, A., 2017. Protein interface prediction using graph convolutional networks. Adv. Neural Inf. Process. Syst. 2017, 6531–6540.

Gao, M., Chen, L., He, X., Zhou, A., 2018. BiNE: bipartite network embedding. In: 41st International ACM SIGIR Conference on Research and Development in Information Retrieval, SIGIR 2018. Association for Computing Machinery, Inc, China, pp. 715–724. https://doi.org/10.1145/3209978.3209987.

Grover, A., Leskovec, J., 2016. Node2vec: scalable feature learning for networks. In: Proceedings of the ACM SIGKDD International Conference on Knowledge Discovery and Data Mining. Association for Computing Machinery, United States, pp. 855–864. https://doi.org/10.1145/2939672.2939754.

Gysi, D.M., Do Valle, Í., Zitnik, M., Ameli, A., Gan, X., Varol, O., Ghiassian, S.D., Patten, J.J., Davey, R.A., Loscalzo, J., Barabási, A.L., 2021. Network medicine framework for identifying drug-repurposing opportunities for COVID-19. Proc. Natl. Acad. Sci. U. S. A 118 (19). https://doi.org/10.1073/pnas.2025581118.

Hamilton, W.L., Ying, R., Leskovec, J., 2017. Inductive representation learning on large graphs. Adv. Neural Inf. Process. Syst. 2017, 1025–1035.

Huang, K., Xiao, C., Glass, L.M., Zitnik, M., Sun, J., 2021. SkipGNN: predicting molecular interactions with skip-graph networks. Sci. Rep. 11 (1), 1–16.

Kipf, T.N., Welling, M., 2016., Variational graph auto-encoders arXiv preprint arXiv:1611.07308.

Kuchaiev, O., Milenković, T., Memišević, V., Hayes, W., Pržulj, N., 2010. Topological network alignment uncovers biological function and phylogeny. J. R. Soc. Interface 7 (50), 1341–1354. https://doi.org/10.1098/rsif.2010.0063.

Li, M.M., Huang, K., Zitnik, M., 2021., Graph representation learning in biomedicine. arXiv, United States arXiv. 23318422.https://doi.org/10.48550/arXiv.2104.04883.

Liu, H., 2021. Network embedding approaches for HIV drug resistance prediction. PLoS Comput. Biol. 17 (3), e1008831.

Luck, K., Kim, D.K., Lambourne, L., Spirohn, K., Begg, B.E., Bian, W., Brignall, R., Cafarelli, T., Campos-Laborie, F.J., Charloteaux, B., Choi, D., Coté, A.G., Daley, M., Deimling, S., Desbuleux, A., Dricot, A., Gebbia, M., Hardy, M.F., Kishore, N., Knapp, J.J., Kovács, I.A., Lemmens, I., Mee, M.W., Mellor, J.C., Pollis, C., Pons, C., Richardson, A.D., Schlabach, S., Teeking, B., Yadav, A., Babor, M., Balcha, D., Basha, O., Bowman-Colin, C., Chin, S.F., Choi, S.G., Colabella, C., Coppin, G., D'Amata, C., De Ridder, D., De Rouck, S., Duran-Frigola, M., Ennajdaoui, H., Goebels, F., Goehring, L., Gopal, A., Haddad, G., Hatchi, E., Helmy, M., Jacob, Y., Kassa, Y., Landini, S., Li, R., van Lieshout, N., MacWilliams, A., Markey, D., Paulson, J.N., Rangarajan, S., Rasla, J., Rayhan, A., Rolland, T., San-Miguel, A., Shen, Y., Sheykhkarimli, D., Sheynkman, G.M., Simonovsky, E., Taşan, M., Tejeda, A., Tropepe, V., Twizere, J.C., Wang, Y., Weatheritt, R.J., Weile, J., Xia, Y., Yang, X., Yeger-Lotem, E., Zhong, Q., Aloy, P., Bader, G.D., De Las Rivas, J., Gaudet, S., Hao, T., Rak, J., Tavernier, J., Hill, D.E., Vidal, M., Roth, F.P., Calderwood, M.A., 2020. A reference map of the human binary protein interactome. Nature 580 (7803), 402–408. https://doi.org/10.1038/s41586-020-2188-x.

Luo, Y., Zhao, X., Zhou, J., Yang, J., Zhang, Y., Kuang, W., Peng, J., Chen, L., Zeng, J., 2017. A network integration approach for drug-target interaction prediction and computational drug repositioning from heterogeneous information. Nat. Commun. 8 (1). https://doi.org/10.1038/s41467-017-00680-8.

Ou, M., Cui, P., Pei, J., Zhang, Z., Zhu, W., 2016. Asymmetric transitivity preserving graph embedding. In: Proceedings of the ACM SIGKDD International Conference on Knowledge Discovery and Data Mining. Association for Computing Machinery, China, pp. 1105–1114. https://doi.org/10.1145/2939672.2939751.

Oughtred, R., Rust, J., Chang, C., Breitkreutz, B.J., Stark, C., Willems, A., Boucher, L., Leung, G., Kolas, N., Zhang, F., Dolma, S., Coulombe-Huntington, J., Chatr-aryamontri, A., Dolinski, K., Tyers, M., 2021. The BioGRID database: a comprehensive biomedical resource of curated protein, genetic, and chemical interactions. Protein Sci. 30 (1), 187–200. https://doi.org/10.1002/pro.3978.

Pal, S., Wu, B., Li, Y., 2021. Reproducibility in network biology: lessons and best practices. Briefings Bioinf. 22 (3), bbaa236.

Perez-Riverol, Y., Xu, Q.W., Wang, R., Uszkoreit, J., 2017. OmicsDI: integrating multi-omics datasets into a single resource. Nat. Biotechnol. 35 (4), 316–319.

Perozzi, B., Al-Rfou, R., Skiena, S., 2014. DeepWalk: online learning of social representations. In: Proceedings of the ACM SIGKDD International Conference on Knowledge Discovery and Data Mining. Association for Computing Machinery, United States, pp. 701–710. https://doi.org/10.1145/2623330.2623732.

Piñero, J., Ramírez-Anguita, J.M., Saüch-Pitarch, J., Ronzano, F., Centeno, E., Sanz, F., Furlong, L.I., 2020. The DisGeNET knowledge platform for disease genomics: 2019 update. Nucleic Acids Res. 48 (1), D845. https://doi.org/10.1093/nar/gkz1021.

Su, C., Tong, J., Zhu, Y., Cui, P., Wang, F., 2020. Network embedding in biomedical data science. Briefings Bioinf. 1 (1), 182–197. https://doi.org/10.1093/bib/bby117.

Szklarczyk, D., Gable, A.L., Nastou, K.C., Lyon, D., Kirsch, R., Pyysalo, S., Doncheva, N.T., Legeay, M., Fang, T., Bork, P., Jensen, L.J., von Mering, C., 2021. The STRING database in 2021: customizable protein-protein networks, and functional characterization of user-uploaded gene/measurement sets. Nucleic Acids Res. 49 (1), D605. https://doi.org/10.1093/nar/gkaa1074.

Tang, J., Qu, M., Wang, M., Zhang, M., Yan, J., Mei, Q., 2015. LINE: large-scale information network embedding. In: WWW 2015 - Proceedings of the 24th International Conference on World Wide Web. Association for Computing Machinery, Inc, China, pp. 1067–1077. https://doi.org/10.1145/2736277.2741093.

Tang, J., Qu, M., Wang, M., Zhang, M., Yan, J., Mei, Q., 2019. OpenNE: an open-source toolkit for network embedding. In: Proceedings of the 27th ACM International Conference on Information and Knowledge Management, pp. 2493–2496.

Veličković, P., Casanova, A., Liò, P., Cucurull, G., Romero, A., Bengio, Y., 2018. Graph attention networks. In: 6th International Conference on Learning Representations, ICLR 2018 - Conference Track Proceedings International Conference on Learning Representations. ICLR, United Kingdom.

Wang, D., Cui, P., Zhu, W., 2016. Structural deep network embedding. In: Proceedings of the 22nd ACM SIGKDD International Conference on Knowledge Discovery and Data Mining, pp. 1225–1234. https://doi.org/10.1145/2939672.2939753.

Wang, H., Huang, H., Ding, C., 2017. Simultaneous clustering of multi-type relational data via symmetric nonnegative matrix tri-factorization. In: Proceedings of the 20th ACM SIGKDD International Conference on Knowledge Discovery and Data Mining, pp. 1247–1256.

Wang, S., Ma, J., Xu, J., 2021. AUC-maximized deep graph embedding for disease gene prediction. Bioinformatics 7, 939–946.

Wilkinson, M.D., Dumontier, M., Aalbersberg, I.J.J., Appleton, G., Axton, M., Baak, A., Blomberg, N., Boiten, J.-W., Bonino da Silva Santos, L., Bourne, P.E., Bouwman, J., Brookes, A.J., Clark, T., Crosas, M., Dillo, I., Dumon, O., Edmunds, S., Evelo, C.T., Finkers, R., Gonzalez-Beltran, A., Gray, A.J. G., Paul, G., Goble, C., Grethe, J.S., Heringa, J., C 't Hoen, P.A., Hooft, R., Kuhn, T., Kok, R., Kok, J., Lusher, S.J., Martone, M.E., Mons, A., Packer, A.L., Persson, B., Rocca-Serra, P., Roos, M., van Schaik, R., Sansone, S.-A., Schultes, E., Sengstag, T., Slater, T., George, S., Swertz, M.A., Thompson, M., Lei, J.van der, Mulligen, E.van, Velterop, J., Waagmeester, A., Wittenburg, P., Wolstencroft, K., Zhao, J., Mons, B., 2016. The FAIR Guiding Principles for scientific data management and stewardship. Sci. Data 3 (1). https://doi.org/10.1038/sdata.2016.18.

Wu, Z., Pan, S., Chen, F., Long, G., Zhang, C., Yu, P.S., 2021. A comprehensive survey on graph neural networks. IEEE Transact. Neural Networks Learn. Syst. 32 (1), 4–24. https://doi.org/10.1109/TNNLS.2020.2978386.

Xu, M., Wang, H., Jin, H., 2021. Generative adversarial networks in bioinformatics: applications and perspectives. Briefings Bioinf. 22 (3), bbaa219.

Yue, X., Wang, Z., Huang, J., Parthasarathy, S., Moosavinasab, S., Huang, Y., Lin, S.M., Wen, Z., Zhang, P., Sun, H., Cowen, L., 2020. Graph embedding on biomedical networks: methods, applications and evaluations. Bioinformatics 36 (4), 1241–1251. https://doi.org/10.1093/bioinformatics/btz718.

Zeng, X., 2019. Deep learning for predicting disease–gene associations. Nat. Biotechnol. 37 (9), 1033–1040.

Zhang, M., Song, Y., 2021. Graph representation learning for multi-omics data integration and disease prediction. Briefings Bioinf. 22 (3), bbaa320.

Zhang, Y., Wang, S., Zhou, Y., 2019. Predicting protein–protein interactions using DeepWalk and node2vec. BMC Bioinf. 20 (1), 1–12.

Zhou, Y., Hou, Y., Shen, J., Huang, Y., Martin, W., Cheng, F., 2020. Network-based drug repurposing for novel coronavirus 2019-nCoV/SARS-CoV-2. Cell Discov. 6 (1). https://doi.org/10.1038/s41421-020-0153-3.

Zitnik, M., Agrawal, M., Leskovec, J., 2018. Modeling polypharmacy side effects with graph convolutional networks. Bioinformatics 34 (13), i457. https://doi.org/10.1093/bioinformatics/bty294.

13

AI use in clinical prediction

Iqra Khan[1], Ananta Dutta Choudhury[1], Iqubal Alam[7], Mohhamed Salahuddin[2], Sushim Harish Mathur[3], Farheen Siddiqui[4] and Mohammad "Sufian" Badar[5,6,8,9]

[1]DEPARTMENT OF COMPUTER SCIENCE, JAMIA MILLIA ISLAMIA, NEW DELHI, INDIA; [2]INFORMATION SPECIALIST, CORE42, ABU DHABI, UNITED ARAB EMIRATES; [3]NEW DELHI MUNICIPAL COUNCIL, DR RAM MANOHAR LOHIA HOSPITAL, NEW DELHI, INDIA; [4]DEPARTMENT OF COMPUTER SCIENCE AND ENGINEERING, SCHOOL OF ENGINEERING SCIENCES AND TECHNOLOGY (SEST), JAMIA HAMDARD UNIVERSITY, NEW DELHI, INDIA; [5]DEPARTMENT OF BIOENGINEERING, UNIVERSITY OF CALIFORNIA, RIVERSIDE, CA, UNITED STATES; [6]UNIVERSAL SCIENTIFIC EDUCATION AND RESEARCH NETWORK (USERN), TEHRAN, IRAN; [7]EMERGENCY MEDICINE DEPARTMENT, DR SULAIMAN AL HABIB HOSPITAL, RIYADH, SAUDI ARABIA; [8]SPI DARBHANGA, DARBHANGA, BIHAR, INDIA; [9]DEPARTMENT OF COMPUTER SCIENCE AND ENGINEERING (BIOINFORMATICS), SCHOOL OF ENGINEERING SCIENCES AND TECHNOLOGY (SEST), JAMIA HAMDARD UNIVERSITY, NEW DELHI, INDIA

What you will learn

After studying this chapter, the reader will be able to:

- Understand the concept of **clinical prediction** and its significance in modern healthcare
- Differentiate between **traditional statistical models** and **AI-based predictive approaches**
- Identify key **domains of clinical prediction**, including diagnosis, prognosis, treatment optimization, and disease monitoring
- Explain the role of **machine learning (ML), deep learning, and natural language processing** in clinical prediction
- Recognize the importance of **multi-modal clinical data sources** such as EHRs, medical imaging, genomics, and wearable devices
- Appreciate the role of **AI-driven clinical decision support systems (CDSS)** in real-world healthcare workflows
- Understand how **explainable AI (XAI)** enhances transparency, trust, and clinician adoption

Applying Artificial Intelligence to Computational Biology and Medical Informatics. https://doi.org/10.1016/B978-0-443-44915-4.00013-7

- Evaluate **ethical, legal, and social implications** associated with AI-based clinical prediction
- Gain insight into **future directions**, including federated learning, continual learning, and human–AI collaboration

13.1 Introduction to AI in clinical prediction

13.1.1 Definition and overview

Clinical Prediction refers to the process of predicting health outcomes based on the current and historical data of the patient. This involves using artificial intelligence (AI), ML, and deep learning (DL) algorithms to develop models that can forecast events that can occur in the future in real time, possibly so that individual health can be improved.

Early detection and prevention of diseases, improving diagnosis accuracy, and treatment plan improvement are some of the use cases where clinical prediction can be applied. Through the use of a clinical prediction healthcare system, efficiency can be improved resulting in better patient outcomes.

13.1.2 Role of AI in clinical decision-making

With the advancement of technology, data has become widely and easily available. Extracting meaning and patterns from this huge amount of data (also termed Big Data) is an impossible task for human beings. AI has made a significant contribution in dealing with this problem (Khalifa and Albadawy, 2024). By processing medical data quickly, AI has made it easier to identify patterns and correlations, also with high precision. AI can identify occurrences of diseases in advance through the analysis of medical images, patient history, and lab results.

Advancements in AI have also made it possible to tailor specific treatment to individual patients to provide personalized medicine instead of generic treatments. It can help medical practitioners to understand the progression of different diseases in different individuals and to predict how various treatments will affect them.

With AI, patient readmission risk and potential complications can also be predicted, which ensures the safety of the patient. Therefore, with AI in Clinical Prediction, better healthcare services can be ensured. AI's ability to deal with huge amounts of data and analyze and interpret in a short amount of time helps in improving prediction accuracy, giving personalized treatment, promoting preventive healthcare, and increasing the overall efficiency of the treatments (Khalifa and Albadawy, 2024).

13.1.3 Traditional versus AI-based models

Predicting health outcomes accurately is a long-standing goal in clinical medicine, whether predicting the progression of a disease, patient's readmission, or the efficiency of a treatment.

Traditionally, statistical models that relied on linearity and interoperability were used to determine future outcomes, such as logistic regression, Cox proportional hazard models, and linear discriminant analysis. These were widely adopted by clinicians, but they fell short in capturing[Rephrase] nonlinear interaction and high-dimensional dependencies present in modern EHRs (Samson, 2025). The more complex clinical cases were unable to fit the linearly derived prediction model criteria; hence, they could not be solved by such models (Luo et al., 2020).

With the advent of AI, especially ML and DL algorithms, the accuracy of predictions has improved tremendously. Algorithms such as K-Nearest Neighbor (KNN), Support Vector Machine (SVM), Gradient Boosting (XGBoost), Random Forest, and Neural Network take in large amounts of data and can predict the outcome very accurately. They can learn nonlinearity without being explicitly programmed and are well suited for complex and unstructured data from which deriving patters is not humanly possible (Samson, 2025; Luo et al., 2020).

13.2 Key domains of clinical prediction

13.2.1 Diagnosis and early detection

Diagnosis refers to the process of detecting a fault, identifying its type, and determining its severity or degree. This process requires the direct assessment of the patients physiology to understand the underlying illness or condition of the patient (Machine learning for patient risk stratification).

With the help of AI, diagnostic accuracy can be significantly improved. DL, a subfield of AI has collected remarkable articulation due to its ability to diagnose a disease with exceptional accuracy, surpassing, in some cases, that of human experts.

In the field of medical imaging, several DL algorithms are being used, such as Convolution Neural Network (CNNs), Deep Neural Networks (DNNs), Deep Belief Networks (DBNs), Deep Autoencoders, Deep Boltzmann Machines (DBMs), Deep Extreme Learning Machines (DC-ELM), and Recurrent Neural Networks (RNNs), along with their variants such as Bidirectional Long Short-Term Memory (BLSTM) and MDLATM (Aggarwal, 2018).

Region Aggregation Graph Convolutional Networks (RAGCNs) represent an advanced DL approach for medical data analysis. RAGCNs leverage Graph Convolutional Networks (GCNs) to aggregate information from different regions of an image. This technique is particularly suited for medical images such as CT and MRI scans, which often contain multiple Regions of Interest (ROIs) requiring independent analysis. RAGCN employs a graph-based framework to segment an image into distinct regions and applies GCNs to each region for feature extraction and prediction.

These examples highlight the diverse range of DL techniques currently used for disease prediction and diagnosis. It is important to note that DL is a rapidly evolving field, with new models and hybrid approaches continually being developed.

13.2.2 Prognosis and risk stratification

Prognosis is the determination of the disease trajectory of a patient using predictive tools to plan necessary treatments and care. ML algorithms have proved to be equivalent to a skilled dermatologist in the prognosis of skin cancer (Dermatologist-level classification of skin cancer with deep neural networks) and to predict prediabetes type 2 using routinely obtained electronic health record data of patients (Anderson et al., 2015; Full article).

AI also has widespread potential in risk stratification, where models are required to predict the future disease state of a patient based on their current health state and available history (Machine learning for patient risk stratification).

AI is being used in identifying precisely and risk-stratifying individuals with "coronary artery disease." AI also demonstrated potential as an early triage tool in the diagnostic accuracy of coronary artery disease (Full article).

13.2.3 Treatment response and optimization

The treatment response is an essential facet of modern healthcare, focusing on predicting how patients will respond to specific treatments or therapies (Turki and Wang, 2019).

AI-driven predictive outcome proves valuable in optimizing patient treatment outcomes. ML and DL techniques optimize an objective function (e.g., medication dosage) based on complex and multidimensional patient data (such as patient history from EHRs) (Frontiers).

Reinforcement Learning models are dynamically adapted to schedule therapy based on patient response, which results in 22% improvement of treatment efficacy and drug toxicity reduction (View of The role of artificial intelligence in personalized oncology).

In oncology, the prediction of cancer patient's response to chemotherapy regimen based on patient's genetic profile and characteristics of their tumor through AI models is also done (Liao et al., 2023).

AI can detect subtle changes in brain imaging or cognitive function tests for neurodegenerative diseases like Alzheimer's and predict the progression of disease (Revolutionizing the Early Detection of Alzheimer's Disease).

13.2.4 Disease monitoring and adverse event prediction

Acute decompensation occurs rarely but is most commonly accompanied by increasing physiological derangements and worse outcomes. Early intervention may prevent poor outcomes, though oftentimes it is difficult to identify before significant hemodynamic compromise with our traditional standard monitoring and early warning scores used commonly. Intervention can be made for such a patient as six to 8 hours precede acute patient decompensation (Taenzer et al., 2010).

To detect the subtle trends in physiologic parameters over time to provide precision and reliability, multivariate regression-based models or AI-based early warning systems can be used(Predicting vital sign deterioration, 2019).

13.3 AI techniques in clinical prediction

13.3.1 Machine learning and deep learning

ML methods are a subset of AI that are used in clinical prediction to estimate the risk of a disease, its outcomes, and treatment response using patient data that includes labs, clinical history, vitals, and imaging history.

An example of a ML model developed by Lundberg et al. is TreeExplainer, which is an interpretability algorithm that improves transparency in tree-based models. It, in turn, helps in enabling more reliable and explainable medical predictions ([1905.04610] Explainable AI for Trees).

DL is a subset of ML that uses neural network architecture to learn the underlying features and patterns in the data to make predictions without hardcoding the rules explicitly. Convolution Neural Network (CNNs) are a type of DL architecture that is used for image processing tasks such as medical imaging.

Esteva et al. developed a DL model for classifying skin cancer with accuracy comparable to that of board-certified dermatologists. Using CNN, the model was able to distinguish between malignant and benign skin lesions with high specificity and sensitivity (Aravazhi et al., 2025).

13.3.2 Natural language processing

A study developed a hybrid fusion method combining structured EHR data with unstructured clinical narratives using pretrained language models to predict injury diagnoses from the National Electronic Injury Surveillance System (NEISS) dataset. The RoBERTa-based hybrid fusion model achieved 75% accuracy for top injury prediction and 93.54% for top three injuries, significantly outperforming both single-modality models and traditional fusion approaches. The research demonstrates that integrating clinical text with structured data through advanced natural language processing can substantially improve clinical prediction model performance (Elhaddad and Hamam, 2024).

13.4 Clinical data and challenges

13.4.1 Data sources—EHRs, imaging, genomics, wearables

Clinical prediction models utilize data from diverse sources across the healthcare ecosystem, each contributing unique information that enhances the accuracy of prediction and clinical utility. As healthcare has become increasingly digitized, the volume, variety, and velocity of clinical data have expanded exponentially. To understand the nature of the data better and develop robust, reliable, and clinically useful prediction models.

Healthcare generates enormous volumes of diverse data types, each with unique characteristics and applications:

i. *EHRs*: Electronic health records (EHRs), often also referred to as electronic medical records (EMRs), are a major source of clinical data. EHRs are a digital record of health information of the patient. EHRs not only contain data but also deal with how data is shared efficiently with clinicians and organizations involved in caring for patients, such as labs, specialists, imaging facilities, pharmacies, emergency facilities, workplace clinics, etc.
ii. *Medical Imaging*: Refers to the techniques and processes used to create images of human body or specific parts for various clinical purposes such as medical procedures and diagnosis or medical science including study of normal anatomy and function. It includes techniques such as X-ray radiology, which, when combined with computed temography generated three-dimensional cross-sectional images of the body. Magnetic Resonance Imaging (MRI), Ultrasound Imaging are also notable techniques,
iii. *Genomics*: Genomics is the detailed study of an organism's complete genetic material. The analysis of gene or allele identification could help in the diagnosis of disease and in designing targeted therapies. The prediction of high-dimensional, heterogeneous genetic datasets is a complex task, which is why ML and DL models are being used to solve this problem.
iv. *Wearables*: Biosensors are wearable, implantable, and ambient devices that convert biological responses into electro-optical signals and make continuous monitoring of health and well-being possible, even with a variety of mobile apps. Since EHRs often lack patients' self-reported experiences, human activities, and vital signs outside of clinical settings, tracking those continuously is expected to improve treatment outcomes by closely analyzing patient's condition.

13.4.2 Data quality, bias, and integration issues

AI models are remarkably capable of analyzing complex clinical data, but their real-world effectiveness is bound by the quality of data, biases embedded in data and algorithms, and practical difficulties in integrating AI in existing healthcare workflows.

Data quality issues—The data is the backbone for any kind of AI model, without which the model will not work as expected. Real-world data is often messy, incomplete, or missing. EHR data frequently has missingness due to variation in documentation practices. Most real-world data is also heterogeneous in nature and inconsistent because of a lack of standard documentation, terminology usage, and measurement protocols.

Data bias issues—As AI models are hugely reliant on data, it becomes largely difficult to mitigate the unexpected learning of patterns from the data instead of useful information. Bias is introduced in the model when the data used for training is not properly representative of a diverse population or when the data is imbalanced, which makes the model predict everything in favor of the dominating class.

Data integration issues—Interoperability and standardization prove to be an obstacle to integrating data from multiple sources and healthcare systems. The modality of data is also a challenge when integrating heterogeneous data types—such as EHR data, medical images, genomic sequences, etc.—into a unified prediction model.

13.5 Clinical implementation

13.5.1 AI decision support tools

In acute care environments, AI-based tools are needed to aid in the early detection and assessment of diseases since the amount of data that is generated exceeds the capability of human processing. ML helps Clinical Decision Support Systems (CDSS) in processing vast volumes of data, such as medical imaging, genomic information, and EHRs, to extract meaningful insights and to make informed clinical decisions.

Studies have demonstrated the efficacy of CNN-based CDSS in improving diagnostic accuracy and facilitating early disease detection (Elhaddad and Hamam, 2024). CNNs, originally developed for image processing tasks, have been adapted for various medical imaging applications, including lesion detection, organ segmentation, and disease classification.

NLP algorithms enable CDSS to parse, interpret, and extract relevant information from clinical narratives, facilitating automated clinical documentation, coding, and information retrieval. These algorithms leverage linguistic analysis techniques to identify clinical concepts, extract structured data elements, and populate structured databases, thereby streamlining clinical workflows and enhancing data interoperability.

13.5.2 Workflow integration and user trust

The need for clinicians is more than the healthcare industry can provide (A review of augmented reality systems). Clinicians who are overwhelmed with difficult tasks and limited time experience a decline in efficient clinical decision-making.

According to the cognitive load theory, an excessive workload can negatively impact decision-making. In other words, clinicians who are overwhelmed with complex tasks and time constraints often experience a decline in their clinical decision-making efficiency (A review of augmented reality systems).

AI can assist clinicians by automating the mundane repetitive tasks, thereby reducing their workload.

A report published by McKinsey and Company in 2023 reported that AI-driven automation cut down the administrative time of healthcare professionals by 34% and saw a 20% decline in burnout rates among clinicians (Aravazhi et al., 2025).

Even though AI has profound significance in clinical practice, its black box nature leads to a level of ambiguity that raises concerns about the predictions. Thus, to increase transparency and consequently the trust in AI, explainable AI is being used to understand how the models use data to make predictions (JMIR AI).

Explainability can ensure that clinicians are able to discern the model's logic and ensure the patients understand the findings that will build trust and help in informed decision-making (Fostering trust and interpretability).

SHapley Additive exPlanations (SHAP), together with Local Interpretable Model-agnostic Explanations (LIME) adds functionality to the proposed system by displaying important features contributing to each prediction.

LIME, as demonstrated in multiple studies (Exploring Multimodal Framework, 2025), is a model-agnostic explainability technique that can enhance the interpretability of ML models.

Band et al. (2023) also reviewed XAI methods on healthcare datasets, highlighting their strengths in transparency but noting challenges like resource intensity and data biases.

13.6 Personalized and precision medicine

Personalized and precision medicine focus on delivering healthcare interventions that are tailored to individual patient characteristics rather than relying on generalized treatment protocols. By incorporating genetic, molecular, clinical, and environmental information, this approach aims to improve treatment effectiveness while minimizing adverse outcomes. AI has emerged as a critical enabler of precision medicine by allowing the analysis of large, complex datasets that exceed human cognitive capacity and traditional statistical techniques (Topol, 2019).

AI-based clinical prediction systems are particularly valuable in identifying subtle interactions among biological variables, supporting clinicians in making data-driven decisions that reflect patient-specific variability. As healthcare moves toward individualized care, AI-driven models are increasingly embedded within precision medicine frameworks to support diagnosis, prognosis, and therapeutic decision-making.

13.6.1 Genetic data and drug response prediction

Genetic differences among individuals significantly influence how drugs are absorbed, metabolized, and eliminated. Variations in genes encoding drug-metabolizing enzymes, transporters, and receptors can lead to differences in drug efficacy and toxicity. AI techniques, including supervised ML and deep neural networks, are widely applied in pharmacogenomics to model these complex genotype–phenotype relationships (Chen and Snyder, 2013).

By integrating genomic data with clinical and demographic information, AI models can predict patient-specific drug responses and identify individuals at higher risk of adverse drug reactions. These predictive capabilities support optimized drug selection and dosing, reducing reliance on trial-and-error prescribing practices. As a result, AI-enhanced pharmacogenomic prediction contributes to safer, more effective, and cost-efficient healthcare delivery.

13.6.2 Precision oncology and treatment planning

Precision oncology represents one of the most impactful applications of AI in personalized medicine. Cancer is a highly heterogeneous disease, with significant variability in tumor genetics and patient response to treatment. AI systems can analyze multi-modal oncology data, including genomic sequencing results, histopathological images, radiological scans, and longitudinal clinical records, to support cancer diagnosis and treatment planning (Esteva et al., 2019).

DL models have demonstrated strong performance in tumor classification, survival prediction, and therapy response assessment. AI-driven clinical decision support tools assist oncologists by synthesizing patient-specific molecular profiles with evidence-based treatment guidelines. While these systems enhance consistency and personalization in oncology care, clinical judgment remains essential to interpret model outputs and ensure patient safety.

13.7 Applications and case studies

The application of AI in clinical prediction has expanded from experimental research to real-world healthcare environments. Numerous case studies demonstrate that AI-based systems can enhance diagnostic accuracy, improve workflow efficiency, and support timely clinical interventions across diverse medical specialties.

13.7.1 Real-world examples—Radiology, ICU, oncology, etc

In radiology, AI-based image analysis systems have achieved performance comparable to that of experienced radiologists in detecting abnormalities across a range of imaging modalities. DL models are commonly used for tasks such as tumor detection, fracture identification, and disease classification in X-rays, CT scans, and MRI images. These systems help reduce diagnostic delays, prioritize urgent cases, and alleviate clinician workload (A survey on deep learning).

In intensive care units, AI-driven predictive models analyze real-time physiological signals and electronic health record data to identify patients at risk of deterioration. Applications include early sepsis detection, mortality prediction, and optimization of treatment strategies. Early warning systems based on AI have demonstrated potential to improve clinical outcomes by enabling faster and more targeted interventions (Komorowski et al., 2018).

In oncology practice, AI systems support pathology analysis, treatment selection, and outcome prediction. Clinical case studies indicate that AI can identify clinically relevant patterns within large-scale datasets that may not be immediately apparent to human clinicians, thereby enhancing decision-making in complex care settings (Topol, 2019).

13.8 Ethical, legal, and social considerations

The increasing reliance on AI for clinical prediction introduces a range of ethical, legal, and social challenges that must be carefully managed to ensure responsible deployment in healthcare.

13.8.1 Fairness and bias

Bias in AI systems can arise when training data are unrepresentative of the broader patient population. Historical inequalities, incomplete datasets, and systemic biases embedded within healthcare systems may be reflected in AI predictions, potentially leading to unequal care outcomes. Ensuring fairness requires continuous monitoring of model performance across demographic groups, diverse data collection, and transparent evaluation processes (Obermeyer et al., 2019).

13.8.2 Privacy and consent

Clinical prediction models depend on access to sensitive personal health information, raising concerns about patient privacy, data security, and informed consent. Safeguarding patient data requires strict adherence to data protection regulations and the use of technical safeguards such as encryption, anonymization, and privacy-preserving learning methods. Federated learning approaches further enhance privacy by enabling collaborative model training without sharing raw patient data.

13.8.3 Regulation and accountability

The deployment of AI in clinical settings challenges existing regulatory and legal frameworks, particularly in determining accountability when AI-supported decisions contribute to adverse outcomes. Clear governance structures are necessary to define the responsibilities of clinicians, healthcare institutions, and AI developers. While regulatory bodies have begun developing guidelines for AI-based medical systems, standardized global regulations remain an ongoing challenge (Topol, 2019).

13.9 Challenges and future directions

Despite significant advances, multiple challenges continue to limit the widespread adoption of AI-based clinical prediction systems.

13.9.1 Data generalizability and trust

Many AI models demonstrate high performance in development environments but struggle to generalize across institutions, populations, and clinical workflows. Differences in data collection practices, patient demographics, and healthcare infrastructure can reduce model reliability. Addressing these limitations requires robust external validation, multi-center datasets, and continuous model evaluation to build clinician trust and ensure safe deployment (Kelly et al., 2019).

13.9.2 Federated and continual learning

Federated learning has emerged as a promising solution to data-sharing barriers by enabling collaborative model training across institutions while preserving patient privacy. Continual learning approaches allow AI systems to adapt to evolving clinical practices and population changes over time. Although these methods offer significant advantages, they introduce new technical and governance challenges related to model stability, validation, and oversight (The future of digital health with federated learning).

13.9.3 Human-AI collaboration

Future clinical prediction systems are increasingly designed to function as decision-support tools rather than autonomous decision-makers. Effective human–AI collaboration emphasizes transparency, interpretability, and clinician involvement throughout the AI lifecycle. Explainable AI approaches improve user trust and enable clinicians to critically assess model recommendations, ensuring AI systems complement human expertise rather than replace it (Explainability for artificial intelligence in healthcare, 2020).

13.10 Conclusion

AI has become a transformative force in clinical prediction, enabling more accurate diagnosis, improved prognostic assessment, and personalized treatment strategies. Advances in ML, DL, and data integration have expanded the scope of precision medicine and enhanced clinical decision support across multiple domains.

However, challenges related to data quality, generalizability, ethical concerns, and regulatory uncertainty remain significant. The successful integration of AI into healthcare depends on responsible development practices, rigorous validation, ethical governance, and strong collaboration between clinicians and AI systems. When thoughtfully implemented, AI-driven clinical prediction has the potential to substantially improve patient outcomes and reshape the future of healthcare.

Take-home messages

- AI has significantly transformed clinical prediction by enabling **data-driven, personalized, and proactive healthcare decision-making.**
- Advanced **ML, DL, and natural language processing** techniques allow AI systems to identify complex patterns in diverse clinical data that traditional models cannot capture.
- AI-based clinical prediction improves **diagnostic accuracy, prognostic assessment, and treatment optimization** across multiple healthcare domains.
- The effective adoption of AI in clinical practice depends on:
- High-quality and representative data.
- Ethical and regulatory governance.
- Model transparency and explainability.
- Seamless integration into existing clinical workflows.
- AI systems are best positioned as **decision-support tools**, complementing rather than replacing clinical expertise.
- The future of clinical prediction lies in **effective human–AI collaboration**, enabling safer, more efficient, and patient-centered healthcare delivery.

References

A Review of Augmented Reality Systems and their Effects on Mental Workload and Task Performance: Heliyon. Available from: www.cell.com/heliyon/fulltext/S2405-8440(21)00382-0.

A Survey on Deep Learning in Medical Image Analysis - PubMed. Available from: https://pubmed.ncbi.nlm.nih.gov/28778026/.

Aggarwal, C.C., 2018. Neural Networks and Deep Learning: A Textbook Neural Networks and Deep Learning: A Textbook. Springer International Publishing, United States, pp. 1–497. https://doi.org/10.1007/978-3-319-94463-0. Available from: https://link.springer.com/book/10.1007/978-3-319-94463-0.

Anderson, P., Shenfeld, I., Marks, C., et al., 2015. Reverse engineering and evaluation of prediction models for progression to type 2 diabetes: an application of machine learning using electronic health records. J. Diabetes Sci. Technol. 10 (1), 6–18.

Aravazhi, P.S., Gunasekaran, P., Benjamin, N.Z.Y., Thai, A., Chandrasekar, K.K., Kolanu, N.D., Prajjwal, P., Tekuru, Y., Brito, L.V., Inban, P., 2025. The integration of artificial intelligence into clinical medicine: trends, challenges, and future directions. Disease-a-Month 71 (6). https://doi.org/10.1016/j.disamonth.2025.101882. Available from: https://www.sciencedirect.com/science/journal/00115029.

Band, S.S., Yarahmadi, A., Hsu, C.C., Biyari, M., Sookhak, M., Ameri, R., Dehzangi, I., Chronopoulos, A.T., Liang, H.W., 2023. Application of explainable artificial intelligence in medical health: a systematic review of interpretability methods. Inform. Med. Unlocked 40. https://doi.org/10.1016/j.imu.2023.101286. Available from: https://www.sciencedirect.com/science/journal/23529148.

Chen, R., Snyder, M., 2013. Promise of personalized omics to precision medicine. WIREs Syst. Biol. Med. 5 (1), 73–82. https://doi.org/10.1002/wsbm.1198.

Dermatologist-level Classification of Skin Cancer with Deep Neural Networks | Nature. Available from: https://www.nature.com/articles/nature21056.

Elhaddad, M., Hamam, S., 2024. AI-Driven Clinical Decision Support Systems: An Ongoing Pursuit of Potential. Cureus 6. Available from: https://cureus.com/articles/244463-ai-driven-clinical-decision-support-systems-an-ongoing-pursuit-of-potential.

Esteva, A., Robicquet, A., Ramsundar, B., Kuleshov, V., DePristo, M., Chou, K., Cui, C., Corrado, G., Thrun, S., Dean, J., 2019. A guide to deep learning in healthcare. Nat. Med. 25 (1), 24–29. https://doi.org/10.1038/s41591-018-0316-z.

Explainability for artificial intelligence in healthcare: a multidisciplinary perspective. BMC Med. Inf. Decis. Making 20 (1), 2020, 310. Available from: https://link.springer.com/article/10.1186/s12911-020-01332-6.

(1905.04610) Explainable AI for Trees: From Local Explanations to Global Understanding. Available from: https://arxiv.org/abs/1905.04610.

Exploring Multimodal Framework of Optimized Feature-Based Machine Learning to Revolutionize the Diagnosis of Parkinson's Disease: AI-Driven Insights | Request PDF, 2025. ResearchGate. Available from: https://www.researchgate.net/publication/389305343_Exploring_Multimodal_Framework_of_Optimized_Feature-Based_Machine_Learning_to_Revolutionize_the_Diagnosis_of_Parkinson's_Disease_AI-Driven_Insights.

Fostering Trust and Interpretability: Integrating Explainable AI (XAI) with Machine Learning for Enhanced Disease Prediction and Decision Transparency | Diagnostic Pathology | Springer Nature Link. Available from: https://link.springer.com/article/10.1186/s13000-025-01686-3.

Frontiers | Accessing Artificial Intelligence for Clinical Decision-Making. Available from: https://www.frontiersin.org/journals/digital-health/articles/10.3389/fdgth.2021.645232/full.

Full article: Machine Learning and AI in Cancer Prognosis, Prediction, and Treatment Selection: A Critical Approach. Available from: https://www.tandfonline.com/doi/full/10.2147/JMDH.S410301#references-Section1.

JMIR AI - How Explainable Artificial Intelligence Can Increase or Decrease Clinicians' Trust in AI Applications in Health Care: Systematic Review. Available from: https://ai.jmir.org/2024/1/e53207.

Kelly, C.J., Karthikesalingam, A., Suleyman, M., Corrado, G., King, D., 2019. Key challenges for delivering clinical impact with artificial intelligence. BMC Med. 17 (1), 17417015. https://doi.org/10.1186/s12916-019-1426-2.

Khalifa, M., Albadawy, M., 2024. Artificial intelligence for clinical prediction: exploring key domains and essential functions. Comp. Methods Prog. Biomed. Update 5, 100148. https://doi.org/10.1016/j.cmpbup.2024.100148.

Komorowski, M., Celi, L.A., Badawi, O., Gordon, A.C., Faisal, A.A., 2018. The Artificial Intelligence Clinician learns optimal treatment strategies for sepsis in intensive care. Nat. Med. 24 (11), 1716–1720. https://doi.org/10.1038/s41591-018-0213-5.

Liao, J., Li, X., Gan, Y., Han, S., Rong, P., Wang, W., et al., 2023. Artificial intelligence assists precision medicine in cancer treatment. Front. Oncol. 12. Available from: https://www.frontiersin.org/journals/oncology/articles/10.3389/fonc.2022.998222/full.

Luo, J.-C., Zhao, Q.-Y., Tu, G.-W., 2020. Clinical prediction models in the precision medicine era: old and new algorithms. Ann. Transl. Med. 8 (6). https://doi.org/10.21037/atm.2020.02.63, 274–274.

Machine Learning for Patient Risk Stratification: Standing on, or Looking Over, the Shoulders of Clinicians? | npj Digital Medicine. Available from: https://www.nature.com/articles/s41746-021-00426-3.

Obermeyer, Z., Powers, B., Vogeli, C., Mullainathan, S., 2019. Dissecting racial bias in an algorithm used to manage the health of populations. Science 366 (6464), 447–453. https://doi.org/10.1126/science.aax2342.

Predicting vital sign deterioration with artificial intelligence or machine learning. J. Clin. Monit. Comput. 33 (6), 2019, 949–951. Available from: https://link.springer.com/article/10.1007/s10877-019-00343-7.

Revolutionizing the Early Detection of Alzheimer's Disease through Non-Invasive Biomarkers: The Role of Artificial Intelligence and Deep Learning | MDPI. Available from: https://www.mdpi.com/1424-8220/23/9/4184.

Samson, O., 2025. A comparative study of traditional statistical models vs. AI. In: Predicting Health Outcomes.

Taenzer, A.H., Pyke, J.B., McGrath, S.P., Blike, G.T., 2010. Impact of pulse oximetry surveillance on rescue events and intensive care unit transfers: a before-and-after concurrence study. Anesthesiology 112 (2), 282–287. https://doi.org/10.1097/ALN.0b013e3181ca7a9b.

The future of digital health with federated learning | npj Digital Medicine. Available from: https://www.nature.com/articles/s41746-020-00323-1.

Topol, E.J., 2019. Deep Medicine How Artificial Intelligence Can Make Healthcare Human Again. Basic Books. - References - Scientific Research Publishing. Available from: https://www.scirp.org/reference/referencespapers?referenceid=3758241.

Turki, T., Wang, J.T.L., 2019. Clinical intelligence: new machine learning techniques for predicting clinical drug response. Comput. Biol. Med. 107, 302–322. https://doi.org/10.1016/j.compbiomed.2018.12.017.

View of the role of artificial intelligence in personalized oncology: predictive models and treatment optimization | Academic Journal of Science, Technology and Education. Available from: http://integrumpublication.org/index.php/ajste/article/view/65/67.

14

AI/ML for medical informatics and public health

Mohammad "Sufian" Badar[1, 2, 3, 4], Asma Maheen[5], Asma Orooj[6], Shazmeen Shamsi[7], Syed Fawaz Ayazi[8], Faiq Alam[8], Md Kaleem[9], Mohd Shahzad[10], Shaista Parveen[11], Mohd Maksuf Ul Haque[12] and Mohd Afaque[13]

[1]*DEPARTMENT OF BIOENGINEERING, UNIVERSITY OF CALIFORNIA, RIVERSIDE, CA, UNITED STATES;* [2]*UNIVERSAL SCIENTIFIC EDUCATION AND RESEARCH NETWORK (USERN), TEHRAN, IRAN;* [3]*SPI DARBHANGA, DARBHANGA, BIHAR, INDIA;* [4]*DEPARTMENT OF COMPUTER SCIENCE AND ENGINEERING (BIOINFORMATICS), SCHOOL OF ENGINEERING SCIENCES AND TECHNOLOGY (SEST), JAMIA HAMDARD UNIVERSITY, NEW DELHI, INDIA;* [5]*NEW DELHI, INDIA;* [6]*DEPARTMENT OF COMMUNITY MEDICINE, AGARTALA GOVERNMENT MEDICAL COLLEGE AND GBP HOSPITAL, AGARTALA, TRIPURA, INDIA;* [7]*DEPARTMENT OF COMPUTER SCIENCE, JMI, NEW DELHI, INDIA;* [8]*BACHELOR'S OF TECHNOLOGY, ZAKIR HUSAIN COLLEGE OF ENGINEERING AND TECHNOLOGY, ALIGARH MUSLIM UNIVERSITY (AMU), UTTAR PRADESH, INDIA;* [9]*DEPARTMENT OF PHYSICS, JAHANGIRABAD INSTITUTE OF TECHNOLOGY, BARABANKI, UTTAR PRADESH, INDIA;* [10]*FACULTY OF LAW, HAMDARD INSTITUTE OF LEGAL STUDIES AND RESEARCH (HILSR), JAMIA HAMDARD UNIVERSITY, NEW DELHI, INDIA;* [11]*DENTAL CONSULTANT, PATNA, INDIA;* [12]*DEPARTMENT OF BIOTECHNOLOGY, JAYPEE INSTITUTE OF INFORMATION TECHNOLOGY, NOIDA, UTTAR PRADESH, INDIA;* [13]*DEPARTMENT OF RESPIRATORY MEDICINE, HAMDARD INSTITUTE OF MEDICAL SCIENCES AND HAHC HOSPITAL, NEW DELHI, INDIA*

What you will learn

- Overview of AI/ML concepts and how medical informatics differs from public health
- Role of AI/ML in EHRS, CDSS, Medical imaging, genomics, and related areas of medical informatics
- How AI/ML supports public health through disease surveillance and population health management
- Exploring the role of AI/ML in COVID-19 and TB patients
- Existing concerns and limitations of AI/ML implementation in healthcare

Applying Artificial Intelligence to Computational Biology and Medical Informatics. https://doi.org/10.1016/B978-0-443-44915-4.00014-9

14.1 Introduction

14.1.1 Brief introduction to AI/ML and its history

We have officially entered into a period of AI/ML innovation in many domains and disciplines, including healthcare systems. The development and use of AI has been fueled in various ways to analyze a huge volume of data emerging from healthcare system sources. AI, in simpler terms, can be understood as machines with the ability to execute tasks that necessitate the use of human intelligence, that is, analyzing and reasoning, learning, and solving problems (Dabla et al., 2021). AI has been built to mimic human cognitive functions and match its capabilities, and it can be broadly classified as Narrow AI or General AI. While the narrow AI performs limited tasks, the general AI can perform more intellectual tasks. Narrow AI, also acknowledged as Artificial Narrow Intelligence (ANI), is trained for particular activities or domains, lacking human multipurpose proficiency, but its ability to perform specific tasks swiftly and precisely is strikingly remarkable, through specialized algorithms enabling the task (Babu and Banana, 2024). On the other hand, General Artificial Intelligence (AGI) is made to have cognitive capabilities similar to a human being, excelling in various domains (Latif et al., 2023). ML, or machine learning, often used together with AI, is actually a subset of AI. Both AI and ML are currently redefining healthcare in many ways. Some of them include applications in diagnostics and clinical decision making, or on a more individual level, personal health monitoring. Personal health monitoring devices are wearable tools that integrate AI/ML algorithms with devices, for instance, Samsung Health. To support clinicians, AI can employ different methods such as image recognition, decision-making, and disease diagnosis (Mandala et al., 2023). For example, BotkinAI, a Russian company, enables AI-driven analysis through a software platform that can process medical images for cancer and COVID-19 predictions. When startup companies collaborate globally, they can drive innovations and better healthcare delivery.

The current state of AI innovation has been paved by preliminary developments from the past. The origins of AI began around the 20th century, when mathematical logics, together with computational advancement, made the thought of machines that could stimulate human thought actually possible (Masic, 2022). The rule-based system, often called symbolic AI, was the earliest form of AI that, just like its name, used predefined rules to enable problem-solving (Muthukrishnan et al., 2020). Some of its limitations are rigid sets and rules that cannot deal with uncertainty. Another one is the lack of applicability to the real world. This led to the transition toward more scalable and dynamic approaches, that is, machine learning. Machine learning, as the name suggests, can learn patterns and make decisions based on the fed data. This learning feature from data, as more data is exposed to it, escalated to another subset of ML called Deep learning (DL) (El Naqa and Murphy, 2015). DL utilizes neural networks comprised of many layers, allowing tasks like image and speech recognition, and natural language processing (NLP). More characteristically, it can leverage large data and computational capabilities to understand hierarchical data, related to each other through the principle of containment.

14.1.2 AI/ML in medical informatics and public health

The adoption of AI in healthcare is gaining popularity due to quick decision-making leveraged through data obtained from sources such as Internet of Things (IoT) devices, clinical decision support systems (CDSS) used by physicians, and electronic health records (EHRs) generated from the healthcare sector, and is analyzed using big data technologies (Yao et al., 2025). For example, IoT and big data are applied to image processing to create structured information utilized in remote diagnostics.

Beyond applications in medical informatics, there are many broader implications of AI for public health. Medical informatics can be understood as a cross-sectional discipline that allows efficient and innovative research in biomedicine and computer sciences, for the purpose of advancements in human health. Essentially, the use of computers for medicine marked the beginning of medical informatics (Haux, 2010). Efforts by researchers in medical informatics may make it possible in the future to automate data capture and storage through diagnostics and therapeutics. The clear distinction between medical informatics and public health is due to the emphasis of the latter on population-level disease prevention, health surveillance, preparedness in times of health calamities and outbreaks, often involving governmental and public sector support. While public health professionals are more inclined toward analyzing and optimizing systems to enhance surveillance systems for planning, data collection, analysis, and implementation phases, medical informatics professionals support patient care through technologies like EHRs, decision support systems, and data analytics (Aziz, 2017; Savel, Foldy and Control, 2012).

To facilitate the understanding of AI/ML in medical informatics and public health, this chapter includes a brief review of the foundations of AI/ML in healthcare, discussing the purpose of CDSS, EHRS, medical imaging, and genomics in medical informatics. Then, following that, the context of AI/ML in public health, its principles, and population health management will be reviewed. Overall, this chapter aims to comprehend the current landscape of AI/ML in the field of medical informatics and public health. Navigation through it will provide insights into the same.

14.2 Foundational concepts of AI/ML in healthcare

14.2.1 Types of machine learning

Machine learning is alternatively known as data mining and is based on the cross-section of statistics and engineering. Its goal is to extract useful patterns in the data to give solutions. Broadly, ML algorithms are divided into four types: 1. Supervised learning, 2. Unsupervised learning, 3. Semi-supervised learning, and 4. Reinforcement learning (Morales and Escalante, 2021). The difference between them is mentioned below.

Supervised learning: In this type, algorithms are trained on labeled data along with a corresponding known outcome to uncover underlying patterns and enable the model to predict outcomes for new, unknown inputs. Techniques such as Support Vector Machine, Logistic Regression, and decision trees are commonly used in this.

Unsupervised learning: In this type, algorithms are trained on unlabeled data and focus on discovering hidden structures, anomalies, or groupings without any predefined outcomes. It includes various algorithms like clustering, KNN, and Apriori ((Alloghani et al., 2019).

Semi-supervised learning: It combines both supervised and unsupervised learning techniques, that is., both labeled and unlabeled data are used to predict outcomes.

Reinforcement learning (RL): It comprises an agent, an external environment, and a trial-and-error process. The agent learns by interacting with the environment through a trial-and-error approach. It also involves strategies to maximize a cumulative reward over time. Algorithms used are Q-learning, SARSA, DQN, and double DQN (Shethiya, 2024).

Application of these algorithms can be found in various domains. Supervised learning is commonly used to forecast outbreaks of diseases on the basis of epidemiological past data, climate patterns, and mobility data patterns, or even to assign regions into high-risk or low-risk. Unsupervised learning can be seen in early warning systems for anomaly detection (e.g., symptom-related social media activity), clustering cases or outbreaks to reveal transmission dynamics, and to detect emerging public health concerns. Although RL is complex to design and implement and it can be used to optimize public health interventions, such as determining when and where to deploy vaccines or allocate resources, and in robotics and gaming. The uses of all four types of machine learning are shown in Fig. 14.1, shown below.

Given the unprecedented complexity of machine learning and its types, another tool known as natural language processing (NLP) is utilized in a variety of language-related activities in several domains to bridge the gap between human language and computer comprehension. Human language contains nuances such as sentiment, context, and structure which is not easily comprehensible by computers. NLP tools bridge these intricacies, enabling communication across various applications. The application of NLP can be seen in virtual assistants and chatbots commonly used today (Shrivastava et al., 2025).

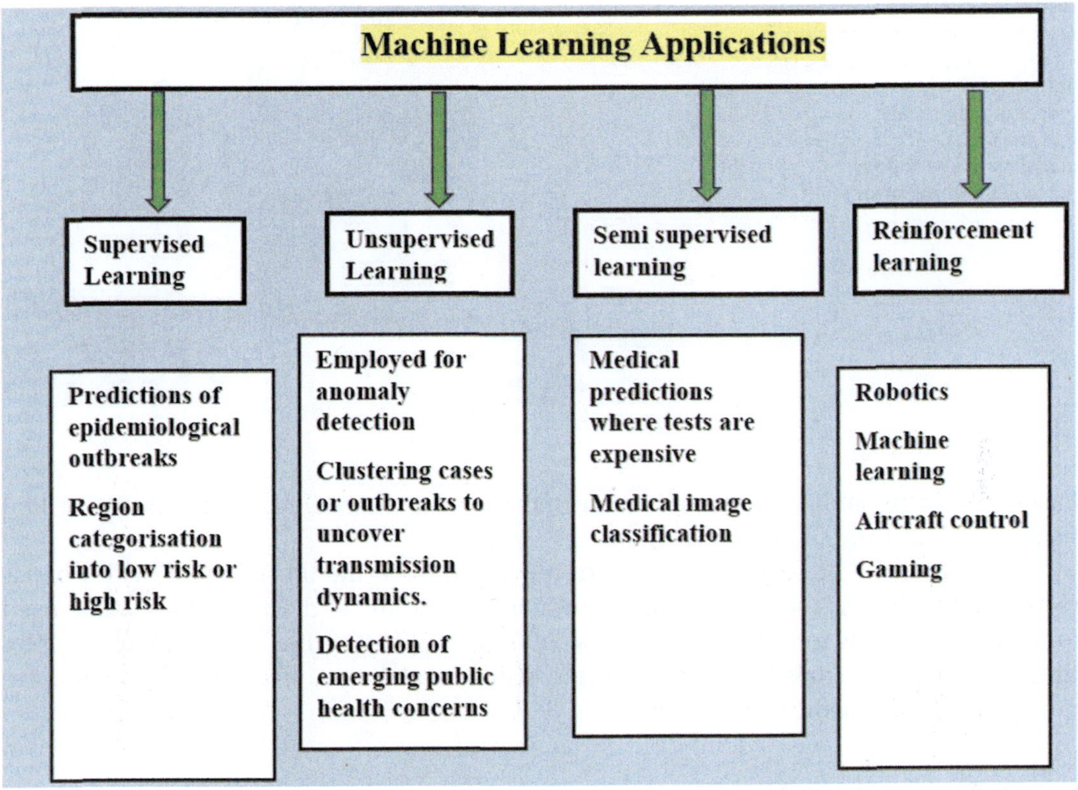

FIGURE 14.1 Types of machine learning and applications in various fields.

14.2.2 Various sources of data

Machine learning requires huge data sets, and in the converging sphere of AI and healthcare systems, a huge amount of structured and unstructured data that is heterogeneous in nature arises. The vast data can be accessed through sources like EHRs, CDSS, medical imaging, genetic data, and client data that is often organized but unstructured. This data goes into machine learning pipelines for procedures like risk assessment and diagnosis. A feedback system makes iterative model development possible to enhance accurate diagnosis and interpretation. Data sources can be classified into two types: 1. Qualitative and 2. Quantitative. Qualitative data includes laboratory tests, gene arrays, and sensor data, while qualitative data includes text and demographics (Javier (Andreu-Perez et al., 2015).

In the healthcare sector, "big data" refers to all types of data connected to patient care and wellbeing. It can be clinical data, patient data, machine-generated data, and less patient-specific data. Data collected from prescriptions, medical imaging, pharmacy, laboratory tests, etc, comes under clinical data. Patient data is driven through electronic health records, while machine-generated data can include vital sign monitoring.

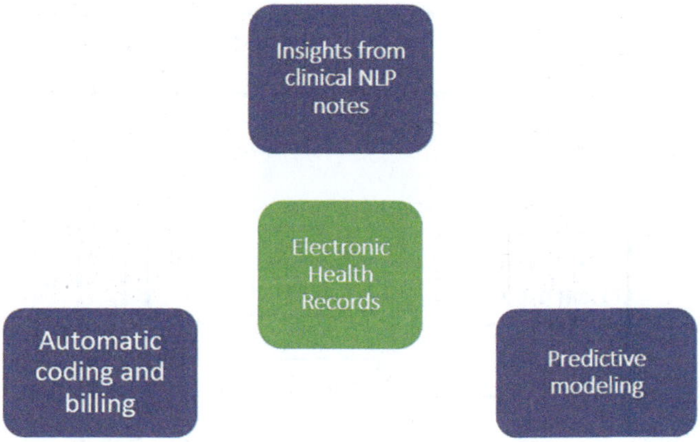

FIGURE 14.2 Uses of electronic health records in healthcare- meaningful information can be extracted from unstructured text using natural language processing; automatic coding and billing reduce manual effort; and predictive modeling allows the forecast of health outcomes and risk assessment.

Facebook status and Twitter feeds also make up social media data (Raghupathi and Raghupathi, 2014).

Big data can be beneficial to detect trends to improve the lifestyle and health of society, predict disease incidences, and detect diseases at the earliest stage. It can be used for genome analysis for personalized treatment in cancer or other similar diseases. New drugs and clinical trials can be supported using modeling and predictive analysis to design improved medications as well.

14.3 Role of AI/ML in medical informatics

14.3.1 Applications of electronic health records (EHRs)

Electronic health records (EHRs) describing patient treatments and outcomes are rich but underused information, especially in free-text form. The medical records of patients are electronically stored, thereby opening new avenues for research that could improve the healthcare industry, as depicted in Fig. 14.2. Physicians, nurses, and caregivers keep track of records in free text format, enabling nuanced reflection. Free text data is not similar to organized data, which can be used directly. Free text data naturally calls for a data mining process that is ultimately difficult to process. Efficient use of algorithms makes this task easier and scalable to extract useful information. A study has successfully demonstrated the use of statistical natural language processing in extracting geriatric patients' records through EHRs (Chen et al., 2019).

Another area where EHRs can be employed is predictive modeling. In traditional studies, data is often collected at a scheduled time, whereas data contained in EHRs is collected in an unscheduled manner due to the patient requiring treatment as and when necessary (Wu et al., 2010). Eventually, researchers do not have hold of what and when information is gathered. A typical paradigm would consider the majority of patients to have missing data since researchers do not control what information is gathered or when it is gathered. But now computational breakthroughs have made data analysis easier, and newer machine learning algorithms can easily exploit EHRs for medical diagnosis. These predictive models essentially make patient identification, risk assessment much easier, which leads to better patient outcomes. A recent study on 423 patients used predictor variables from clinical, financial, and geographic data obtained from EHRs to build models. Findings revealed that out of the 423 patients, 223 saw an improvement in their clinical state, 101 saw a decline, and 99 had no change at all. The models examined had the best accuracy in predicting individual patient response, ranging from 70% to 72%, based on baseline modeling of many clinical markers (Bennett et al., 2012).

Another interesting application of EHRs can be found in automated coding and billing. EHRs contain patients' data in the form of treatment, lab tests, prescriptions, and diagnoses. Clinical coding involves ascribing standardized codes to this information. Automated coding and billing can scan such data and assign codes, thus reducing intense manual labor.

14.3.2 Clinical decision support systems (CDSS) for assisting physicians

The idea of computers assisting clinicians in supporting and making crucial therapeutic decisions has always been the main research topic and dominated the medical informatics research. It is established that healthcare hazards can be caused due to medication mistakes and negligence. The traditional CDSS system uses static and established criteria, often limiting its general applicability. But with the advent of AI and ML next generation of CDSS may be transformed to optimize and improve accuracy. More fair and informed clinical judgments and lower costs can be possible due to this (Khude and Shende, 2025). Predictive analytics, real-time monitoring and alerts, medication interactions and adverse event identification, patient data analysis, diagnoses and treatment recommendations, and continuous learning and improvement are just a few of the characteristics and advantages of the AI-CDSS that are examined by this system. Some CDSS can be automated others may require input and annotation. However, it should be noted that AI-assisted decision support systems are not made to take over physicians and their sound judgment, that is, the autonomy will and should be with the physicians, since CDSS are made as supplementary tools. The adoption of CDSS systems also depends upon the interpretability of AI-CDSS's diagnostic findings and its ease of integration into intricate clinical processes (Castaneda et al., 2015).

The last 20 years have seen a growing research based on CDSS, and mostly they account for non-comparative studies (46.1%). The studies comprise systematic reviews (3.4%) and randomized trials (3.6%). One study, involving 42 doctors and nurses, explored CDSS's role in reducing medical errors. Physicians noted benefits like fewer allergic reactions and drug interactions (60%), quicker access to patient data (50%), and computerized orders (30%). Nurses highlighted improved legibility (25%), fewer missed tests (18.8%), faster info access (12.5%), and reduced embolism risk (9.4%) (Hajieslam and Javanmard, 2023).

14.3.3 AI/ML-based medical imaging & diagnosis

The potential of AI/ML has been applied to various facets of the medical field. The medical field often requires timely intervention and test results, which may be time-consuming and lacking efficiency. In the current medical landscape, there is a surge of research to integrate AI/ML in the diagnosis process, especially in the radiology field, which requires specialized equipment. Some of the applications of AI/ML in the different medical domains are mentioned below.

14.3.3.1 Radiology (CT, MRI, X-ray interpretation)

Advanced diagnostic techniques like computed tomography (CT), magnetic resonance imaging (MRI), positron emission tomography (PET), ultrasound, and X-rays help to document a patient's health visually and guide prompt clinical interventions and treatment monitoring. These images provide a complex understanding of anatomical, physiological, and molecular disease. An accurate and swift image report speeds up the diagnosis process and improves communication between experts. Apart from decision making a radiologists can contribute toward AI development through means such as collecting, selecting, cleaning, and organizing the data that is needed for the training and validation of the algorithm, and act as domain experts on labeling the medical data, and are to be consulted for establishing the ground truth (Najjar, 2023).

14.3.3.2 Cardiology

In the field of cardiology, AI/ML is being applied to read and interpret echocardiograms to predict conditions like myocardial infarction and atrial fibrillation. Detection of heart anomalies via ML algorithms is also a possibility, with the benefit of reducing interobserver variability. Apart from ECG and image processing, EHRs are being used to predict mortality rates due to adverse events commonly encountered. There are other promising AI/ML applications in clinical cardiology that have not been implemented yet, but hold promise for both patients and physicians (Stamate et al., 2024).

14.3.3.3 Digital pathology

AI application in pathology can be found in automated tissue image analysis and digital scanning of glass slides for remote use, which allows the pathologists to examine minute histopathological characteristics invisible to the human eye. Digital pathology allows collaboration between experts and enables accurate and swift diagnosis in telemedicine (Shen and Zhang, 2025; Shafi and Parwani, 2023).

14.3.3.4 Oncology

AI-based analysis of cancer imaging, CT scan, MRI, tumor classification example, breast, lung, prostate, radiomic data modeling, and multi-omics integration are some of the applications of AI in oncology. It can differentiate benign tumors from malignant tumors and enhance diagnosis, prognosis, and treatment planning. It also helps automate tumor measurement and supports precision oncology (Scalia et al., 2025).

14.3.4 AI/ML in genomics & drug discovery

The promise of AI and machine learning cannot be understated in the field of genomics, especially pharmacogenomics and clinical genomics, both of which produce a huge volume of data. But what drives AI integration in pharmacogenomics and precision medicine? The answer lies in better patient care and treatment responses. This happens when healthcare workers can predict reactions and tailor them according to each patient's genetic composition. Treatments' results are enhanced when there is a decrease in adverse drug responses and an increase in drug efficiency. Additionally, AI in pharmacogenomics can aid in the development of new medicines as well (Taherdoost and Ghofrani, 2024).

Development of new drugs is time-consuming and demands labor-intensive techniques. AI can invariably expedite the drug development process by zeroing in on the most appropriate therapeutic compound. Moreover, when novel compounds are made, AI can predict their pharmacokinetic and pharmacodynamic properties and even stimulate clinical trials and discover new targets. This leads to better time and cost management.

In the advancing field of precision medicine, biomarkers and patient profiling are two key aspects that need the most attention, as biomarkers provide valuable information about drugs. AI systems can predict drug effectiveness and find response-related biomarkers, and select the right group of patients for treatment. Insights from AI algorithms enable physicians to choose medication, dose, and safety, thus optimizing the whole process (Abdelhalim et al., 2022).

14.3.5 Wearables & remote monitoring

Healthcare facilities often face constraints and issues when diseases that could have been detected at an earlier stage escalate beyond control. In order to prevent such cases, the use of AI has been introduced in wearable devices with machine learning and predictive analytics capabilities that track and forecast developments in health. Some of the examples of wearable devices include smartwatches, fitness trackers, specialized biosensors, and continuous glucose monitoring (CGM) devices for collecting and analyzing data (Paul, 2025). Wearable technologies have made real-time monitoring of health data in real time. Over time, it can provide accurate results as it adjusts to a person's health data. The AI-enabled monitoring devices can predict future health issues like diabetes or a heart attack to reduce emergency health occurrences (Jafleh et al., 2024).

The most interesting aspect of wearable devices can be seen in remote patient care and health monitoring. The state of health of a patient is monitored by healthcare workers and improve adherence to treatment plans.

14.4 Role of AI/ML in public health

14.4.1 Epidemiology & disease surveillance

Epidemiology deals with the effective control and prevention of diseases within a specific population. It helps in navigating how a disease outbreak is impacting our society and economy, often measured in DALYs (Disability- Adjusted Life Years). Effective surveillance systems are often needed to detect harmful diseases at early stages with tools to understand how a disease is spreading and evolving in a specific group of population group. Often, data collection forms the core of a monitoring and surveillance system. A surveillance system works only if at least these three components are visible: a monitoring system to detect patterns, an action plan when disease cases rise above normal levels, and a defined threshold of disease frequency giving clear indications when to start implementing the action plan (Chow and Leo, 2016; Akindahunsi et al., 2024).

AI/ML in public health has been increasingly dominating epidemiology surveillance, management of population health, and certain aspects of health behavior and promotion. A recent study showed the development of ML models combining spatio-temporal data to predict and improve an early warning system for dengue outbreaks (Anggraini Ningrum et al., 2024). Dengue is a mosquito-borne viral infectious disease whose symptoms are similar to influenza and the common cold. Another study used a Long Short-term Memory (LSTM) model, which is a recurrent neural network model, to organize the symptoms of patients as text (Nadda et al., 2022). There were a lot of drawbacks in the traditional surveillance methods, often due to delayed reporting of cases and untimely action response. Ahead of the traditional surveillance, a SmartHealth-Track achieved precise accuracy in the detection of outbreaks 92%, fever detection, 93.5%, and wastewater pathogen classification 94%. The SmartHealth-Track integrated various data from IoT networks, wearable data, environmental sensors, and pharmacy data, with experimental validation through real-world datasets (Alwakeel, 2025).

During the COVID-19 pandemic, credit card and mobile phone data were used to track people with COVID-19 exposure (Park et al., 2020). It has been reported that a Canadian AI-based platform called BlueDot detected the COVID-19 outbreak much before the official international authorities. A variety of approaches known as epidemic modeling that uses statistical analysis, mathematical tools, and computer simulations, helped to understand how COVID-19 was able to spread and evolve (Villanueva-Miranda et al., 2025).

14.4.2 Population health management

The main goal of population health management is to optimize and maximize population health, taking each individual's health into consideration, according to the Institute for Healthcare Improvement (IHI). Population health management is highly people-centric, driven by data analysis and social determinants of health (Swarthout and

Bishop, 2017). Population health management essentially helps to identify opportunities to make interventions more effective based on both social and clinical needs, and not just on demand. The aims of population health management include but are not limited to reducing healthcare costs, addressing inequalities, and enhancing patient care. The working model of population health management is based on targeting a population or community, assessing the health disparities within that population, assigning an individual a score to classify into high, medium, or low risk groups, proactive intervention delivery among different groups, and monitoring and evaluating the intervention.

In a particular population, there are social and economic variations influencing health outcomes, such as housing, food security, education, and income. AI at a population level can assess these disparities and inequities, called social determinants of health (SDOH). Integrating and evaluating datasets, mapping risk factors, predicting health policy effects, and optimizing health services are some of the goals of AI. At a deeper level, AI can assist in targeting underlying causes of health disparities with qualitative health data.

In the 21st century, both computers and smartphones provide a means for the population to actively take part in a surveillance system. For example, surveillance systems for infectious diseases like Influenzanet, FluTracking, Reporta, etc (Smolinski et al., 2017). Notable digital platforms like Facebook, Google Trends, and Twitter have been used for public health surveillance, and their rise has been incremental since 2005. Google Trends and Twitter accounted for communicable outbreak prediction, while Facebook, Instagram, and YouTube were used for behavioral risk factors that included smoking patterns, substance abuse, or unhealthy lifestyle patterns. A popular means of health promotion and intervention is through AI-powered chatbots. They have become popular because they provide a means to improve adherence to treatment while maintaining anonymity. A chatbot is essentially a computer program with deep learning and NLP technologies incorporated in it (Denecke et al., 2021). AI chatbots can be classified into three types: chatbots with menus and buttons, chatbots based on keyword recognition, and AI chatbots with an understanding of attitudes and intents of users due to speech recognition technologies. People engaging with chatbots are much more likely than healthcare professionals due to the nature of diseases like mental health disorders, dementia, substance abuse, etc. Chatbots can determine a user's symptoms, make a diagnosis, and suggest treatment plans (Balcombe, 2023).

Our smartphones are also a great way to promote health and wellness. It has been reported that smokers are more likely to quit smoking when using apps, especially when they are adhering to the complete app's features and plans (Lee et al., 2018).

14.4.3 AI in infectious diseases management: TB and COVID-19

Tuberculosis (TB) is an airborne, infectious disease caused by Mycobacterium tuberculosis and is a key concern for developing countries. The estimated mortality rate is 40 per 100,000 in low-income countries compared to 3 per 100,000 in the USA or Europe

(Glaziou et al., 2018). TB is known to cause mortality more than any other infectious disease. Diagnostic tests, such as the Mantoux skin test, sputum smear microscopy, culture tests, and chest X-rays, are used for detection (Etim et al., 2023). These tests are time-consuming, less sensitive, with low accuracy. In recent years, deep learning neural networks have been employed to interpret chest radiography to screen people for pulmonary tuberculosis. An interesting study comparing the TB detection accuracy with and without AI assistance revealed that detection accuracy increases with AI assistance (Qin et al., 2021; Memon et al., 2025).

The recent COVID-19 pandemic has been one of the most challenging due to the evolving nature of the SARS-CoV-2 virus. Drug repurposing allows the exploration of previously approved drugs as an alternative to novel drug discovery. The drugs remdesivir and dexamethasone were used as credible treatments through AI-powered drug repurposing for the COVID-19 pandemic (Gandla et al., 2022). For this purpose, proteomic and genomic datasets are examined by an AI system, while genomic scanning is performed to identify viral and bacterial proteins as drug targets by DL models. For example, baricitinib, a rheumatoid arthritis drug, was discovered as a prospective COVID-19 treatment through NLP and ML technologies (Mohanty et al., 2020). Similarly, a DL-based prediction model called Molecule Transformer Drug-Target Interaction was proposed by Beck et al., to list out commercially available anti-viral drugs for COVID-19. AlphaFold, an AI-based protein structure prediction tool by DeepMind, accurately predicted the structures of SARS-CoV-2 proteins, accelerating drug development efforts (Bhattacharya and Mitra, 2023).

14.5 Challenges of AI in healthcare

The healthcare system contains both structured and unstructured datasets that have the potential to benefit society at large. However, there are several ethical and regulatory concerns that need to be addressed, especially when dealing with such a vast volume of data (Panch et al., 2019). Some of the common challenges encountered are shown in Fig. 14.3.

14.5.1 Algorithmic bias

It should be noted that algorithms are trained on data from the real world and lead to predictions based on prejudice and inequality and resulting in inaccurate outcomes. Notably, access to healthcare data across different platforms and systems is challenging. There may be a lack of data and lead to underestimation and overestimation of risk in a certain population of patients. Clear definitions, sufficient contextual specificity are invariably required for algorithms to work efficiently. ML systems often incorporate deep learning neural networks that transform real-world data into an output. Their internal logic produces an incomprehensible and unexplained output or outcome. The term

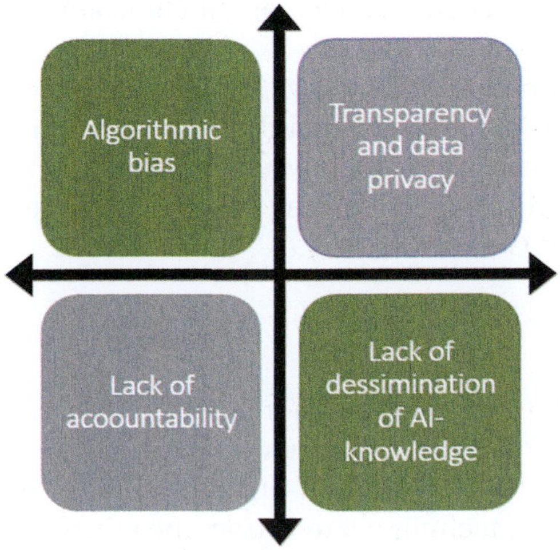

FIGURE 14.3 Common challenges and limitations of ai in medical informatics and public health.

"black box" refers to this deep neural network activity (Adeyinka et al., 2023; Marcus and Teuwen, 2024).

14.5.2 Transparency and ethical data sharing

The public healthcare system requires medical data; consequently, this data is collected, stored, and shared. This raises a lot of questionable aspects related to safety and privacy. The health records of patients may be subject to distribution across AI networks. An ethical question can be raised regarding the informed consent and usage to prevent psychological and reputational harm to patients. Therefore, public norms should be implemented to understand why and how personal data is being shared (Fehr et al., 2024).

14.5.3 Lack of accountability

AI systems are prone to making mistakes that can potentially harm patients during patient treatment and diagnosis. Holding the physician responsible and accountable for misdiagnosis is challenging since the responsibility of misdiagnosis can be leveraged on the developer of the algorithm. Determining accountability during times of failure is another challenge that needs to be addressed (Goktas and Grzybowski, 2025).

14.5.4 Dissemination of knowledge regarding AI

There is a lack of understanding when it comes to AI usage and nuances, both in the public and among physicians. Skepticism and negative perceptions can be easily found regarding AI because of a lack of proper understanding. The interactions between patients and clinicians can become challenging if clinicians lack the skills to use the AI system properly.

While there is a lot of potential for AI/ML in the healthcare industry in terms of accurate and timely diagnosis, reduced expenditure, predictions, and surveillance of diseases, and reducing the workload of physicians, further efforts to demonstrate the benefits in the real world by overcoming these challenges are also required and encouraged (Santosh and Gaur, 2021).

14.6 Conclusion

The implications of AI/ML for medical informatics and public health cannot be overstated. With so many data sources emerging from the healthcare industry, such as electronic health records (EHRs) and AI-enabled clinical decision support systems (CDSS), proper data mining techniques to extract meaningful information are required. The adoption of AI/ML is taking place in the diagnostic processes across the fields of radiology, cardiology, pathology, and oncology. Additionally, AI/ML in public health is dominating through epidemiological surveillance, population health management, and promotion of healthy behavior through machine learning models, mobile apps, and chatbots. Although promising, AI/ML remains a double-edged sword due to growing concern regarding data privacy, transparency, and accountability, so the need for ethical frameworks, fair and equitable access to AI technologies is much needed.

Take-home messages

Artificial intelligence and machine learning in the field of medical informatics and public health are providing insights that were previously thought to be unattainable. They can be used for early disease outbreak prediction, introduce new population health care approaches, and even make use of social media platforms. A notable shift from disease diagnosis to disease detection and prevention is emerging. The global population suffers from numerous infectious and fatal diseases, so the incorporation of AI/ML in scanning images or drug repurposing can bring about significant changes and reduce healthcare burdens. Although revolutionary in many ways, it should be noted that there is a presence of apprehension regarding AI/ML usage in medical informatics and public health among the general public.

Production Note:Please request the author to link the unlinked reference(s) with the corresponding text.

References

Abdelhalim, H., Berber, A., Lodi, M., Jain, R., Nair, A., Pappu, A., Patel, K., Venkat, V., Venkatesan, C., Wable, R., Dinatale, M., Fu, A., Iyer, V., Kalove, I., Kleyman, M., Koutsoutis, J., Menna, D., Paliwal, M.,

Patel, N., Patel, T., Rafique, Z., Samadi, R., Varadhan, R., Bolla, S., Vadapalli, S., Ahmed, Z., 2022. Artificial Intelligence, healthcare, clinical genomics, and pharmacogenomics approaches in precision medicine. Front. Genet. 13. https://doi.org/10.3389/fgene.2022.929736.

Adeyinka, A., Lamina, Y., Adeyeye, Y.I. Yaw, A. Minkah, O.E.T., AI in healthcare frameworks that prioritize fairness, security, and accountability. (2023).

Akindahunsi, T., Olulaja, O., Ajayi, O., Prisca Onyenegecha, I., Hanson, U., Fadojutimi, B., 2024. Analytical tools in diseases epidemiology and surveillance: a review of literature. International Journal of Applied Research 10 (9), 155–161. https://doi.org/10.22271/allresearch.2024.v10.i9c.12018.

Alloghani, M., Al-Jumeily, D., Mustafina, J., Hussain, A., Ahmed, J., Aljaaf, 2019. A Systematic Review on Supervised and Unsupervised Machine Learning Algorithms for Data Science. Springer Science and Business Media LLC, pp. 3–21. https://doi.org/10.1007/978-3-030-22475-2_1.

Alwakeel, M.M., 2025. AI-assisted real-time monitoring of infectious diseases in urban areas. Mathematics 13 (12), 1911. https://doi.org/10.3390/math13121911.

Andreu-Perez, J., Poon, C.C.Y., Merrifield, R.D., Wong, S.T.C., Yang, G.Z., 2015. Big data for health. IEEE J. Biomed. Health Inform. 19 (4), 1193–1208. https://doi.org/10.1109/JBHI.2015.2450362.

Anggraini Ningrum, D.N., Li, Y.C.J., Hsu, C.Y., Solihuddin Muhtar, M., Pandu Suhito, H., 2024. Artificial intelligence approach for severe dengue early warning system. Stud. Health Technol. Inf. 310, 881–885. https://doi.org/10.3233/SHTI231091.

Aziz, H.A., 2017. A review of the role of public health informatics in healthcare. J. Taibah Univ. Med. Sci. 12 (1), 78–81. https://doi.org/10.1016/j.jtumed.2016.08.011.

Babu, M.V.S., Banana, K., 2024. A study on narrow artificial intelligence—an overview. Int. J. Eng. Sci. Adv. Technol. 24, 210–219.

Balcombe, L., 2023. AI chatbots in digital mental health. Informatics 10 (4), 82. https://doi.org/10.3390/informatics10040082.

Bennett, C.C., Doub, T.W., Selove, R., 2012. EHRs connect research and practice: where predictive modeling, artificial intelligence, and clinical decision support intersect. Health Policy and Technology 1 (2), 105–114. https://doi.org/10.1016/j.hlpt.2012.03.001.

Bhattacharya, A., Mitra, A., 2023. Development and Repurposing of Drugs against COVID-19 Using Artificial Intelligence Computer Intelligence against Pandemics: Tools and Methods to Face New Strains of COVID-19. De Gruyter, India De Gruyter, India, pp. 139–152. https://doi.org/10.1515/9783110767681-007.

Castaneda, C., Nalley, K., Mannion, C., Bhattacharyya, P., Blake, P., Pecora, A., Goy, A., Suh, K.S., 2015. Clinical decision support systems for improving diagnostic accuracy and achieving precision medicine. J. Clin. Bioinf. 5 (1). https://doi.org/10.1186/s13336-015-0019-3.

Chen, T., Dredze, M., Weiner, J.P., Hernandez, L., Kimura, J., Kharrazi, H., 2019. Extraction of geriatric syndromes from electronic health record clinical notes: assessment of statistical natural language processing methods. JMIR Med. Inform. 7 (1), e13039. https://doi.org/10.2196/13039.

Chow, A., Leo, Y.S., 2016. Surveillance of Disease: Overview International Encyclopedia of Public Health. Elsevier Inc., Singapore, pp. 124–138. https://doi.org/10.1016/B978-0-12-803678-5.00439-2.

Dabla, P.K., Gruson, D., Gouget, B., Bernardini, S., Homsak, E., 2021. Lessons learned from the COVID-19 pandemic: emphasizing the emerging role and perspectives from artificial intelligence, mobile health, and digital laboratory medicine. Electronic Journal of the International Federation of Clinical Chemistry and Laboratory Medicine 32 (2), 224–243. https://www.ifcc.org/ifcc-communications-publications-division-cpd/ifcc-publications/ejifcc-journal/e-journal-volumes/ejifcc-2021-vol-32/ejifcc-vol-32-no-2/.

Denecke, K., Abd-Alrazaq, A., Househ, M., 2021. Artificial Intelligence for Chatbots in Mental Health: Opportunities and Challenges. Springer Science and Business Media LLC, pp. 115–128. https://doi.org/10.1007/978-3-030-67303-1_10.

El Naqa, I., Murphy, M.J., 2015. What Is Machine Learning? Machine Learning in Radiation Oncology: Theory and Applications. Springer International Publishing, Canada, pp. 3–11. https://doi.org/10.1007/978-3-319-18305-3_1.

Etim, N.G., Mirabeau, T.Y., Olorode, O.A., Nwodo, M.U., Izah, S.C., 2023. Current Diagnostics Tools of Tuberculosis: Challenges and Opportunities. ES General. https://doi.org/10.30919/esg1059.

Fehr, J., Citro, B., Malpani, R., Lippert, C., Madai, V.I., 2024. A trustworthy AI reality-check: the lack of transparency of artificial intelligence products in healthcare. Front. Digit. Health 6. https://doi.org/10.3389/fdgth.2024.1267290.

Gandla, K., Reddy, K.T.K., Babu, P.V., Sagapola, R., Sudhakar, P., 2022. A review of artificial intelligence in treatment of COVID-19. J. Pharm. Negat. Results 13, 254–264.

Glaziou, P., Floyd, K., Raviglione, M.C., 2018. Global epidemiology of tuberculosis. Semin. Respir. Crit. Care Med. 39 (3), 271–285. https://doi.org/10.1055/s-0038-1651492.

Goktas, P., Grzybowski, A., 2025. Shaping the future of healthcare: ethical clinical challenges and pathways to trustworthy AI. J. Clin. Med. 14 (5), 1605. https://doi.org/10.3390/jcm14051605.

Hajieslam, F., Javanmard, Z., 2023. Investigating the role of clinical decision support systems in reducing medical errors. Journal of Health and Biomedical Informatics 10 (1), 82–90. https://doi.org/10.34172/jhbmi.2023.16.

Haux, R., 2010. Medical informatics: past, present, future. Int. J. Med. Inf. 79 (9), 599–610. https://doi.org/10.1016/j.ijmedinf.2010.06.003.

Jafleh, E.A., Alnaqbi, F.A., Almaeeni, H.A., Faqeeh, S., Alzaabi, M.A., Zaman, K.Al, 2024. The role of wearable devices in chronic disease monitoring and patient care: a comprehensive review. Cureus 16 (9). https://doi.org/10.7759/cureus.68921.

Khude, H., Shende, P., 2025. AI-driven clinical decision support systems: revolutionizing medication selection and personalized drug therapy. Advances in Integrative Medicine 12 (4), 100529. https://doi.org/10.1016/j.aimed.2025.100529.

Latif, E., Mai, G., Nyaaba, M., Wu, X., Liu, N., Lu, G., Li, S., Liu, T., Zhai, X., 2023. Artificial General Intelligence (AGI) for Education. arXiv, United States.

Lee, M., Lee, H., Kim, Y., Kim, J., Cho, M., Jang, J., Jang, H., 2018. Mobile app-based health promotion programs: a systematic review of the literature. Int. J. Environ. Res. Publ. Health 15 (12), 2838. https://doi.org/10.3390/ijerph15122838.

Mandala, G., Reddy, R., Nishanth, A., Yasmeen, Z., Maguluri, K.K., 2023. AI and ML in healthcare: redefining diagnostics, treatment, and personalized medicine. Int. J. Adv. Eng. Technol. 5.

Marcus, E., Teuwen, J., 2024. Artificial intelligence and explanation: how, why, and when to explain black boxes. Eur. J. Radiol. 173, 111393. https://doi.org/10.1016/j.ejrad.2024.111393.

Masic, I., 2022. Artificial intelligence in health care: past, present and future. Acta Inf. Med. 30 (1), 61–67. https://doi.org/10.5455/aim.2022.30.61-67.

Memon, S., Bibi, S., He, G., 2025. Integration of AI and ML in tuberculosis (TB) management: from diagnosis to drug discovery. Diseases 13 (6), 184. https://doi.org/10.3390/diseases13060184.

Mohanty, S., Harun AI Rashid, M., Mridul, M., Mohanty, C., Swayamsiddha, S., 2020. Application of artificial intelligence in COVID-19 drug repurposing. Diabetes Metabol. Syndr.: Clin. Res. Rev. 14 (5), 1027–1031. https://doi.org/10.1016/j.dsx.2020.06.068.

Morales, E.F., Escalante, H.J., 2021. A Brief Introduction to Supervised, Unsupervised, and Reinforcement Learning Biosignal Processing and Classification Using Computational Learning and Intelligence: Principles, Algorithms, and Applications. Elsevier, Mexico, pp. 111–129. https://doi.org/10.1016/B978-0-12-820125-1.00017-8.

Muthukrishnan, N., Maleki, F., Ovens, K., Reinhold, C., Forghani, B., Forghani, R., 2020. Brief history of artificial intelligence. Neuroimaging Clin. 30 (4), 393–399. https://doi.org/10.1016/j.nic.2020.07.004.

Nadda, W., Boonchieng, W., Boonchieng, E., 2022. Influenza, dengue and common cold detection using LSTM with fully connected neural network and keywords selection. BioData Min. 15 (1). https://doi.org/10.1186/s13040-022-00288-9.

Najjar, R., 2023. Redefining radiology: a review of artificial intelligence integration in medical imaging. Diagnostics 13 (17), 2760. https://doi.org/10.3390/diagnostics13172760.

Panch, T., Mattie, H., Atun, R., 2019. Artificial intelligence and algorithmic bias: implications for health systems. J. Glob. Health 9 (2). https://doi.org/10.7189/jogh.09.020318.

Park, Y.J., Cho, S.Y., Lee, J., Lee, I., Park, W.H., Jeong, S., Kim, S., Lee, S., Kim, J., Park, O., 2020. Development and utilization of a rapid and accurate epidemic investigation support system for covid-19. Osong Public Health and Research Perspectives 11 (3), 118–127. https://doi.org/10.24171/j.phrp.2020.11.3.06.

Paul, J., Real-time predictive health monitoring using AI-driven wearable sensors: enhancing early detection and personalized interventions in chronic disease management. (2025).

Qin, Z.Z., Ahmed, S., Sarker, M.S., Paul, K., Adel, A.S.S., Naheyan, T., Barrett, R., Banu, S., Creswell, J., 2021. Tuberculosis detection from chest x-rays for triaging in a high tuberculosis-burden setting: an evaluation of five artificial intelligence algorithms. Lancet Digit. Health 3 (9), e543. https://doi.org/10.1016/S2589-7500(21)00116-3.

Raghupathi, W., Raghupathi, V., 2014. Big data analytics in healthcare: promise and potential. Health Inf. Sci. Syst. 2 (1). https://doi.org/10.1186/2047-2501-2-3.

Santosh, K., Gaur, L., 2021. Introduction to AI in Public Health SpringerBriefs in Applied Sciences and Technology. Springer Science and Business Media Deutschland GmbH, United States, pp. 1–10. https://doi.org/10.1007/978-981-16-6768-8_1.

Savel, T.G., Foldy, S., Centers for Disease Control, 2012. The Role of Public Health Informatics in Enhancing Public Health Surveillance. Morbidity and mortality weekly report. Surveillance summaries, Washington, D.C, pp. 20–24.

Scalia, I.G., Pathangey, G., Abdelnabi, M., Ibrahim, O.H., Abdelfattah, F.E., Pietri, M.P., Ibrahim, R., Farina, J.M., Banerjee, I., Tamarappoo, B.K., Arsanjani, R., Ayoub, C., 2025. Applications of artificial intelligence for the prediction and diagnosis of cancer therapy-related cardiac dysfunction in oncology patients. Cancers 17 (4). https://doi.org/10.3390/cancers17040605.

Shafi, S., Parwani, A.V., 2023. Artificial intelligence in diagnostic pathology. Diagn. Pathol. 18 (1). https://doi.org/10.1186/s13000-023-01375-z.

Shen, I.Z., Zhang, L., 2025. Digital and artificial intelligence-based pathology: not for every laboratory – a mini-review on the benefits and pitfalls of its implementation. Journal of Clinical and Translational Pathology 5 (2), 79–85. https://doi.org/10.14218/JCTP.2025.00007.

Shethiya, A.S., 2024. Decoding intelligence: a comprehensive study on machine learning algorithms and applications. Academia Nexus Journal 3.

Shrivastava, N., Tewari, P., Sujatha, S., 2025. Natural language processing for conversational AI: chatbots and virtual assistants. In: IEEE International Conference on Interdisciplinary Approaches in Technology and Management for Social Innovation, IATMSI 2025. Institute of Electrical and Electronics Engineers Inc, India. https://doi.org/10.1109/IATMSI64286.2025.10984818.

Smolinski, M.S., Crawley, A.W., Olsen, J.M., Jayaraman, T., Libel, M., 2017. Participatory disease surveillance: engaging communities directly in reporting, monitoring, and responding to health threats. JMIR Public Health and Surveillance 3 (4). https://doi.org/10.2196/publichealth.7540.

Stamate, E., Piraianu, A.-I., Roxana Ciobotaru, O., Crassas, R., Duca, O., Fulga, A., Grigore, I., Vintila, V., Fulga, I., Ciobotaru, O.C., 2024. Revolutionizing cardiology through artificial intelligence—big data from proactive prevention to precise diagnostics and cutting-edge treatment—a comprehensive review of the past 5 years. Diagnostics 14 (11), 1103. https://doi.org/10.3390/diagnostics14111103.

Swarthout, M., Bishop, M.A., 2017. Population health management: review of concepts and definitions. Am. J. Health Syst. Pharm. 74 (18), 1405–1411. https://doi.org/10.2146/ajhp170025.

Taherdoost, H., Ghofrani, A., 2024. AI's role in revolutionizing personalized medicine by reshaping pharmacogenomics and drug therapy. Intelligent Pharmacy 2 (5), 643–650. https://doi.org/10.1016/j.ipha.2024.08.005.

Villanueva-Miranda, I., Xiao, G., Xie, Y., 2025. Artificial intelligence in early warning systems for infectious disease surveillance: a systematic review. Front. Public Health 13. https://doi.org/10.3389/fpubh.2025.1609615.

Wu, J., Roy, J., Stewart, W.F., 2010. Prediction modeling using EHR data: challenges, strategies, and a comparison of machine learning approaches. Med. Care 48 (6), S106. https://doi.org/10.1097/MLR.0b013e3181de9e17.

Yao, L., Liu, Y., Wang, T., Han, C., Li, Q., Li, Q., You, X., Ren, T., Wang, Y., 2025. Global trends of big data analytics in health research: a bibliometric study. Front. Med. 12. https://doi.org/10.3389/fmed.2025.1456286.

15

AI in medical imaging for developing countries: Challenges and opportunities

Sobia Naaz[1], Marya Saeed[2], Rehan Saeed Khan[3], Akhtar Ali[4], Azhar Latif[5], Shabbir Ahmad Warsi[6], Labeebah Rizwan Badar[7], Aamna Hassan[8], Arif Wahab[9], Abul Kalam Azad[10] and Mohammad "Sufian" Badar[1, 11, 12, 13]

[1]*DEPARTMENT OF COMPUTER SCIENCE AND ENGINEERING (BIOINFORMATICS), SCHOOL OF ENGINEERING SCIENCES AND TECHNOLOGY (SEST), JAMIA HAMDARD UNIVERSITY, NEW DELHI, INDIA;* [2]*CARDIOTHORACIC AND VASCULAR SURGERY (CTVS) AND CARDIAC CRITICAL CARE UNIT, MEDANTA HOSPITAL, NOIDA, UTTAR PRADESH, INDIA;* [3]*DEPARTMENT OF COMPUTER SCIENCE AND ENGINEERING, SEST, JAMIA HAMDARD, NEW DELHI, INDIA;* [4]*CENTRE FOR FEDERAL STUDIES, PUBLIC POLICY AND GOVERNANCE, JAMIA HAMDARD UNIVERSITY, NEW DELHI, INDIA;* [5]*META, MENLO PARK, CA, UNITED STATES;* [6]*AURA COSMETIC SURGERY KLINIK, PARAS HMRI HOSPITAL, PATNA, BIHAR, INDIA;* [7]*BADAR MEDICAL CENTRE (BMC), NEW DELHI, INDIA;* [8]*DEPARTMENT OF BOTANY, JAMIA HAMDARD, NEW DELHI, INDIA;* [9]*INDRAPRASTHA APOLLO HOSPITAL, SARITA VIHAR, NEW DELHI, INDIA;* [10]*MAX HEALTH CARE, MAX SUPERSPECIALTY HOSPITAL, NEW DELHI, INDIA;* [11]*DEPARTMENT OF BIOENGINEERING, UNIVERSITY OF CALIFORNIA, RIVERSIDE, CA, UNITED STATES;* [12]*UNIVERSAL SCIENTIFIC EDUCATION AND RESEARCH NETWORK (USERN), TEHRAN, IRAN;* [13]*SPI DARBHANGA, DARBHANGA, BIHAR, INDIA*

What you will learn

- Understand the application of medical imaging modalities such as X-rays, ultrasounds, CT scans, and MRI in clinical environments with constrained resources, as well as their fundamental principles of operation.
- Elucidate the techniques employed by AI systems to analyze medical images through deep learning, machine learning, and intricate structures.
- Identify the primary imaging AI tasks, such as classification, detection, and segmentation, and select the appropriate assessment metrics, including AUC,

Applying Artificial Intelligence to Computational Biology and Medical Informatics. https://doi.org/10.1016/B978-0-443-44915-4.00015-0

sensitivity, specificity, and calibration. Utilize explicable AI methodologies, such as Grad-CAM, LIME, and SHAP, to enhance clarity, foster trust in healthcare environments, and assist in identifying errors.

- Understand the unique constraints faced by developing nations, such as insufficient infrastructure, a shortage of labor, and inferior data quality.
- Develop deployment protocols encompassing mobile/edge inference, radiology integration, and mechanisms to provide quality assurance.
- Evaluate clinical readiness through external validation, subgroup analysis, prospective studies, and postdeployment surveillance. Summarize the critical regulations and guidelines (consistent with FDA/CE/SaMD principles) and ethical standards necessary for successful implementation.

15.1 Introduction

Medical imaging is a crucial component of contemporary clinical practice, as it enables physicians to visualize interior anatomical structures and their functions, hence facilitating diagnosis, risk evaluation, treatment strategies, and continuous monitoring. Imaging modalities such as X-rays, ultrasounds, CT scans, and MRI assist physicians in decision-making across various medical fields, including emergency medicine, cardiology, infectious disease management, and maternal and neonatal care.

In underdeveloped nations, imaging systems are typically organized in tiers: fundamental imaging is employed at primary and local hospitals, whereas sophisticated imaging is predominantly conducted at regional referral centers. Medical imaging has improved due to advancements in artificial intelligence (AI), enhanced deep learning frameworks, increased availability of digital imaging datasets, and superior computing infrastructure. For successful clinical application, the algorithm must perform effectively on internal datasets. Developing nations encounter challenges such as inadequate connectivity, restricted access to specialists, varying imaging technologies, and disparate regulations that influence operational efficacy, safety, and durability.

When evaluating image AI in resource-constrained environments, it is crucial to consider its practical efficacy, user-friendliness, and compliance with safety regulations. This chapter addresses the fundamental concepts of medical imaging and artificial intelligence techniques, beginning with the most elementary principles and progressing to more intricate ones. It emphasizes the critical necessity for impoverished nations to utilize this technology judiciously. It provides practical examples and valuable guidance for developing scalable, secure, and equitable AI-driven image applications (Dey et al., 2019).

15.2 AI applications and their tangible benefits

As medical images and AI get better, they lay the groundwork for flexible, clinically relevant decision support in healthcare systems that don't have a lot of resources. By using the right imaging methods along with AI techniques, doctors can make better decisions about which diagnoses to focus on, make the most of the experts they have access to, and move things along faster without losing their clinical sense. In this case, it is easiest to see how AI can really help when systems are built to improve current processes, work within clear governance and validation frameworks, and connect easily with referral pathways.

15.2.1 Fundamentals and significance of medical imaging

Medical imaging converts the physical characteristics of tissue into comprehensible information. The imaging physics, therapeutic applications, and organizational requirements vary for each modality.

- X-ray imaging detects variations in tissue attenuation and is extensively utilized for screening programs, thoracic imaging, and injury assessment due to its cost-effectiveness and capacity for simultaneous scans.
- Ultrasound employs sound waves for imaging purposes. It is portable, non-radiative, and effective for rapid assessments at the point of care, particularly in maternal and neonatal evaluations in emergency situations.
- CT scans are essential for assessing the lungs, brain injuries, and cancer due to their ability to produce high-resolution cross-sectional images. Nonetheless, they require reliable infrastructure and the capacity for consistent maintenance.
- MRI presents numerous operational and financial challenges yet, it offers superior contrast for soft tissues and is applicable in neurological, musculoskeletal, and certain cardiovascular imaging.

Imaging typically serves two critical functions in underdeveloped nations: (i) promptly categorizing patients for urgent treatment, and (ii) screening large populations for prevalent diseases. When AI technologies are designed to assist physicians in their duties rather than usurp medical decision-making, these functions are rational (Moore and Copel, 2011).

15.2.2 Definition of AI and its function in healthcare

Artificial intelligence (AI) is a domain within computer science that develops algorithms capable of doing tasks often requiring human intelligence, such as pattern recognition, predictive analysis, and assisting medical professionals in decision-making. These are many of the most prevalent applications of AI in healthcare:

- In machine learning (ML), models are constructed using structured clinical data such as demographics, laboratory values, and comorbidity to assess disease risk or forecast patient progression.
- Deep learning (DL) uses neural networks to develop models capable of learning intricate representations directly from extensive datasets. Convolutional neural networks (CNN) are extensively employed in medical imaging due to their capability to extract valuable features from image inputs for tasks such as segmentation, detection, and classification.

- Multimodal AI integrates imaging data with additional clinical variables, such as laboratory biomarkers, vital signs, and electronic health record (EHR) information, to enhance predictive accuracy and facilitate risk stratification.
- Explainable AI (XAI) comprises strategies designed to elucidate the salient components of an image or input data that significantly influence a model's prediction. This aids in clinical interpretation and auditing.

Artificial intelligence is frequently employed in medical imaging to classify images (indicating the presence of disease), identify anomalies, segment images (distinguishing anatomical features or lesions), and enhance workflow by automating processes such as quality control and prioritization. When appropriately utilized in healthcare environments, AI could facilitate more consistent and efficient decision-making while enabling physicians to monitor patient safety effectively (Beam and Kohane, 2018).

Fig. 15.1 illustrates a generalized AI workflow for medical imaging, beginning with image acquisition and reprocessing, followed by automated quality control and model inference. The pipeline culminates in clinically actionable outputs—such as triage scores or localization maps—supported by continuous monitoring to ensure performance stability, safety, and fairness over time.

15.2.3 The necessity of AI solutions for developing nations

Infectious diseases, health problems between mothers and children, and late-stage discovery of chronic diseases like cancer and heart disease affect developing countries more than other countries. Imaging services are often held up because there aren't

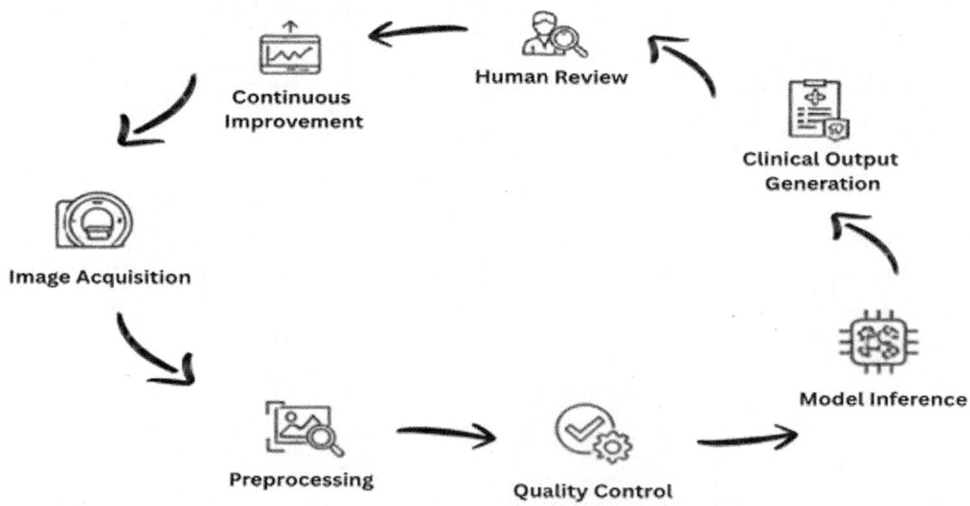

FIGURE 15.1 General AI pipeline for medical imaging, from image acquisition and prepossessing to model inference, clinical output generation, and monitoring.

enough workers, and it is hard for rural clinics to connect with expert referral centers. These delays may make early intervention pathways less successful and cause illness that could have been avoided.

AI could help people in real life by:

- It is possible for high-sensitivity screening tools to find cases that need more tests or an immediate referral.
- Sharing tasks can help healthcare workers who aren't as skilled by making decision support outputs more consistent.
- Quality assurance: detecting poor-quality scans to avoid repeated imaging and improve interpretability.
- Workforce amplification: using AI to speed up reporting through hub-and-spoke models. But adoption should be based on facts, since wrong classification could mean missing diagnoses, getting the wrong help, or not having equal access if performance changes between different patient and demographic groups (117).

Fig. 15.2 depicts common diagnostic pathway delays in developing countries, including limited access to imaging, workforce shortages, and prolonged referral timelines. It highlights points where AI-supported imaging may reduce time-to-triage and time-to-referral by enabling earlier prioritization, automated quality checks, and structured escalation within hub-and-spoke care models.

15.2.4 Chapter learning pathway

This chapter follows a structured learning pathway:

- Section 15.2 introduces imaging modality fundamentals and challenges in low-resource settings. Section 15.3 explains AI methods from basic ML to deep learning, uncertainty estimation, interpretability, and lightweight deployment.
- Section 15.4 examines health-system barriers, including infrastructure gaps, workforce shortages, and regulatory constraints.
- Section 15.5 describes practical AI applications and measurable operational benefits in imaging workflows.
- Section 15.6 outlines case studies and scaling lessons, emphasizing evidence requirements and implementation design.
- Section 15.7 highlights ethical considerations, privacy, policy, training, and next-generation systems.

15.3 Fundamentals of medical imaging

Medical imaging is a way for doctors to see inside the body without hurting the patient. This helps with drug monitoring, diagnosis, triage, and risk assessment. Lack of infrastructure, high prices, a lack of staff, and an uneven distribution of imaging technologies make it hard for poor countries to use other imaging methods. So, chest X-rays and

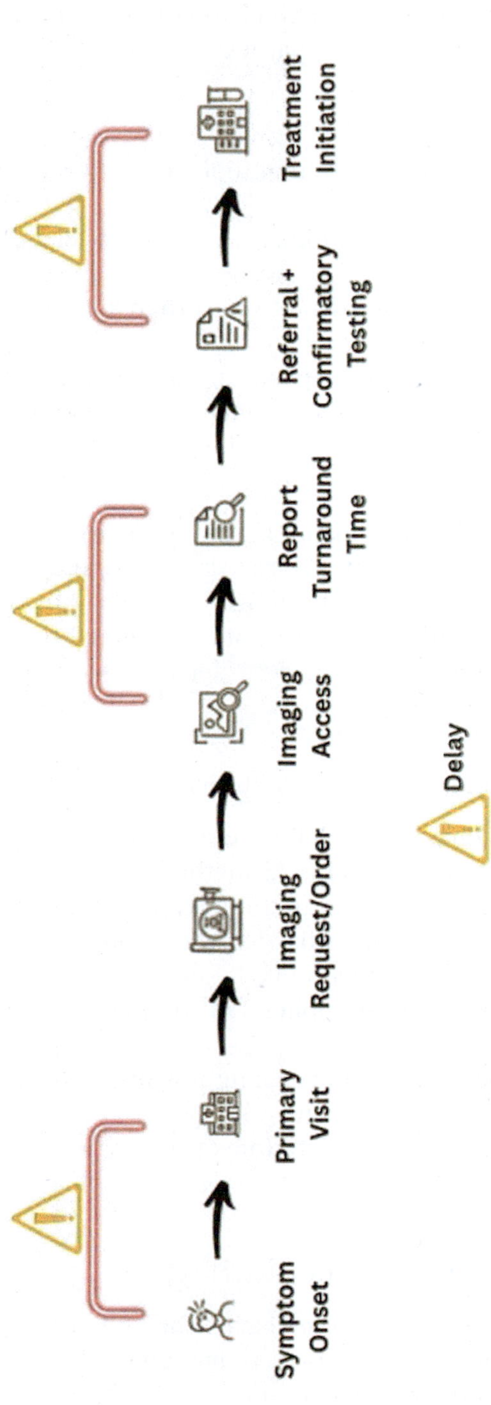

FIGURE 15.2 Common diagnostic pathway delays in developing countries and the potential roles of AI-supported imaging in reducing time-to-triage and time-to-referral.

point-of-care ultrasound are two examples of methods that can be used by a lot of people. On the contrary, advanced cross-sectional imaging methods like CT and MRI are mostly only available in major referral hospitals. This part talks about the basics of a few common imaging methods and focuses on issues of practicality that are important in places with limited resources.

15.3.1 Radiographic and thoracic imaging (tuberculosis emphasis)

Radiographic imaging is one of the most popular ways to take pictures in low- and middle-income countries (LMIC) because it is cheap, rapid, and works well for high-throughput procedures. The chest X-ray (CXR) is a very significant aspect of systems for screening and antiaging respiratory problems. People often use it to treat tuberculosis (TB), pneumonia, chronic lung illness, and heart and lung problems. It is particularly beneficial when integrated into organized referral pathways. Tuberculosis is still a major public health issue in LMIC. It is typical for early diagnosis and treatment to be delayed because it is hard to receive confirmatory testing, specialists take too long to assess cases, and referral systems are not properly organized. CXR is often used to classify people based on how important they are for more molecular testing. That so, a chest X-ray alone can't identify for sure if someone has tuberculosis; the scan results could also show symptoms of other lung diseases. AI-enabled systems should not be employed as self-diagnosing tools in TB programs; instead, they should be used to help make decisions about screening and escalation.

In regions with little resources, many things affect how well people learn CXR in real life. There can be differences because of different types of detectors, portable versus fixed devices, exposure settings that don't always function well, motion artifacts, positioning problems, and radiographers who don't get adequate training or help. Before AI can be used on a broad scale, there should be standard acquisition techniques, automated quality control, and external validation across sites and technologies. This is because these things can make it tougher to understand and use AI in other circumstances (Zech et al., 2018) (Table 15.1).

Fig. 15.3 illustrates an AI-assisted tuberculosis screening workflow in resource-limited settings, beginning with community or facility-based chest X-ray acquisition followed by automated AI triage. High-risk cases are prioritized for confirmatory molecular testing, while low-risk cases are managed through routine follow-up, enabling efficient use of diagnostic resources and supporting timely treatment initiation.

15.3.2 Ultrasound for point-of-care applications

Ultrasound is a key imaging modality in developing countries because it is radiation-free, portable, and capable of supporting rapid bedside clinical decisions. Point-of-care ultrasound (POCUS) is increasingly used in maternal–fetal medicine, emergency care, trauma triage, and limited cardiovascular assessment, particularly where access to advanced imaging is delayed or unavailable (Moore and Copel, 2011).

Table 15.1 Medical imaging modalities commonly used in developing countries: typical applications and feasibility constraints.

Modality	Typical clinical use cases	Key strengths	Major constraints in low resource settings
X-ray (CXR)	TB screening, Pneumonia triage, trauma evaluation	Low cost, rapid, scalable, high throughput	Variable image quality, limited digital infrastructure, reader variability
Ultrasound (POCUS)	Obstetric care, Trauma triage (FAST), bedside Cardiac/abdominal assessment	Portable, radiation-free, real-time decisions	Operator dependent, training needs, variable protocols
CT	Trauma, stroke Assessment, cancer staging	High-resolution cross sectional imaging, rapid acquisition	High cost, radiation safety requirements, stable Power/maintenance needed
MRI	Neuron-imaging, soft-tissue tumors, selected cardiac Imaging	Excellent soft-tissue Contrast, no ionizing Radiation	Very high cost, long scan times, infrastructure and specialist demands

Despite its clinical value, ultrasound is highly operator-dependent. Image quality depends on correct probe positioning, acquisition technique, and machine settings. Inconsistent protocols and variable training levels may lead to heterogeneous diagnostic accuracy across settings, especially in rural facilities or outreach programs. These constraints underscore the importance of structured training programs and competency-based skill assessment, supported by quality assurance mechanisms (Oktay et al., 2018).

AI can improve the usability and consistency of POCUS through functions such as automated view recognition, acquisition guidance, real-time quality scoring, and automated measurement support. These capabilities may strengthen task-sharing models by enabling frontline healthcare workers to capture diagnostically meaningful scans and escalate uncertain cases to expert review. However, safe implementation requires uncertainty-aware decision support and clearly defined escalation thresholds to prevent overconfidence in automated outputs (Steyerberg, 2019).

Fig. 15.4 depicts an AI-supported point-of-care ultrasound workflow in which real-time acquisition guidance assists operators during image capture, followed by automated image quality assessment and preliminary triage support. This approach facilitates consistent image acquisition by nonexpert users and enables timely escalation of clinically relevant findings in resource-constrained clinical settings.

15.3.3 Fundamentals of CT and MRI

Cross-sectional imaging modalities, such as CT and MRI, are highly effective for diagnostic purposes but require substantial infrastructure and proficient personnel. Consequently, the majority of CT and MRI procedures are located in tertiary referral hospitals within developing nations.

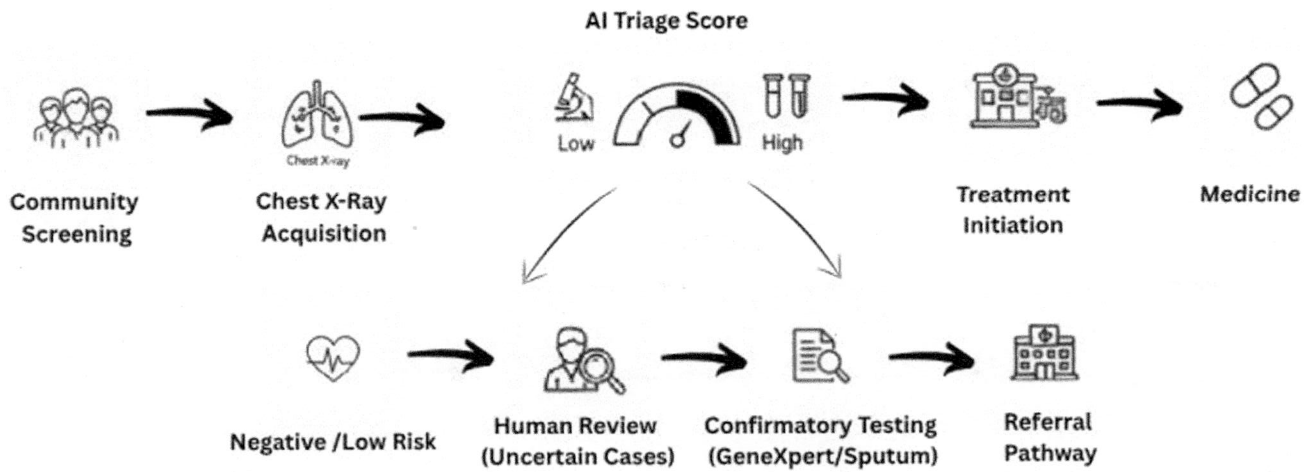

FIGURE 15.3 AI-assisted TB screening workflow using chest X-ray triage and confirmatory testing pathways in resource-limited settings.

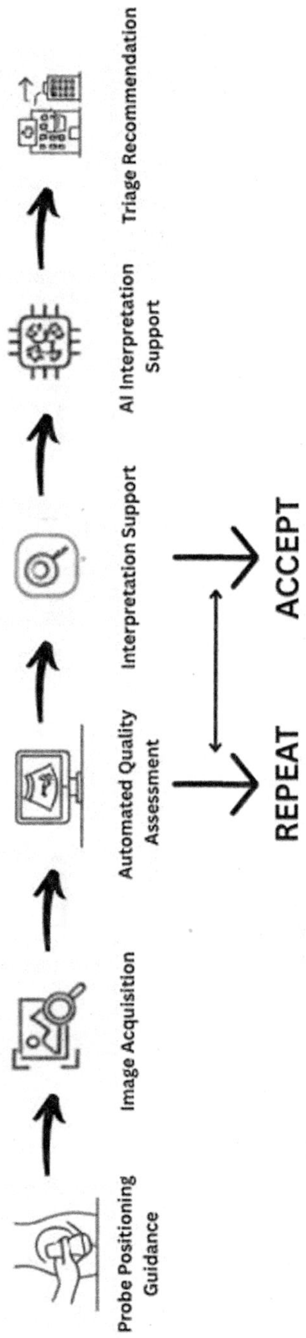

FIGURE 15.4 AI-supported point-of-care ultrasound workflow illustrating acquisition guidance, image quality assessment, and triage support.

Computed tomography (CT) utilizes spinning X-ray beams to generate cross-sectional images. Trauma assessment, stroke triage, pulmonary embolism evaluation, complex chest imaging, and oncology staging all require it. CT scans are rapid and provide high-resolution images; however, they require a reliable power supply, stringent radiation safety protocols, the capability to administer contrast agents, and routine maintenance. AI applications in CT are frequently employed to identify lesions, categorize them, quantify their volume, and prioritize urgent cases within reporting procedures. Nonetheless, performance may vary among scanners and reconstruction techniques, underscoring the necessity of conducting external validation at many locations.

Magnetic resonance imaging (MRI) employs robust magnetic fields and radiofrequency radiation to enhance the visualization of soft tissues. MRI is particularly advantageous for examining the brain, assessing the musculoskeletal system, and certain cardiovascular applications. MRI systems require substantial financial investment, extended scanning durations, secure environments, and proficient operators, hence complicating their widespread utilization in numerous low- and middle-income countries (LMIC). AI methodologies are frequently employed in MRI to expedite reconstruction, detect artifacts, and perform automated segmentation. While these tools may expedite processes and reduce scanning duration, they require meticulous testing, as modifications to the reconstruction could influence diagnostic outcomes (Zech et al., 2018) Table 15.2.

15.3.4 Prevailing issues in low-resource environments

Imaging systems in resource-constrained settings are occasionally hindered by operational constraints that diminish diagnostic efficacy and obstruct the secure implementation of AI techniques. The constraints encompass an unreliable power supply, uneven connectivity, restricted human resources, and delays in maintenance and equipment repair. Notwithstanding the availability of imaging technology, an inadequate PACS/RIS infrastructure may obstruct access to longitudinal data, delay auditing processes, and restrict integrated clinical decision support (Collins and Moons, 2019).

Data constraints are significant obstacles for visual artificial intelligence. Numerous datasets in developing nations exhibit a deficiency of labels, insufficient follow-up, inconsistent metadata, and heterogeneity in annotations between different assessors. These issues may introduce bias, exacerbate internal performance assessments, and diminish generalization when models are applied across varied contexts or populations. Moreover, screening algorithms often exhibit significant class imbalance, necessitating meticulous threshold selection and assessment through sensitivity/specificity at operational cutoffs, predictive values, and calibration (Steyerberg, 2019). To address these limitations, effective imaging programs should emphasize standardized acquisition protocols, automated quality control, external validation, fairness-aware subgroup analysis, and ongoing monitoring for dataset drift and calibration modifications

Table 15.2 Compares CT versus MRI in low-resource settings based on utility, cost, infrastructure, power, maintenance, scan time, and feasibility.

Criteria	CT (computed tomography)	MRI (magnetic resonance imaging)
Clinical utility	Strong for **trauma**, emergency imaging, lung pathology, stroke (hemorrhage), oncology staging, fractures	Strong for **brain**, spine, soft tissue, tumors, neuron-imaging, musculoskeletal injuries, cardiac imaging
Cost	**Moderate–High** (lower than MRI in most settings)	**High–Very High** (equipment + installation + operations)
Infrastructure requirements	Requires **shielded room (basic), stable power, cooling, radiation safety**, trained staff	Requires **RF-shielded suite, controlled environment, strong safety protocols**, specialized site preparation
Power needs	**High**, but generally more feasible than MRI; can operate with stable electrical backup	**Very high** and continuous; sensitive to fluctuations; requires robust backup power
Maintenance complexity	**Moderate** (service contracts + spare parts needed); downtime possible but manageable	**High** (magnet safety, hardware maintenance, specialized servicing); longer downtime risks
Scan time	**Fast** (seconds to minutes), supports high patient throughput	**Longer** (minutes to 30+ minutes), limited throughput in busy settings
Feasibility in low resource settings	**Higher feasibility** due to speed and broader emergency utility, but still constrained by costs and maintenance	**Lower feasibility** due to high costs, infrastructure demands, and specialist requirements

postdeployment. AI systems must be incorporated into therapy protocols with clear escalation procedures and accountability structures to guarantee patient safety and foster clinical acceptance (Zech et al., 2018) (Table 15.3).

15.4 AI techniques: From novice to expert

15.4.1 Fundamental concepts of machine learning

Machine learning (ML) models use data to find statistical links and make predictions. In medical imaging, ML can work with engineered features, such as radiomics texture descriptors, histogram statistics, shape measurements, and metadata deep representations acquired directly from pixels via neural networks (Lundervold and Lundervold, 2019; Moore and Copel, 2011; Rajpurkar et al).

In clinical AI, the design of the dataset and the way it is labeled often set the highest level of performance that can be reached. Radiology data, expert adjudication, or clinical outcomes can all be used to make labels, but each method of labeling has its own level of uncertainty and bias. For instance, labels based on reports may show differences in reporting rather than actual pathology, whereas labels based on outcomes may include the effects of treatment or restrictions in follow-up (Collins and Moons, 2019).

Table 15.3 Key technical and clinical considerations for AI deployment across imaging modalities in developing countries.

Consideration	Chest X-ray (CXR)	Ultrasound (POCUS)	CT	MRI
Data variability risk	Moderate–High	High	Moderate	Moderate
Need for quality control	Very high	Very high	High	High
Typical AI use cases	TB triage, Abnormality Screening	Guidance, quality scoring, measurement	Detection, Segmentation, urgent triage	Reconstruction, Segmentation, artifact correction
Key safety requirement	High-sensitivity triage + escalation	Uncertainty gating + expert fallback	Protocol-aware validation	Sequence-aware validation
External validation priority	Multisite, multi device	Multi-user, multi device	Cross-scanner generalization	Cross-vendor, cross sequence validation

In developing countries, label quality might be further affected by incomplete longitudinal follow-up and restricted access to confirmatory tests. Some of the most important ideas are as follows:

- Train/validation/test divides to stop data leaking: Data leakage happens when knowledge about the test set is accidently used during training. In imaging, leakage can happen when images from the same patient, device, or site show up in more than one split. Strategies for patient-level splitting and site-aware validation help lower this risk.
- Controlling over-fitting via regularization and external testing: Over-fitting makes the model work well on the training data but not on new hospitals or devices. This is especially important for LMIC deployments, as different regions may have different image capture methodologies and machine manufacturers.
- Cross-validation when the dataset is small: Cross-validation gives more reliable estimates when there isn't much labeled data, but it shouldn't be used instead of external validation, especially when making decisions on policy or clinical adoption.
- Managing class imbalance: Most screening datasets are not balanced, with fewer positive instances. When choosing a threshold, you should think about clinical goals (such maximum sensitivity for triage) and the capacity of the health system.

In developing countries, disease prevalence differs by location; thus, predictive values (PPV/NPV) may fluctuate even when sensitivity and specificity are constant. For instance, a model with high sensitivity and specificity may nevertheless create operational problems if the condition is rare, resulting in a low PPV and more effort for confirmatory testing. This is especially essential for TB screening programs and cancer triage pipelines, where referral paths may already be short on resources (Zech et al., 2018).

Along with discrimination, model evaluation should also incorporate calibration, which shows how well the projected probabilities match the observed event rates. A well-calibrated model makes clinical safety better by providing clear risk assessments and allowing for uniform triage threshold across contexts. Decision-curve analysis may further delineate therapeutic utility by measuring net benefit at various thresholds; however, its application necessitates meticulous interpretation in public health screening contexts (Collins and Moons, 2019).

15.4.2 An explanation of convolutional neural networks

Deep learning systems called convolutional neural networks (CNNs) use convolutional filters to find patterns in pictures that show up in space. In chest imaging triage jobs like pneumonia and TB screening, CNNs have made things better. They can learn hierarchical features, starting with low-level edges and textures and working up to higher-level patterns that match to anatomical structures and pathological signatures, and they are useful in clinical settings.

What CNNs usually have:

- Convolution layers with activations that aren't linear
- Strides or pooled convolution for down-sampling
- Normalization for stable training
- Heads of classification that make chance outputs

One important thing to keep in mind is that CNN's outputs are usually continuous probability scores rather than clear "truth" in diagnostics. Because of this, the clinical use case must be used to choose the limits. In triage workflows, thresholds are usually set to get the most sensitive results to make sure no cases are lost while still getting enough specific results to keep confirmatory services from getting too busy. For TB screening, this method fits with using AI as a way to classify risks instead of a self-diagnosing tool (Steyerberg, 2019).

When transfer learning is used, models are usually trained on big datasets first, and then they are fine-tuned on medical imaging datasets. When there aren't many resources available, transfer learning needs to be carefully checked because device and population differences can cause the domain to change. Domain shift can happen when the contrast, resolution, or types of patients are different between the training environment and the release environment. For instance, models that were trained on carefully chosen academic datasets might not do well in mobile screening programs, where portable radiography adds motion artifacts and different exposure settings. Because of this, external validation should be done across sample sites and equipment as part of the deployment process (Zech et al., 2018).

Lastly, CNN distribution should think about the needs for interpretability. CNNs are very accurate in many situations, but the fact that they are black boxes may make clinicians less likely to believe them. This is especially important in task-sharing models where direct healthcare workers depend on decision support to find ways to escalate situations. Explainability tools, uncertainty outputs, and formal reporting suggestions can help make things safer and more widely used (Tjoa and Guan, 2021).

Fig. 15.5 Simplified convolutional neural network (CNN) architecture for chest X-ray classification, illustrating input image reprocessing, hierarchical feature extraction through convolutional layers, and generation of probability-based outputs for disease prediction.

15.4.3 Techniques for segmentation and detection

Beyond classification, clinically actionable outputs often require localization. In clinical practice, localization improves interpretability and allows quantitative assessment, which may support follow-up decisions and treatment monitoring. In LMIC settings, localization can also support training by illustrating why a case was escalated for specialist review.

15.4.3.1 Detection

Detection systems identify regions of interest using bounding boxes. Applications include:

- Pulmonary abnormality localization
- Nodule/mass detection
- Multifinding triage in screening workflows

Detection models are useful when the clinical objective is to flag abnormal regions rather than quantify precise boundaries. In chest imaging, detection may support triage

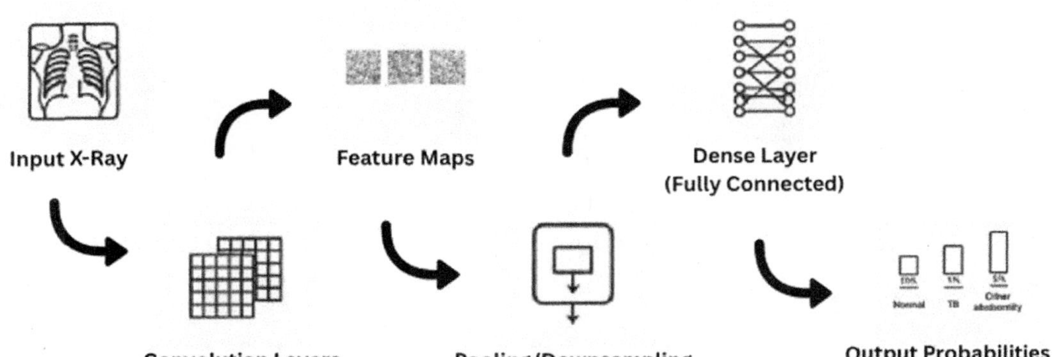

FIGURE 15.5 Simplified CNN architecture for chest X-ray classification showing input image processing, feature extraction layers, and probability-based output.

for suspicious lung lesions, pleural effusions, or consolidations. A key operational consideration is false positives: overly sensitive detectors can reduce specificity and increase clinical workload. Therefore, the evaluation should report sensitivity at clinically meaningful false positive rates.

15.4.3.2 Segmentation

Segmentation creates lines at the pixel level that can help with quantification:

- Figuring out the amount of cavitation in TB images
- Estimating the limits of cardiomegaly
- Figuring out the size of a tumor for oncology tracking

The results of segmentation can be very useful for long-term monitoring, like keeping track of the number of lesions or figuring out the sizes of chambers in cardiac imaging. In real life, the quality of segmentation rests a lot on how consistent the annotations are. It might be hard to get expert pixel-level labels in LMICs, so weak supervision or semiautomated annotation processes might be needed (Oktay et al., 2018).

Some common ways to rate things are the Dice similarity coefficient, the intersection over union (IOU) measure, and boundary-based metrics like the Hausdorff distance. Criteria, on the contrary, should be linked to clinical importance. Small mistakes near boundaries might not have a big effect on clinical decisions, but mistakes close to important structures (like pleural borders) might. Because of this, doctors should do a qualitative review as part of the evaluation, especially before the system is put into use. In Fig. 15.6 conceptual comparison of classification, detection, and segmentation tasks in medical imaging, highlighting differences in output granularity, localization capability, and clinical interpretability across common diagnostic workflows (Collins and Moons, 2019).

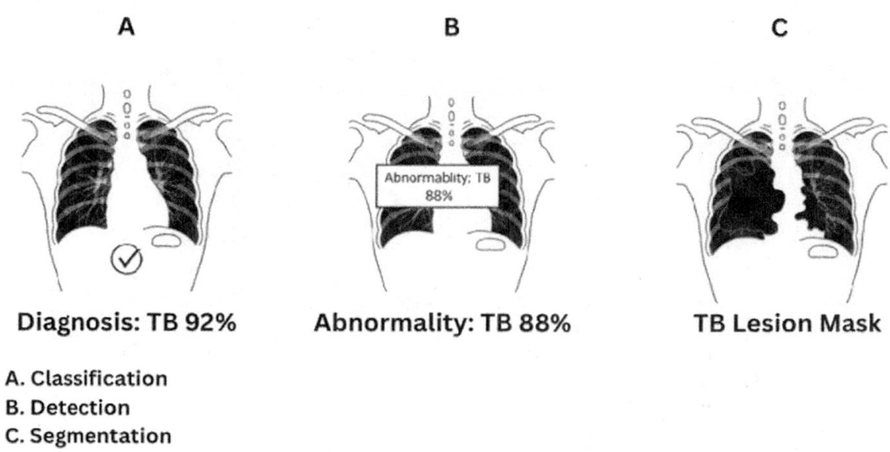

FIGURE 15.6 Differences between classification, detection, and segmentation tasks in medical imaging, demonstrating the clinical interpret-ability of each output type.

15.4.4 Explainable artificial intelligence for clinical trust

Explainable AI (XAI) improves transparency and supports clinical oversight, particularly in screening and triage systems. The objective of XAI is not to replace clinical judgment, but to provide clinically meaningful context for model decisions and facilitate error auditing, bias detection, and workflow acceptance (Tjoa and Guan, 2021).

15.4.4.1 Grad-CAM

Grad-CAM generates heat maps highlighting image regions contributing most strongly to a CNN prediction. It can help confirm whether the model focuses on clinically plausible regions. In chest X-ray triage, Grad-CAM overlays can help clinicians identify whether the model relies on lung parenchyma patterns, hilar regions, or nonanatomical artifacts such as embedded text markers.

15.4.4.2 LIME

LIME approximates local decision boundaries via perturbations and surrogate models. Outputs may vary by parameter selection and require careful interpretation. LIME is useful for localized reasoning, but its stability can be limited in high-dimensional image spaces (Tjoa and Guan, 2021).

15.4.4.3 SHAP

SHAP estimates feature contributions, especially useful in multimodal models combining imaging with structured clinical variables. Pixel-level SHAP may be less stable than structured-variable explanations. In LMIC implementations where EHR data may be incomplete, SHAP-based explanations should be interpreted cautiously, since missingness itself may bias predictions (Tjoa and Guan, 2021).

15.4.4.4 Practical guidance for XAI in developing countries

In resource-constrained settings, XAI should support:

- Clinician trust and auditability
- Structured escalation decisions
- Error analysis and training for task-sharing models
- Detection of spurious correlations or artifacts

XAI outputs should be treated as supportive evidence rather than proof of correctness. A practical recommendation is to present explainability alongside uncertainty estimation and structured triage outputs, enabling users to understand both "why" and "how confident" the system is (Steyerberg, 2019).

In Fig. 15.7 Example of explainable artificial intelligence (XAI) visualization using a Gradient-weighted Class Activation Mapping (Grad-CAM) heat-map overlaid on a chest X-ray, highlighting image regions that most strongly contribute to the model's prediction and supporting clinical interpretability.

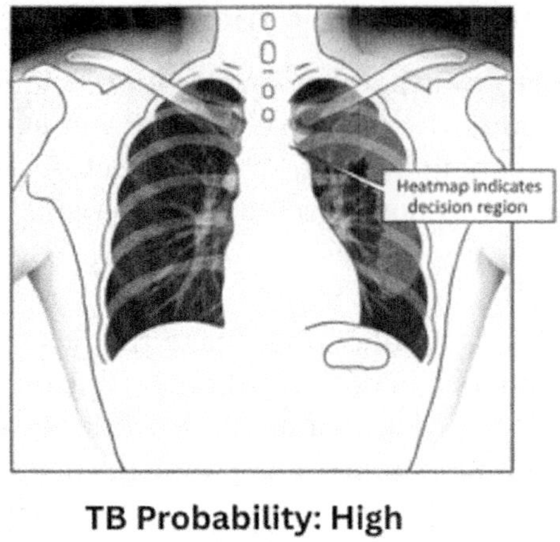

FIGURE 15.7 Explainable AI visualization using Grad-CAM heat-map overlay on chest X-ray to highlight regions contributing to model predictions.

15.4.5 Ensemble techniques and uncertainty assessment

Uncertainty estimation is essential for safe AI-supported triage and referral decisions. In real-world deployment, uncertainty may arise from poor image quality, out-of-distribution inputs, and demographic or device shift relative to training data.

15.4.5.1 Ensembles

Ensembles combine predictions from multiple models to improve robustness and reduce variance. Benefits may include improved reliability under a moderate distribution shift. In practice, ensembles may reduce the risk of unstable predictions that occur when a single model encounters an unfamiliar acquisition pattern.

15.4.5.2 Uncertainty quantification

Approaches include Monte Carlo dropout, deep ensembles, and Bayesian approximation. Uncertainty can support safety-gated workflows:

- Low uncertainty → automated routing
- Moderate uncertainty → human review
- High uncertainty → repeat imaging or specialist referral

In LMIC settings, uncertainty-based escalation can reduce the likelihood of harm by ensuring ambiguous cases receive expert review. This also aligns with clinical governance requirements by ensuring AI does not operate beyond its validated scope.

15.4.5.3 Calibration

AUC measures discrimination (ranking ability) but does not ensure probability accuracy. Calibration should be evaluated using calibration plots, Brier score, and expected calibration error (ECE). Calibration is particularly critical when models are used for risk-based referral decisions rather than binary classification. In Fig. 15.8. Schematic representation of an uncertainty-aware decision-gating framework for safe AI deployment, demonstrating how model confidence levels guide automated actions, human review, or referral escalation to ensure clinical safety and accountability (Steyerberg, 2019).

15.4.6 Optimization of lightweight deployment

Developing countries often require AI systems that work under limited computing and intermittent internet access. Lightweight deployment strategies include:

- Quantization (e.g., INT8 inference)
- Pruning and knowledge distillation
- Efficient architectures optimized for edge devices

Deployment testing should include inference latency, offline fallback, device compatibility, cybersecurity safeguards, and audit logging. For clinical safety, offline deployment must include mechanisms for secure local storage, delayed upload when connectivity resumes, and clear labeling of model versions used for each prediction (Tjoa and Guan, 2021).

15.4.7 Clinical validation and regulatory pathways

Clinical readiness requires evidence beyond internal testing.

15.4.7.1 External validation and generalization

External validation across multiple sites, devices, and populations is essential to evaluate domain shift. For LMIC deployment, validation should include representative rural and urban sites and heterogeneous device manufacturers where applicable.

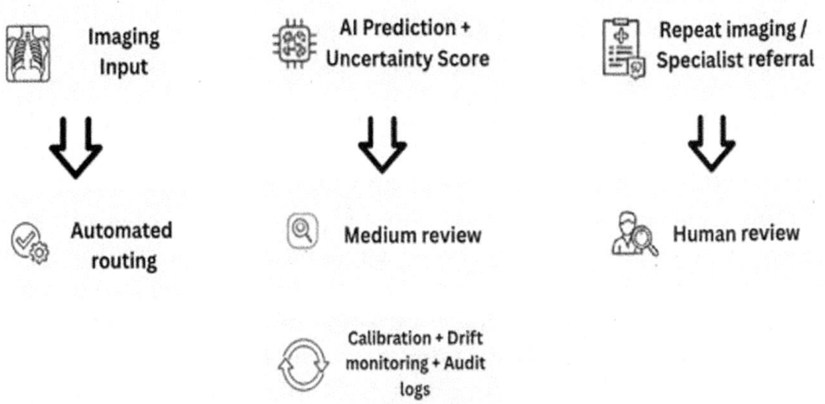

FIGURE 15.8 Uncertainty-aware decision gating model for safe AI deployment, illustrating escalation pathways based on confidence levels.

15.4.7.2 Subgroup evaluation and equity

Fairness evaluation should report performance stratified by clinically relevant subgroups, including:

- Gender
- Age groups
- Geographic region
- Hospital type (primary vs. tertiary)

15.4.7.3 Prospective evaluation

Prospective approaches include silent trials and pragmatic trials assessing workflow endpoints. Clinical outcome studies are recommended when feasible. Even when outcome trials are not feasible, workflow outcomes such as turnaround time, referral completion, and diagnostic yield can provide meaningful evidence of utility.

15.4.7.4 Regulatory pathways

AI systems used for clinical decision support may be classified as Software as a Medical Device (SaMD). Core regulatory principles include risk-based classification, clinical evidence, usability testing, cybersecurity, and postmarket surveillance. Procurement decisions should require transparent documentation of model performance, limitations, and monitoring plans (Rajkomar et al., 2018; Kelly et al., 2019).

15.5 Healthcare challenges in developing nations

15.5.1 Deficiencies in infrastructure and equipment

Significant structural challenges include:

- Limited availability of gadgets for individuals.
- Intermittent electrical supply.
- Insufficient space exists for PACS and workstations.
- Delayed maintenance and machine downtime.

AI models may prove less advantageous in practical applications, despite doing well in controlled trials. Access to imaging in numerous low- and middle-income countries (LMICs) is impeded by a scarcity of devices and their inequitable distribution. Modern imaging techniques are predominantly situated in tertiary centers in urban areas, while rural locations have restricted diagnostic capabilities. This complicates the early detection of conditions that respond favorably to timely intervention, such as pulmonary tuberculosis, malignancies, obstetric complications, and cardiovascular diseases. Infrastructure has significant challenges due to unpredictable energy, including voltage fluctuations, power outages, and insufficient backup systems. This directly influences the acquisition of images (such as ensuring uniform radiographic exposure), maintaining device connectivity, and the reliability of digital operations. Despite the availability of

AI software, unstable power can disrupt model reasoning and impede case escalation. When considering the utilization of AI, one should contemplate power resilience strategies such as uninterruptible power supplies (UPS), battery-operated mobile devices, or inference platforms capable of functioning offline when feasible.

The velocity of data processing and connectivity constitutes significant limitations. In a hub-and-spoke model, photographs are transmitted to regional centers for analysis. However, if the bandwidth is inadequate, the transfer may be prolonged, and the images may deteriorate due to compression problems. Compression can reduce file size; nevertheless, excessive compression may alter minor visual characteristics, thereby affecting AI performance in tasks such as recognition and segmentation. Quality assurance protocols must delineate the criteria for acceptable compression and ensure that AI models are evaluated on images processed through the identical pipeline as the deployment.

Equipment commonly malfunctions due to maintenance deficiencies, including insufficient spare parts, delayed vendor assistance, or inadequate professional service coverage. If these constraints persist, it may hinder the advancement of screening systems and diminish public trust in AI-enabled procedures. A fundamental principle for employing AI is that it cannot compensate for ineffective imaging services; it must be founded on solid operational frameworks. Enhancing infrastructure and advancing digital maturity are essential for optimizing AI performance across diverse environments. A further concern is the absence of standardized imaging workstations, high-quality monitors, or ergonomic reporting systems. Subpar display quality of datasets utilized for local validation or fine-tuning may hinder accurate analysis and annotation by individuals. If local labeling is required, minimal workstation characteristics and standardized viewing protocols must be specified to enable consistent evaluations (WHO, 2021a; Zech et al., 2018).

In Fig. 15.9 overview of system-level barriers to the deployment of AI-enabled medical imaging in developing countries is provided, including limitations in infrastructure, workforce capacity, data availability and quality, financial resources, and regulatory frameworks.

15.5.2 Deficiencies in medical personnel and educational requirements

There aren't enough radiology specialists, and imaging knowledge isn't evenly distributed, which are huge difficulties. If there are clear regulations, mechanisms to be monitored, and quality control processes, sharing tasks can help you get more done. AI can aid by arranging cases in order of significance, but it can't replace governance, clinical duty, or training infrastructure. At LMICs, big cities usually have a number of radiologists. In smaller towns and cities, general practitioners, nurses, or radiographers do imaging work at hospitals and community centers. This makes it hard for doctors to understand the results, especially for high-throughput screening systems like tuberculosis triage. AI-based triage may help individuals quickly define priorities, which could

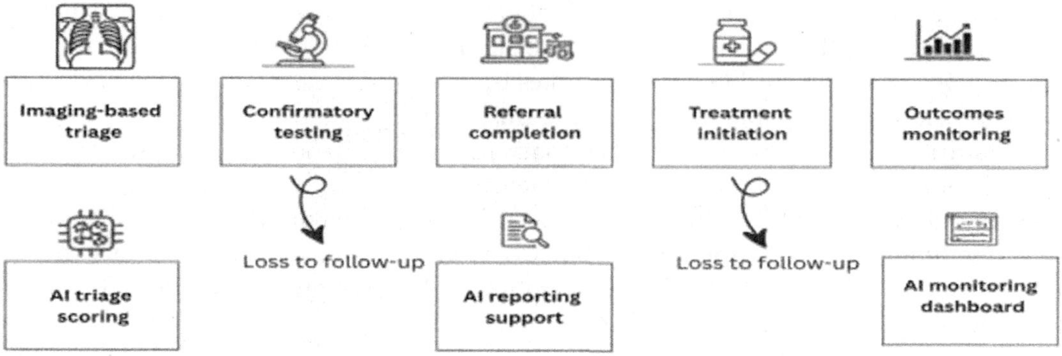

FIGURE 15.9 System-level barriers to AI-enabled imaging deployment in developing countries, including infrastructure, workforce, data, cost, and regulation.

speed up the decision-making process in these instances. But for this to function, there must be clear ways to deal with problems and clinical accountability (Kelly et al., 2019).

Standards for training aren't just for radiologists. AI-based imaging programs depend on:

- Radiographers or technicians who know how to get good pictures on a regular basis. Frontline doctors and nurses who know how to assess triage scores, ambiguity, and demands for escalation.
- Persons who work with health information and can help with data control and digital infrastructure as well as biomedical engineers in the area, who can help keep tools stable and speed up maintenance processes.

One important concern is that what AI says doesn't necessarily match what people know. People who haven't been to school might not know what risk scores, probability values, and model confidence mean. People can put too much faith in model results if they mix up the terms AUC (a measure of discrimination) and clinical threshold performance (sensitivity/specificity at the operational decision point). Training should clearly address how to understand thresholds, how PPV and NPV depend on frequency, and calibration principles so that individuals don't make wrong choices.

A number of personnel are also needed for quality assurance (QA). Standard quality assurance procedures should include image quality testing, reviews of the wrong, and regular inspections to make sure everything is working properly. In practice, LMIC locations may not have enough dedicated QA people, which implies that there isn't a lot of monitoring after deployment. This is a huge danger since AI performance might go down over time if equipment is replaced, protocols are altered, or the number of diseases varies. Because of this, training programs should teach people how to do "AI operations." These abilities should include how to report occurrences, how to watch for drift, and how to escalate problems in a structured fashion.

Lastly, the government needs to be clear about who is in charge of what. AI technologies should help individuals make choices, but licensed healthcare professionals and institutional oversight frameworks should remain in charge of clinical choices. If there is no clear duty, implementation could lead to medical and legal concerns for systems and make doctors less likely to trust them (Steyerberg, 2019; Zech et al., 2018).

15.5.3 Limitations of data and quality concerns

AI requires representative data reflecting local disease patterns and imaging protocols. Common limitations include:

- Missing clinical labels and incomplete follow-up.
- Noisy annotations due to inter-reader variability.
- Limited digitization and inconsistent metadata.
- Class imbalance in screening workflows.

These issues may cause biased performance estimates and reduced generalization if not addressed systematically. Data limitations represent one of the most persistent barriers to LMIC-ready AI deployment, because imaging AI models can fail when the deployment population differs from the training cohort (domain shift) or when label quality is poor (label noise) (Zech et al., 2018).

A frequent limitation is **incomplete ground truth**. For example, TB screening datasets may lack confirmatory molecular test results for all individuals, especially when test availability is constrained. If only high-risk patients receive confirmatory testing, datasets can develop verification bias, inflating apparent model performance. Reporting should clearly describe how labels were obtained and whether verification was uniformly applied. Similarly, oncology datasets may be limited by incomplete pathology confirmation or inconsistent follow-up intervals.

Metadata limitations are also critical. Many datasets lack consistent identifiers for device type, acquisition settings, and site information. Without metadata, it becomes difficult to detect site-level drift, assess subgroup performance, or perform stratified validation. Data governance frameworks should therefore encourage standardized metadata capture as part of digitization initiatives.

Another challenge is image quality variability, including underexposure, positioning errors, and motion artifacts. These factors can degrade both AI performance and human interpretation. Quality assessment models may help reduce nondiagnostic scans, but they must be validated in realistic field conditions rather than curated datasets. Where quality is inconsistent, performance reporting should include stratification by image quality grade, because high performance on high-quality images may not generalize to mobile screening programs.

Bias and fairness are central concerns. Datasets may under-represent certain subgroups such as women, older adults, or rural populations. Disease prevalence may differ across these groups, leading to differences in performance that can reinforce inequity. Therefore, performance should be evaluated across clinically relevant strata, including sex/gender, age bands, site type, and geographic region. When subgroup disparities are observed, mitigation strategies may include re-weighting, targeted data enrichment, or threshold adjustments, but these interventions must be justified and validated externally (Rajkomar et al., 2018; World Health Organization, 2021a, 2021b, 2023).

15.4.4 Financial and regulatory impediments

The costs of digitization, recurring license payments, the requirement for computers, and the cost of training are all barriers to entry. Regulatory issues may encompass the inability to assess all aspects, ambiguity in purchasing procedures, and insufficient postmarket surveillance mechanisms (World Health Organization, 2021a, 2021b, 2023). For long-term success, it is necessary to figure out which policies work best in each case and compare their costs and benefits. It needs a source.

How well it fits into the budget is often what decides whether or not to adopt it. Even while AI tools are cheap per unit, they are nevertheless expensive to own because you have to pay for internet access, hardware updates, integration with PACS/RIS, staff training, and maintenance. For long-term success in LMICs, long-term budgeting, local service support, and integration into existing public health programs may be more significant than short-term experiments.

Adding processes is a big cost driver. Staff members have a tougher time using AI products that involve distinct user interfaces or manual file transfers. Adding AI capabilities directly to imaging work-lists or reporting processes could make a difference in how quickly things get done and how well triage works. On the other side, integration requires technical knowledge and cooperation among vendors, which may not be viable in systems with limited resources. Various countries have varied standards for medical devices, and there aren't enough methods to measure success at the local level. A lot of LMICs employ foreign approval signals like FDA clearance and CE marking to make sure that items are safe and work, however they may not always work for local people or gadgets. So, even if there are approvals from other countries, procurement policies should still need verification that they are legitimate in the area or region.

Finally, tracking after the market has closed is generally not very good. AI models may need to be recalibrated or updated from time to time, but if changes aren't managed, they can fail in new ways. So, policy frameworks should provide basic rules for how to control monitoring, revalidation, and model updates that follow the SaMD life-cycle principles. This is especially critical for screening tools that have a major effect on referral choices, since even modest changes in performance can have a big effect on a wide group of people (WHO, 2021a, 2021b; EU MDR, 2017).

15.6 AI applications and their tangible benefits

15.6.1 Disease identification

AI-assisted chest radiography has been evaluated for TB screening and triage and may support prioritization of high-risk cases in high-burden settings. In practical programs,

AI can be integrated as a triage layer that assigns a risk score to each scan and prioritizes confirmatory testing for individuals most likely to have active disease. This design aligns with WHO-aligned screening strategies, where imaging serves as an intermediate step rather than a final diagnostic outcome.

In oncology, AI applications include lesion detection, structured reporting assistance, and triage support. Potential benefits include prioritizing suspicious cases for expedited referral, standardizing reporting language, and supporting longitudinal monitoring when imaging follow-up is required. However, oncology imaging tasks often require higher-resolution modalities (CT/MRI) and richer clinical context, and therefore require careful validation for each cancer type and clinical setting.

From a clinical safety perspective, it is important to distinguish three different AI use case categories:

1. **Rule-out triage:** high sensitivity is prioritized to reduce missed disease cases.
2. **Rule-in decision support:** higher specificity may be required, but false positives remain a concern.
3. **Structured reporting assistance:** AI supports workflow but does not directly determine diagnosis.

While models may demonstrate strong AUC values in retrospective studies, clinical outcome improvements must be demonstrated through prospective evidence rather than inferred solely from AUC gains. This distinction is particularly important in LMIC programs where unnecessary referrals may overwhelm already limited diagnostic and treatment pathways (Qin et al., 2018).

In Fig. 15.10 AI-assisted screening cascade demonstrating the integration of imaging-based triage with confirmatory testing, referral completion, and downstream monitoring of clinical outcomes, highlighting how AI supports timely decision-making and continuity of care.

15.6.2 Screening initiatives and early detection

When planning a screening workflow, it is important to think about both sensitivity and specificity, as well as the ability to check results later on. The review of deployment should include more than just AUC. It should also include PPV/NPV, calibration, and decision utility when it makes sense. It is common to use AUC as a summary measure of discrimination, but it doesn't directly show how well it works at the operational level used in clinical triage. For instance, even if a model has a high AUC, it may still have too many false-positives at a certain level that makes sure it is very sensitive. Because of this, choosing a threshold should be based on the unique situation and the limited resources in the area, such as the number of available tests and the completion rate of referrals. Early detection programs also need to take spectrum bias into account. People who are

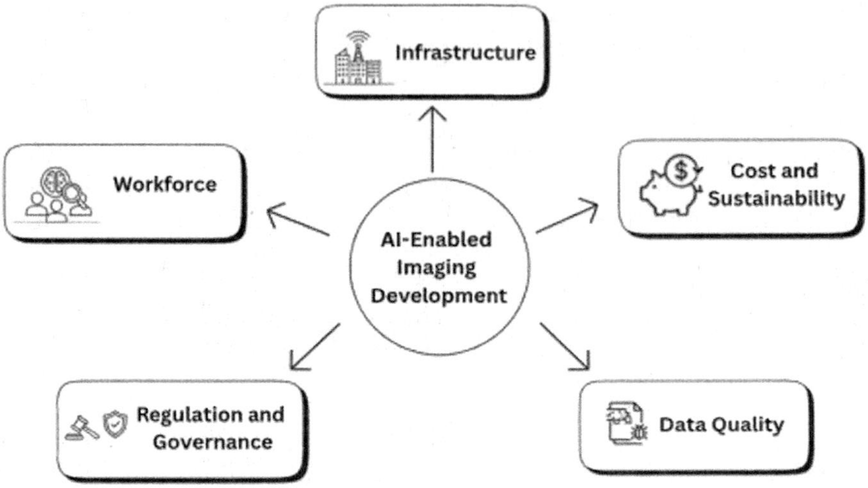

FIGURE 15.10 AI-assisted screening cascade illustrating imaging-based triage, confirmatory testing, referral completion, and downstream clinical outcomes monitoring.

screened may have different patterns of comorbidity and symptoms than people who are admitted to the hospital. If models are built on groups of people who need tertiary care, they might not work well in community-based screening situations. Because of this limitation, it is important to do both local calibration and external evaluation before going big.

Also, the benefits of screening depend on the whole clinical route. If referral systems are weak or there is no way to do confirmatory tests, even a strong AI triage system might not help a lot of people. Because of this, operational measures like time-to-confirmation, referral completion, and time-to-treatment initiation should be part of the evaluation (Collins and Moons, 2019; Steyerberg, 2019).

15.6.3 Enhancements to workflow

Even without complex AI, changes to the way work is done can lead to big improvements in operations. Adopting digital radiography, structured reporting templates, hub-and-spoke radiology, and AI-assisted prioritization within reporting lines are some examples.

Most of the time, process improvement is the main way that AI makes systems work better in LMICs. One useful way to look at AI is as a part of a larger effort to rethink workflow, which should include:

- Standardized procedures for purchase and quality control
- Clear limits for when things get too bad and professional oversight
- Templates for reports that are in line with local clinical standards
- Audit trails that allow review of the past

One benefit that is often mentioned is queue prioritization: AI can change the order of filing queues so that higher-risk cases are looked at first. This can cut down on the time it takes to make a choice for patients who need help right away. But methods for setting priorities should be checked for bias to make sure that some groups of patients aren't automatically lowered in importance.

AI can also improve operational quality by finding low-quality images, recommending retakes when possible, and lowering the amount of work that radiologists have to do on scans that aren't needed for diagnosis. In places with few resources, where it might be hard to do repeat imaging because of travel or equipment issues, checking the quality at the time of gathering can be very helpful (Kelly et al., 2019).

15.6.4 Mobile and remote diagnostics

Coverage can be increased with mobile imaging and remote diagnosis. For safe implementation, you need to be trained on things like data acquisition quality, safe data transfer, reliable referral methods, and the ability to do inference even when you don't have internet access.

Mobile operations are useful for a lot of different reasons, like TB screening, maternal health programs, and outreach services. A lot of the time, these programs work in places where there aren't many specialists or easy access to transfer hospitals. In these situations, AI-based triage can help make quick routing choices and figure out which patients need to be taken to a hospital right away for confirmatory tests.

However, mobile and remote tests add more ways for things to go wrong, such as

- Changes in the conditions of acquisition (lighting, motion, positioning)
- Differences in devices and differences in maintenance
- Clinical metadata that isn't full and has bad follow-up
- Weak spots in cybersecurity during storage or move

Longitudinal evaluation is needed to back up claims of better long-term results. When there aren't any outcome statistics available, programs should report process endpoints like referral completion and confirmed case yield, while being clear about their limitations (World Health Organization, 2021a, 2021b, 2023).

In Fig. 15.11 mobile imaging deployment model illustrates decentralized image acquisition at peripheral sites, AI-supported triage, remote expert interpretation through hub-and-spoke networks, and structured referral escalation for higher-level care.

15.7 Case studies and demonstrated outcomes

15.7.1 Tuberculosis screening in Nigeria (1.75× yield)

Nigeria is still one of the countries with the highest rates of tuberculosis (TB). Delays in detection lead to the disease spreading through the community and causing avoidable

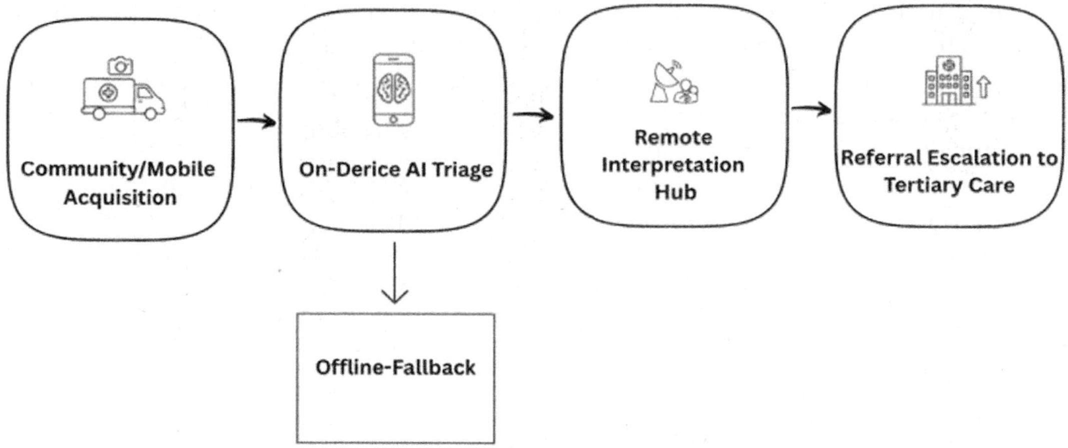

FIGURE 15.11 Mobile imaging deployment model demonstrating decentralized acquisition, AI triage, remote interpretation, and referral escalation.

illness. In many places, screening and diagnosis are hard to do because imaging departments aren't very big, access to confirmatory molecular tests isn't fair, and outreach programs have a lot of patients. To get around these problems, a nationwide active case-finding strategy is using ultraportable digital X-ray systems that are combined with computer-aided detection (CAD) tools more and more to help with screening and processing at the front lines (Useni 2025; WHO, 2023).

In this process, screening in the community or a center usually starts with checking for symptoms and figuring out the risk, then a chest X-ray is taken using portable systems that are moved around by mobile units. The CAD part gives an automated abnormality score to help with triage choices. This helps decide which people should get confirmatory molecular tests and be sent to TB services. This method works especially well in places where radiologists aren't available on-site and where the goal is to get a lot of cases done quickly and accurately so that no cases are lost (World Health Organization, 2021a, 2021b, 2023). An application result from Nigeria shows that using AI to help with CXR triage and improving workflow led to a 1.75 times increase in the number of TB cases found. It needs a reference. In terms of operations, this increase could mean that more cases are found for each number of people screened. This could be because of faster prioritization, more consistent triage, and shorter wait times for diagnoses in high-risk groups. Similar community case-finding methods rest a lot on the confirmatory testing pathway (e.g., the availability of Xpert MTB/RIF), referral completion, and the ability of the program to follow up. For safety reasons, AI results are usually used for triage rather than diagnosis on their own. Confirmatory tests and clinical review are still very important parts of the process (World Health Organization, 2021a, 2021b, 2023).

When these kinds of deployments happen, model validation and monitoring need to take into account different types of devices, different ways of collecting data in the field, and the chance that the training data and operating settings will be different. In real life, programs include quality assurance steps (like checking purchases, performing regular audits, and human control for cases that aren't clear) to keep them reliable over time. Nigeria's experience shows that it is possible to improve screening in places with few resources by using portable imaging, AI triage, and structured referral workflows together. However, it also shows how important it is to keep an eye on things as they change (Zech et al., 2018).

15.7.2 Cancer detection initiatives in India

A lot of people in India have cancer, and getting help for people who are already in late stages is still hard in many places because people don't know about it or can't get to it quickly enough. Breast, cervical, and oral cancers are often the focus of screening and early detection programs. These programs use methods like clinical breast examination, mammography (when possible), visual inspection strategies, and imaging-supported referral routes to find and treat these cancers. In this situation, AI-powered medical imaging has been looked into as a way to make screening more consistent, cut down on reporting delays, and provide decision support in places where there aren't enough specialists.

A good example of a high-quality standard for imaging AI in cancer screening is the evidence from randomized controlled trials in mammography screening. AI-supported screen reading has shown the ability to keep finding cancer while making the job of the radiologist easier. For example, a large trial in Europe reported that AI-supported screening achieved a similar cancer detection rate to standard double reading while greatly reducing the reading workload. Even though this study wasn't done in India, its results are clinically important to LMIC settings because of the same problem: not enough radiologists. Also, mammography workflows need to carefully balance recall rates, detection performance, and the use of human resources.

Several deployments and trials of AI-supported screening have been recorded in India's public and private sectors. These include projects looking into AI-assisted breast screening methods. For example, India's attempts to validate clinical trials have included large, planned screening trials that have been held in public hospitals. It has been said that screening and triage are working better, but there aren't many quantified, peer-reviewed results available for many programs yet. It needs a reference. When AI tools are used together, the best roles they can play are (i) picking out suspicious imaging results to be looked at more quickly; (ii) helping with structured reporting; and (iii) allowing standard triage thresholds that are in line with available confirmatory pathways (Rajkomar et al., 2018).

Workflow integration problems, such as not being able to work with PACS/RIS infrastructure, varying imaging device quality, and different patient populations in different areas, are also things that need to be thought about when implementing in

India. So, validation methods should focus on external testing across multiple datasets and include reporting on subgroup performance to lower the chances of uneven performance between age groups, socioeconomic classes, or urban and rural areas. Also, screening programs shouldn't exaggerate the effects based on AUC alone; for reports to be clinically meaningful, they need to include threshold-based sensitivity/specificity, predictive values in context, and calibration performance where chance estimates inform referral decisions. Overall, India's cancer screening system shows how AI can be useful as a triage support layer. However, it also stresses the need for program-level review and alignment with local diagnostic and treatment capacity (Steyerberg, 2019).

15.7.3 Transformation of workflow in Rwanda

Rwanda is a good example of how redesigning workflows and digitizing data can help health systems with limited resources get better access to diagnostic images. In many LMICs, the main things that make it hard to get imaging interpretations done quickly are a lack of specialists, the fact that imaging acquisition sites and reporting centers are spread out geographically, and a broken system for moving pictures and reports. These problems can be fixed by using digital imaging systems and hub-and-spoke reporting models together. This lets analysis be centralized while acquisition stays decentralized.

A usual method is to take pictures at district hospitals or primary care centers, then send them safely to regional or national reading hubs for reporting. This method can cut down on turnaround time (TAT), make it easier to get expert analysis, and make sure that the quality of all reports is the same (Author, Year). When AI is used in places where it isn't just for diagnosis, it usually makes things easier for people who work there. For example, it can automatically sort through urgent cases, flag problems with quality control, and speed up the review of high-risk studies (Kelly et al., 2019). It is not always possible to find peer-reviewed quantitative results for digitization projects in Rwanda, but similar radiology transformation programs have been said to have a positive effect by reducing the time it takes to report images and making it easier to get expert interpretation. It needs a reference. In real life, shorter wait times may still be useful even if diagnostic accuracy scores don't change. This is because earlier interpretation can lead to faster clinical decisions, referrals, and treatment starts. These changes are especially important in emergency and mother-to-child care routes, where waiting can put people at great risk.

Both technical and clinical indicators should be used for validation and tracking of this kind of workflow change. These include system up-time, the number of successful picture transmissions, the completeness of reports, ways to look over mistakes, and regular checks to see if prioritization tools affect the fairness of access across facilities. So, Rwanda's experience shows that if you want to make big changes to imaging, you should usually start by updating the process and the way services are provided. Then, AI can help with setting priorities and making things run more smoothly once the right structures are in place (World Health Organization, 2021a, 2021b, 2023).

15.7.4 Insights acquired and expansion strategies

Across the above case studies, several consistent lessons emerge regarding safe and scalable adoption of AI-enabled medical imaging in developing countries. First, successful deployments are typically **workflow-first**, meaning that digitization, standardized acquisition protocols, referral pathway design, and confirmatory testing capacity must be addressed alongside AI integration. AI is most effective when it augments a functioning diagnostic pathway rather than attempting to compensate for absent infrastructure or fragmented services (World Health Organization, 2021a, 2021b, 2023).

Second, governance and accountability are essential. AI tools used for triage should operate under clinical oversight, with predefined escalation thresholds and clearly documented responsibility structures for decision-making. In addition, implementation should include monitoring plans that address data drift, device replacement, protocol changes, and subgroup performance differences that may emerge over time. A practical approach is to establish periodic audit cycles in which performance metrics are reviewed against operational outcomes such as referral completion rates and time-to-treatment initiation (Zech et al., 2018; Rajkomar et al., 2018).

Third, model monitoring must extend beyond discrimination metrics such as AUC. Programs should track operational threshold performance, calibration stability, and uncertainty-aware escalation behavior, particularly when AI scores directly influence referral decisions. Continuous monitoring is also needed to detect performance degradation caused by changes in population characteristics or acquisition conditions, which are common in mobile screening and multisite deployments (Zech et al., 2018).

Fourth, training and task-sharing support should be treated as core implementation components rather than optional additions. AI-supported imaging may strengthen task-sharing only when local users are trained to interpret outputs, recognize uncertainty, and apply escalation pathways appropriately. Training programs should be competency-based and incorporate periodic refreshers, error feedback loops, and quality assurance checks to reduce variability across operators (Kelly et al., 2019).

Fifth, sustainable financing and procurement readiness are critical for scale-up. AI tools introduce recurring costs, including licensing, integration, cybersecurity maintenance, and ongoing validation. Procurement decisions should therefore consider the total cost of ownership and long-term sustainability, not only pilot-level feasibility. Where feasible, integration with national TB or cancer programs can support sustainability by aligning AI deployment with existing public health priorities and infrastructure investments (World Health Organization, 2021a, 2021b, 2023).

Finally, prospective evaluation approaches can strengthen evidence for safe adoption. "Silent trials," in which AI runs in the background without influencing decisions, can help quantify real-world performance under operational conditions before clinical integration. Pragmatic evaluations can then assess workflow endpoints such as turnaround time and referral completion, with longer-term clinical outcomes studied where feasible. Collectively, these strategies support responsible scaling of AI-enabled imaging systems that are clinically useful, operationally feasible, and aligned with equity and safety expectations in developing-country health systems (Kelly et al., 2019).

15.8 Future directions and implementation

15.8.1 Ethical artificial intelligence and data privacy

Ethical adoption needs clear information about how AI is used, privacy-protecting rules, fair access, and responsibility for mistakes. Ethical implementation in developing countries needs to be based on real-world workflow conditions, such as limited staffing, limited facilities, and inconsistent patient follow-up.

Imaging data are very revealing, so there are big privacy risks. Role-based access controls, encryption at rest and in transit, audit logs, and secure incident response rules should all be put in place by responsible programs. If cloud services are used, data privacy rules may mean that data needs to be hosted locally or processed in a way that is legal in the region (World Health Organization, 2021a, 2021b, 2023). Federated learning might let people work together without sending data to one place, but it needs to be monitored, gathered, and evaluated securely to see if there are any privacy risks left. Federated approaches also bring up heterogeneity issues: model performance may be different across sites that are involved, and communication problems may make training less stable. If federated learning is used, evaluation should look at how well different sites are doing locally and make sure that gains don't hurt minority groups (Rajkomar et al., 2018).

Transparency with doctors and patients is also an ethical issue. There are different rules about how transparent you have to be with patients depending on where you live. However, clinical governance should make sure that doctors understand AI outputs as decision support rather than absolute truth (Li et al., 2020).

15.8.2 Recommendations for policy

Policy frameworks should spell out the bare minimum of proof needed for purchases, promote and require independent auditing and evaluation, demand safety measures for cyberspace, back up training for workers, and make the government stronger.

A useful guideline for buying things should say:

- External validation data from representative groups of people
- Performance reporting based on thresholds (sensitivity/specificity and predicted values), evaluating accuracy, and dealing with uncertainty
- A plan for tracking drift and an update policy
- Writing down failure modes and ways to get help

Different LMICs have different levels of regulatory power, so policy should encourage the formation of local evaluation partnerships with health ministries and academic organizations. This makes sustainability stronger and lessens reliance on vendor promises that might not hold true in all situations.

A cost-effectiveness analysis should also be added, especially for programs that screen people. If treatment and confirmatory test services can't grow at the same time, AI tools might not be worth the money. These constraints should not be taken for granted. They should be clearly modeled (World Health Organization, 2021a, 2021b, 2023).

15.8.3 Educating local healthcare professionals

Training needs to be built on skills and include learning how to get images, understanding AI (probabilities, thresholds, and limitations), escalation pathways, and regular refresher programs with audit feedback.

It is especially important to know how to use AI because not knowing risk scores can lead to bad clinical decisions. The training should stress:

- Trade-offs between sensitivity and detail
- What PPV and NPV mean when frequency changes
- Limitations on trust and confusion
- The right way to handle cases that aren't in stock

When task-sharing is used, training should make clear who is responsible for what. For example, AI can help with triage, but qualified healthcare workers must still have final clinical responsibility. Structured feedback loops should also be a part of training programs. Frontline users should be able to report failure modes, like bad image quality or wrong predictions, so that governance teams can check performance and change processes (Steyerberg, 2019).

15.8.4 Advanced AI systems of the future

People in developing countries expect the next wave of imaging AI to focus on multimodal risk prediction, being able to adapt to new domains, self-supervised learning, uncertainty-aware triage, and outputs that can be understood to help make decisions.

Using images along with EHR and laboratory data in multimodal systems might help with triage better, but there are times when LMIC EHR infrastructure has gaps that can lead to bias. The multimodal systems need to clearly describe missing data and check for subgroup fairness. Another important area of study is robustness. However, these methods must be carefully controlled to avoid making changes to the model that are unsafe or not recorded. Domain generalization, drift-aware updating, and continual learning frameworks may help keep the model from degrading when devices and populations change.

Following best practices for SaMD, life-cycle governance will still be very important. This includes tracking for drift and postmarket surveillance. For long-term success, developing AI will need to be combined with local efforts to improve the health system. These improvements should include building up referral systems, maintenance facilities, and the skills of health workers (World Health Organization, 2021a, 2021b, 2023).

15.9 Summary

Medical imaging that is powered by AI could help poor countries do better diagnostics by allowing for scalable triage, decision support, and workflow optimization. But for clinical translation to work, model performance, workflow feasibility, and patient safety control must all be carefully aligned. In many low- and middle-income countries, image AI's most important function is not to make diagnoses on its own, but to prioritize high-risk cases, cut down on reporting delays, and help people share tasks when there aren't enough specialists (Kelly et al., 2019).

From a technical point of view, this part stressed that AUC by itself is not enough to determine clinical readiness. For safe implementation, you need to give sensitivity and specificity at operational thresholds, predictive values based on true prevalence, good calibration, and escalation that takes uncertainty into account. Due to domain shift and different acquisition methods, external validation across multiple sites and devices is still needed (Zech et al., 2018).

One big problem with many deployment stories is that there isn't enough information on what will happen in the future with patients. In the past, results may have shown strong discrimination, but in the real world, their usefulness rests on the whole process: getting confirmatory tests, finishing the referral process, and starting treatment. These limits need to be built into review plans to make sure that using AI is safe and useful for the health of the whole population.

Ethics and good government are still very important. Because of the chance of bias and unfairness, performance must be reported and tracked for each grouping. Sensitive patient imaging data must be protected by privacy and cybersecurity measures. Over time, it's likely that LMIC deployment tactics will change to focus on strong but lightweight edge inference, monitoring that is in line with governance, and long-lasting training programs in each country that build up clinical capacity instead of avoiding it.

Overall, AI in medical imaging could help improve diagnostic services in poor countries as long as it is properly integrated into clinical systems, tested by outside experts, and governed in a clear way.

Take-home messages

- Artificial intelligence in medical imaging should be implemented primarily as a tool for triage and clinical decision support, complementing rather than replacing clinician judgment. Robust external validation across multiple sites, imaging devices, and patient populations is essential to minimize generalization failures in real-world deployments (Zech et al., 2018).
- Proper calibration and uncertainty estimation are critical for defining safe referral thresholds, particularly in population-level screening programs (Steyerberg, 2019).

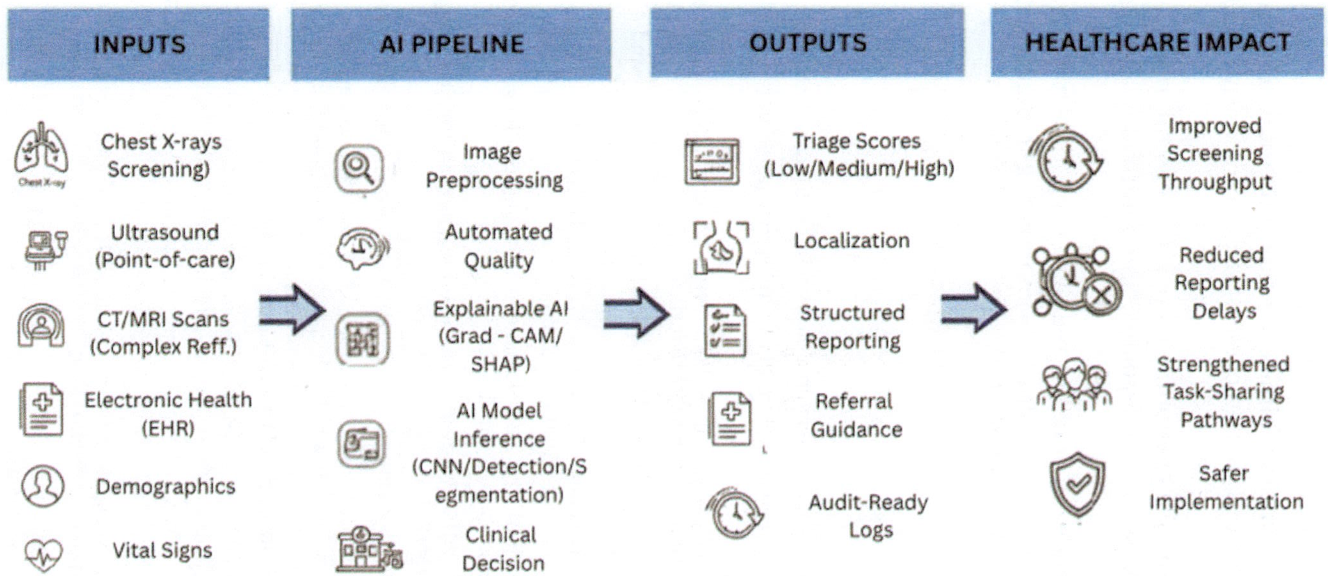

The AI-enabled imaging decision-support pipeline integrates multimodal inputs to generate triage and referral outputs, facilitating scalable screening and workflow improvements in low-resource healthcare settings.

- Explainable AI methods enhance transparency and support clinical auditing; however, interpret-ability alone does not ensure model correctness or clinical safety (Tjoa and Guan, 2021).
- Systematic assessment of bias and fairness through subgroup performance reporting is necessary to prevent inequitable clinical decision pathways (Rajkomar et al., 2018).
- Deployment in edge or mobile environments requires careful model optimization, offline functionality, cybersecurity safeguards, and audit-ready logging mechanisms.
- Sustainable impact depends on workflow redesign, workforce training, governance frameworks, and continuous postdeployment monitoring to detect data drift and performance degradation (Kelly et al., 2019).

References

Beam, A.L., Kohane, I.S., 2018. Big data and machine learning in health care. JAMA 319 (13), 1317–1318.

Collins, G.S., Moons, K.G.M., 2019. Reporting of artificial intelligence prediction models. Br. Med. J. 367, l737.

Dey, D., Slomka, P.J., Leeson, P., Dey, D., Slomka, P.J., Leeson, P., et al., 2019. Artificial intelligence in cardiovascular imaging: JACC state-of the-art review. J. Am. Coll. Cardiol. 73 (11), 1317–1335.

European Union (EU), 2017. Regulation 2017/745 of the European Parliament and of the Council on Medical devices Off J Eur Union.

Kelly, C.J., Karthikesalingam, A., Suleyman, M., Corrado, G., King, D., 2019. Key challenges for delivering clinical impact with artificial intelligence. BMC Med. 17, 195.

Li, T., Sahu, A.K., Talwalkar, A., Smith, V., 2020. Federated learning: challenges, methods, and future directions. IEEE Signal Process. Mag. 37 (3), 50–60.

Lundervold, A.S., Lundervold, A., 2019. An overview of deep learning in medical imaging focusing on MRI. Z. Med. Phys. 29 (2), 102–127119.

Moore, C.L., Copel, J.A., 2011. Point-of-care ultrasonography. N. Engl. J. Med. 364, 749–75713.

Oktay, O., Ferrante, E., Kamnitsas, K., Heinrich, M., Bai, W., Caballero, J., Cook, S.A., De Marvao, A., Dawes, T., O'Regan, D.P., Kainz, B., Glocker, B., Rueckert, D., 2018. Anatomically constrained neural networks (ACNNs): application to cardiac image enhancement and segmentation. IEEE Trans. Med. Imag. 37 (2), 384–395. https://doi.org/10.1109/TMI.2017.2743464.

Qin, Z.Z., Sander, M.S., Rai, B., 2021. et alUsing artificial intelligence to read chest radiographs for tuberculosis detection: a systematic review. Eur. Respir. J. 58 (1), 200317720.

Rajpurkar P, Irvin J, Zhu K, et al CheXNet: Radiologist-Level Pneumonia Detection on Chest X-Rays Using Deep learning. arXiv2017:171105225.

Rajkomar, A., Hardt, M., Howell, M.D., Corrado, G., Chin, M.H., 2018. Ensuring fairness in machine learning to advance health equity. N. Engl. J. Med. 378, 981–983.

Steyerberg, E.W., 2019. Clinical Prediction Models: A Practical Approach to Development, Validation, and Updating, second ed. Springer, Cham.

Tjoa, E., Guan, C.A., 2021. Survey on explainable artificial intelligence (XAI) in healthcare: mechanisms, applications, and challenges. IEEE Transact. Neural Networks Learn. Syst. 32 (11), 4793–481316.

World Health Organization, 2021a. WHO Consolidated Guidelines on Tuberculosis: Module on Screening. WHO, Geneva.

World Health Organization, 2021b. Ethics and Governance of Artificial Intelligence for Health. WHO, Geneva.

World Health Organization, 2023. Global Tuberculosis Report 2023. WHO, Geneva.

Zech, J.R., Badgeley, M.A., Liu, M., Costa, A.B., Titano, J.J., Oermann, E.K., 2018. Variable generalization performance of a deep learning model to detect pneumonia in chest radiographs. PLoS Med. 15 (11), e1002683.

16

AI applications in disease diagnosis and treatment: New applications

Adil Amraz[1], Nida Iftekhar[2], Sadaf Fatima[3], Mohammad Idrees[4], Mohd Afaque[5], Shamsheer Alam[6], Jatin Ahuja[7, 8, 9], Arslan Mansoor[10], Danish Reza[11], Mansaf Alam[12], Bashir Alam[13], Syed Zafer Hashmi[14], Shakeel Ahmad Siddiqui[15], Aasim Zafar[16] and Hatwib Mugasa[17]

[1]DEPARTMENT OF COMPUTER SCIENCE, JAMIA MILLIA ISLAMIA, NEW DELHI, INDIA; [2]DEPARTMENT OF COMPUTER SCIENCE AND ENGINEERING, SCHOOL OF ENGINEERING SCIENCES AND TECHNOLOGY (SEST), JAMIA HAMDARD UNIVERSITY, NEW DELHI, INDIA; [3]DEPARTMENT OF GASTROENTEROLOGY, MEDANTA HOSPITAL, SECTOR 50, NOIDA, UTTAR PRADESH, INDIA; [4]FACULTY OF ENGINEERING & TECHNOLOGY, AMU ALIGARH, ALIGARH, UTTAR PRADESH, INDIA; [5]DEPARTMENT OF RESPIRATORY MEDICINE, HAMDARD INSTITUTE OF MEDICAL SCIENCES AND HAHC HOSPITAL, NEW DELHI, INDIA; [6]ESIC HOSPITAL & PGIMSR, NEW DELHI, INDIA; [7]INFECTIOUS DISEASES & TRAVEL HEALTH SPECIALIST IN INDRAPRASTHA APOLLO HOSPITAL, NEW DELHI, INDIA; [8]AIIMS, NEW DELHI, INDIA; [9]TRAVEL MEDICINE, ISTM, ALPHARETTA, GA, UNITED STATES; [10]AMAZON, BENGALURU, KARNATAKA, INDIA; [11]CENTRE FOR FEDERAL STUDIES, PUBLIC POLICIES AND GOVERNANCE, SCHOOL OF HUMANITIES AND SOCIAL SCIENCES, JAMIA HAMDARD, NEW DELHI, INDIA; [12]DEPARTMENT OF COMPUTER SCIENCE, JAMIA MILLIA ISLAMIA (JMI), NEW DELHI, INDIA; [13]DEPARTMENT OF COMPUTER ENGINEERING, FACULTY OF ENGINEERING AND TECHNOLOGY, JAMIA MILLIA ISLAMIA, NEW DELHI, INDIA; [14]DEPARTMENT OF CRITICAL CARE MEDICINE, MAX NOIDA SECTOR 128, NOIDA, UTTAR PRADESH, INDIA; [15]DEPARTMENT OF COMPUTER SCIENCE AND ENGINEERING, JAMIA HAMDARD, NEW DELHI, INDIA; [16]ALIGARH MUSLIM UNIVERSITY (AMU), ALIGARH, UTTAR PRADESH, INDIA; [17]NATIONAL INFORMATION TECHNOLOGY AUTHORITY, KAMPALA, UGANDA

What you will learn

- **Distinguish between "Narrow AI" and "Generalist AI":** Understand the fundamental shift from task-specific algorithms (e.g., single-image analysis) to multimodal Foundation Models capable of complex clinical reasoning.
- **Explain Multimodal Data Fusion:** Describe how "Deep Fusion" architectures integrate disparate data streams, Electronic Health Records (EHR), medical imaging,

Applying Artificial Intelligence to Computational Biology and Medical Informatics. https://doi.org/10.1016/B978-0-443-44915-4.00016-2

and genomics to create holistic patient profiles that outperform unimodal diagnostics.

- **Analyze the Role of Generative Biology:** Evaluate how generative AI models, such as diffusion models and AlphaFold, are accelerating de novo drug design and enabling the repurposing of existing drugs for rare diseases.
- **Understand Precision Medicine Workflows:** Comprehend the "Digital Twin" concept and how AI-driven pharmacogenomics is moving healthcare from population-based averages to individualized treatment optimization.
- **Navigate Operational and Ethical Challenges:** Identify critical barriers to implementation, including data privacy (Federated Learning), algorithmic bias mitigation, and the evolving regulatory landscape for "Continuous Learning Systems" (e.g., FDA's PCCP).
- **Envision the AI-Augmented Clinician:** Recognize how the future clinical workflow will shift from human-only decision-making to a collaborative "Centaur Model," where AI handles data synthesis, and humans focus on complex judgment and empathy.

16.1 Introduction: The transformative imperative in healthcare

Artificial intelligence (AI) in healthcare is now a practical reality rather than a futuristic theory. But there is currently a sharp division in the story of AI in medicine. We are transitioning from an era of "Narrow AI" characterized by isolated algorithms performing specific tasks like nodule detection on chest X-rays to an era of "Systemic AI," where multimodal, generative, and foundation models begin to address the complexity of human biology and the healthcare workflow itself. This chapter makes the case that the "New Applications" of AI represent a fundamental reorganization of the definition, diagnosis, and treatment of disease rather than just small gains in accuracy.

16.1.1 Defining the scope: Moving beyond conventional AI in medicine

To understand the trajectory of AI in healthcare, one must first delineate the limitations of the "first wave" of medical AI. Historically, Computer-Aided Diagnosis systems and early machine learning models were predicated on **supervised learning** within highly constrained environments. These systems required vast quantities of labeled data, often curated by expensive human experts, to perform a single, narrowly defined task. For example, a Convolutional Neural Network (CNN) trained to identify diabetic retinopathy from fundus photography could not interpret a textual clinical note, nor could it generalize its visual learning to identify melanoma, despite the architectural similarities in image recognition tasks.

This "unimodal" and "task-specific" nature of conventional AI created a fragmented ecosystem. Hospitals deployed disparate algorithms that could not communicate, leading to the "siloization" of clinical intelligence. A radiologist might have an AI tool for lung nodules, a cardiologist for ECG interpretation, and a geneticist for variant calling, but no underlying digital substrate existed to synthesize these inputs into a coherent patient narrative.

The "New Applications" discussed in this chapter refer to the paradigm shift toward **Generalist Medical AI (GMAI)**. These systems differ from their predecessors in three critical ways:

1. **Multimodality:** The ability to ingest and synthesize diverse data types (images, text, omics, and signals) simultaneously.
2. **Self-Supervision:** The capability to learn underlying representations from massive, unlabeled datasets, reducing the dependency on labor-intensive data labeling.
3. **Generativity:** Moving beyond classification (e.g., "Is this cancer? Yes/No") to generation (e.g., "Synthesize a report explaining *why* this region is suspicious and suggest a confirmatory biopsy path").

This shift allows us to move beyond merely automating specific tasks to augmenting the **reasoning** capabilities of the clinician. The scope of this chapter, therefore, encompasses these next-generation technologies that promise to bridge the gap between data accumulation and clinical wisdom.

16.1.2 The need for "new applications": Challenges in current diagnostics and therapeutics

The urgency to adopt these new applications stems from structural failures in the current healthcare delivery model. Despite spending trillions globally, diagnostic errors remain a persistent plague, affecting an estimated 12 million adults in the United States annually. These errors often do not stem from a lack of data, but rather from an excess of unintegrated data. The average hospitalized patient generates gigabytes of data daily from continuous telemetry and labs to imaging and genomic sequencing. The cognitive load required to synthesize this information exceeds unassisted human capacity, leading to "information overload" and burnout.

Furthermore, the therapeutic pipeline is facing a crisis of efficiency, often termed "Eroom's Law", the observation that drug discovery is becoming slower and more expensive over time, despite improvements in technology. The traditional "one-drug-fits-all" blockbuster model is failing to address complex, heterogeneous diseases like Alzheimer's, sepsis, and metastatic cancers.

The "New Applications" of AI address these specific bottlenecks:

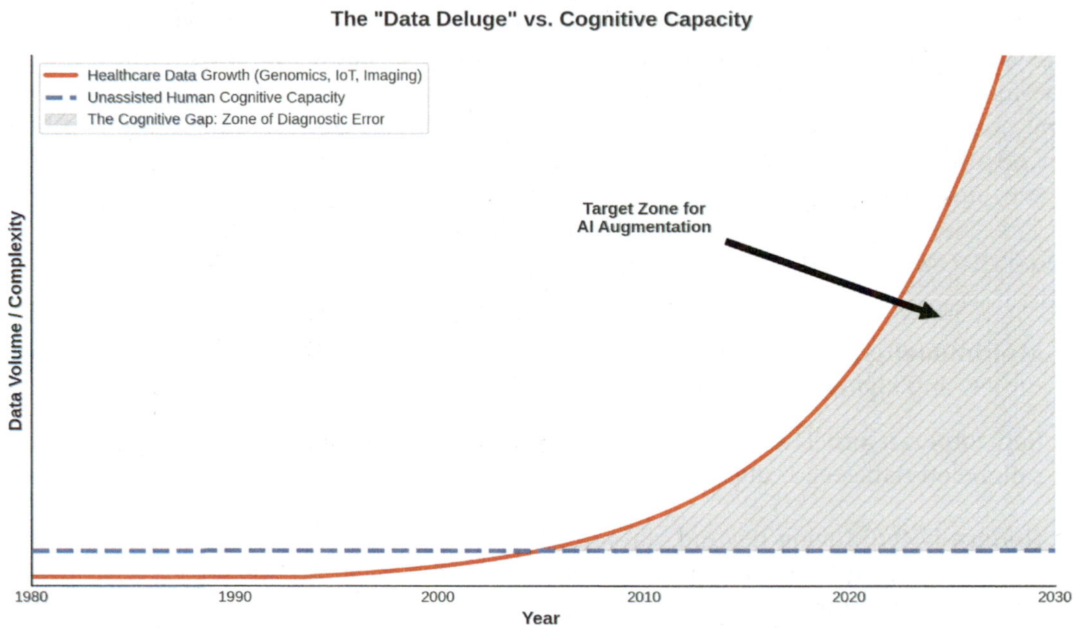

FIGURE 16.1 The widening disparity between the exponential growth of medical data and the fixed cognitive capacity of the human clinician. "New Applications" of AI are required to bridge this "Cognitive Gap" and prevent information overload.

- **The Diagnostic Bottleneck:** By leveraging **Foundation Models**, we can automate the synthesis of fragmented data, turning the EHR from a passive repository into an active diagnostic partner.
- **The Therapeutic Bottleneck:** By utilizing **Generative AI** for de novo molecular design and **Digital Twins** for trial simulation, we can compress the drug discovery timeline and identify patient-specific responders earlier in the development process.
- **The Equity Bottleneck:** By democratizing access to expert-level diagnostics via **Edge AI** (deployable on smartphones in low-resource settings), we can bridge the disparity in global health outcomes.

Thus, the imperative for these new applications is not just technological novelty, but the sustainability and efficacy of modern medicine itself (Fig. 16.1).

16.2 Next-generation diagnostic tools: Leveraging multimodal data

The transition from unimodal data analysis to multimodal data fusion represents the most significant leap in diagnostic AI. While earlier iterations of medical AI excelled at isolated tasks such as detecting pneumonia on a chest X-ray, they lacked the clinical context required for high-stakes decision-making. The "Next-Generation" tools discussed here mimic the cognitive workflow of a human clinician: synthesizing disparate information streams to form a cohesive diagnostic hypothesis.

16.2.1 Multimodal data fusion for holistic patient profiling

Multimodal AI refers to machine learning architectures capable of learning from multiple sensory inputs (modalities) simultaneously. In the clinical setting, these modalities typically include high-dimensional imaging (Radiology/Pathology), unstructured text (EHR), and sequential data (Genomics/Time-series telemetry).

16.2.1.1 Integrating EHR, imaging, and genomic data

The primary challenge in modern diagnostics is the "Heterogeneity Gap" the disconnect between a patient's genotypic profile and their phenotypic presentation. Traditional statistical methods struggle to correlate these distinct data types due to their differing structures (pixels vs. strings vs. numerical arrays).

Deep Fusion Architectures: New applications utilize "Deep Fusion" techniques, specifically **Tensor Fusion Networks**, to bridge this gap.

- **Early Fusion:** This approach concatenates raw features from different modalities (e.g., combining pixel data with genomic variant lists) before feeding them into a neural network. While computationally efficient, it often fails to capture complex interactions between modalities.
- **Late Fusion:** Independent models analyze each modality separately (e.g., a CNN for the X-ray and an Natural Language Processing-NLP model for the notes), and their predictions are averaged at the final layer. This preserves modality-specific features but misses cross-modal correlations.
- **Joint/Hybrid Fusion:** The state-of-the-art approach involves learning a joint representation space. Here, the model learns to align the vector space of an MRI scan with the vector space of a genetic sequence. For example, the model learns that a specific texture in a tumor on an MRI (radiomics) correlates strongly with a specific *EGFR* mutation in the genomic data (Fig. 16.2).

16.2.1.2 Case studies in complex disease: Oncology and neurodegeneration

The utility of multimodal fusion is most evident in complex, systemic diseases where a single test is insufficient for diagnosis.

- **Oncology (The "Virtual Biopsy"):** In nonsmall cell lung cancer, multimodal models integrating CT imaging with clinical demographics have outperformed radiologists in predicting immunotherapy responses. By analyzing the "tumor habitat," the pixel patterns surrounding the tumor alongside the patient's immune profile, these models can identify responders without invasive tissue sampling.

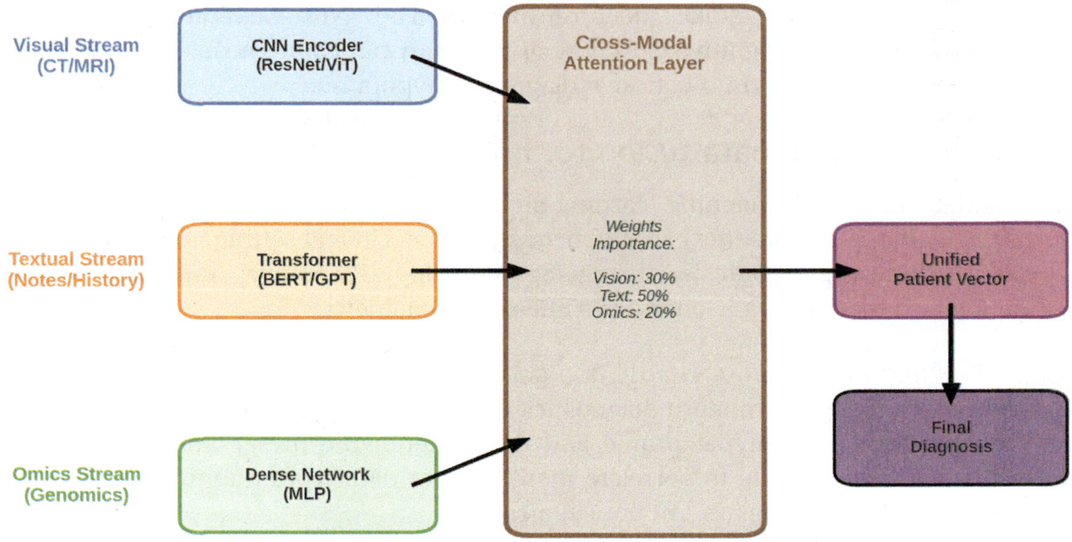

FIGURE 16.2 Multimodal data fusion architecture.

- **Neurodegeneration:** In Alzheimer's disease diagnosis, models that fuse MRI neuroimaging with NLP of speech patterns and cerebrospinal fluid biomarkers detect cognitive decline 3–5 years earlier than standard cognitive testing. The AI identifies subtle linguistic slips in speech that correlate with specific atrophy patterns in the hippocampus, a connection often invisible to human observation.

16.2.2 The role of foundation models in clinical diagnosis

A "Foundation Model" is a large-scale AI model trained on a vast amount of unlabeled data at scale (self-supervised learning) that can be adapted (fine-tuned) to a wide range of downstream tasks. This represents a departure from "task-specific" models trained for a single purpose.

16.2.2.1 Large language models for differential diagnosis and clinical summarization

Large language models (LLMs), such as GPT-4 and Med-PaLM, have demonstrated "emergent abilities" in clinical reasoning. Unlike keyword-search algorithms, these models possess a semantic understanding of medical concepts:

- **Differential Diagnosis:** When presented with a complex list of symptoms, lab values, and patient history, LLMs can generate a ranked list of differential diagnoses with justification. Recent benchmarking shows that medical-grade LLMs perform at or near the level of expert clinicians in USMLE-style diagnostic reasoning questions.

- **Automated Summarization:** One of the most immediate applications is reducing the "data burden." FMs can ingest a patient's entire 500-page medical history and generate a concise "One-Page Summary" highlighting critical events, medication changes, and outstanding risks, thereby reducing cognitive load and preventing missed diagnoses due to information overload.

16.2.2.2 Multimodal FMs for image-text reasoning in pathology and radiology

The frontier of Foundation Models lies in **Vision-Language Models (VLMs)**. These systems are trained on millions of image-text pairs (e.g., medical figures and their captions, or X-rays and their radiology reports).

Visual Question Answering: Clinicians can now interact with diagnostic images using natural language:

- *Pathology*: A pathologist can upload a whole-slide image and ask, *"Identify areas of high mitotic activity and grade the tumor."* The VLM does not just output a classification; it highlights the relevant cells and writes a draft report describing the morphology.
- *Radiology*: In complex trauma cases, VLMs can analyze a full-body CT scan and answer specific queries like, *"Is there evidence of a hairline fracture in the C-spine?"* directing the radiologist's attention to the specific pixels of interest.

16.2.3 Explainable AI for clinical trust and validation

As diagnostic models become more complex (Deep Learning), they become less transparent ("Black Boxes"). For clinical adoption, "interpretability" is not optional; it is a patient safety requirement.

16.2.3.1 Generating diagnostic rationales and "treasure maps"

To trust an AI diagnosis, a clinician must understand *why* the decision was made. Current Explainable AI (XAI) techniques utilize several methods to surface this logic:

- **Saliency Maps (Heatmaps):** These highlight the pixels in an image that most influenced the model's prediction. However, heatmaps are often criticized for being vague.
- **Concept Bottleneck Models:** A superior approach for medicine. Instead of jumping from Image: Diagnosis, these models first predict clinical concepts.
- *Example*: In knee osteoarthritis, the model first detects "Joint Space Narrowing" and "Osteophyte formation" (concepts) and *then* predicts "Severe Arthritis" based on those concepts. This aligns the AI's logic with human medical training.
- **Counterfactual Explanations:** The AI generates an explanation in the form: *"I predicted Malignant. If the border of this lesion were smoother and the intensity lower, I would have predicted Benign."* This helps the clinician visualize the decision boundary (Table 16.1, Fig. 16.3).

Table 16.1 Comparison of explainable AI (XAI) methods in diagnostics.

Method	Mechanism	Clinical utility	Limitation
Saliency maps	Highlights pixels/words driving the prediction.	Quick localization of abnormalities (e.g., identifying a fracture).	Can be noisy; does not explain "why" the feature matters.
Attention mechanisms	Shows how the model links different data parts.	Verifying if the model is looking at the lung versus the background.	Intermediate complexity; requires some technical knowledge to interpret.
Concept bottlenecks	Predicts intermediate medical concepts first.	High validation; mimics human reasoning (e.g., "I see X, therefore Y").	Requires dense annotation of concepts during training.
Counterfactuals	Simulates "what-if" scenarios.	Excellent for understanding decision boundaries and edge cases.	Computationally expensive to generate real-time.

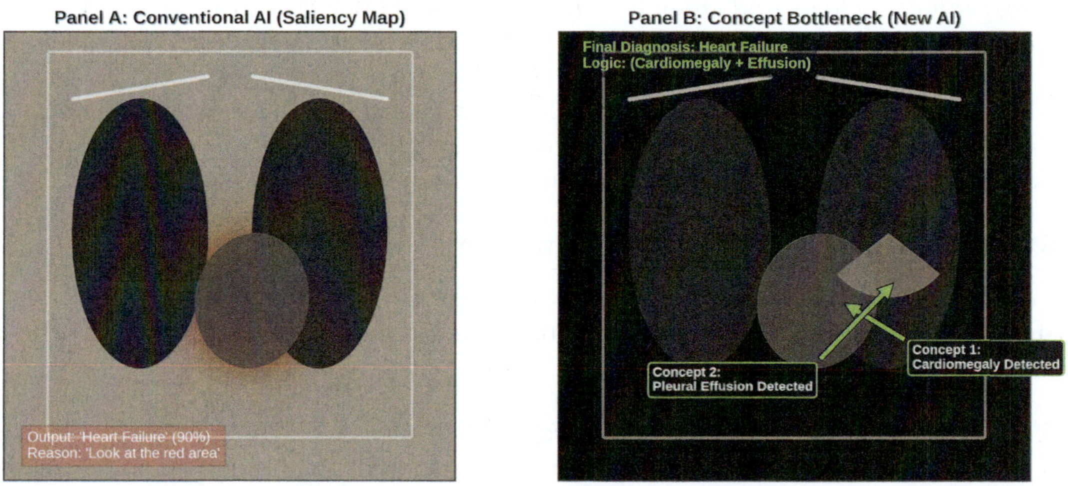

FIGURE 16.3 Evolution of explainability. Panel A demonstrates traditional "Saliency Maps," which highlight regions of interest but lack semantic meaning. Panel B illustrates "Concept Bottleneck Models," which align with clinical reasoning by identifying specific medical features (concepts) before rendering a diagnosis.

16.3 AI-driven innovations in therapeutic design and precision medicine

While the previous section established the role of AI in *identifying* pathology, this section explores its arguably more transformative role: *intervening* in disease biology. The traditional pharmaceutical model is widely considered to be in crisis; the cost of bringing a new drug to market exceeds \$2.5 billion, with a failure rate of over 90% during clinical trials. "New Applications" of AI are disrupting this "Eroom's Law" trajectory by shifting therapeutic design from a process of serendipitous discovery to one of rational, computational engineering.

16.3.1 Accelerating drug discovery and repurposing

The integration of Generative AI and geometric deep learning has birthed the field of "Generative Biology." Rather than screening libraries of existing molecules to see what sticks, algorithms now design novel molecular structures de novo to fit specific biological locks.

16.3.1.1 In silico target identification and molecular optimization

The first step in rational drug design is identifying a "druggable" target, a protein or gene whose modulation impacts disease progression:

- **Target Identification via Knowledge Graphs:** AI models crawl vast repositories of biomedical literature, patent databases, and omics data to construct **Biomedical Knowledge Graphs (BKGs).** These graphs link disparate nodes (e.g., Gene A, Pathway B, and Symptom C) to predict nonobvious therapeutic targets. For instance, in fibrosis, AI models have identified novel kinase targets by analyzing the topology of protein-protein interaction networks, targets that human intuition had overlooked.
- **Generative Chemistry (Inverse Design):** Once a target is validated, the challenge is designing a ligand (drug) that binds to it. New approaches utilize **Diffusion Models** (similar to those used in image generation, like DALL-E) to "denoise" random molecular structures into stable, high-affinity compounds. These models optimize for multiple parameters simultaneously: binding affinity, solubility, toxicity, and synthesis accessibility. This "multiparameter optimization" reduces the lead optimization phase from years to months.

Alpha Fold and Beyond: The release of AlphaFold structural predictions for nearly all known proteins has been a catalyst. Researchers now use these predicted structures to perform virtual screening of billions of compounds (ultralarge library docking) at a scale previously impossible.

16.3.1.2 AI for rare and neglected diseases: A focus on novel repurposing strategies

For rare diseases (orphan diseases), the market size often does not justify the cost of de novo drug development. AI offers a lifeline through **Drug Repurposing,** identifying new uses for already approved drugs:

- **Transcriptomic Signature Matching:** AI algorithms compare the gene expression "signature" of a rare disease against the signature induced by thousands of existing drugs. A negative correlation (where the drug reverses the disease signature) suggests therapeutic potential. This method has successfully identified existing oncology drugs as potential treatments for rare genetic disorders like vascular malformations.

- **Few-Shot Learning:** In neglected tropical diseases where data is scarce, "Few-Shot Learning" algorithms can infer molecular activity from very few data points by transferring knowledge from well-studied pathogens. This allows for the rapid identification of antiparasitic candidates without the need for massive labeled datasets.

16.3.2 Precision treatment planning and optimization

Moving from the molecule to the patient, AI is the engine driving the transition from "One-Size-Fits-All" medicine to "Precision Medicine."

16.3.2.1 Predicting individual treatment response (pharmacogenomics)

Adverse drug reactions are a leading cause of hospital admissions. Pharmacogenomics (PGx) aims to tailor dosing based on genetic variations in drug-metabolizing enzymes (e.g., CYP450 family).

- **Deep PGx Models:** Traditional PGx looks at single gene-drug pairs. New AI models, however, analyze the entire genome (polygenic scores) alongside clinical factors (age, renal function, comedications) to predict drug response.
- *Case Study—Warfarin Dosing*: Neural networks have been trained to predict the optimal dosage of the anticoagulant Warfarin. By integrating genetic variants (*VKORC1*, *CYP2C9*) with clinical variables, these models reduce the risk of bleeding events or thrombotic failures significantly compared to standard dosing algorithms.
- *Oncology*: In cancer care, AI models analyze tumor DNA sequencing data to predict resistance mechanisms before they emerge, allowing oncologists to prescribe combination therapies that preemptively block escape pathways (Fig. 16.4).

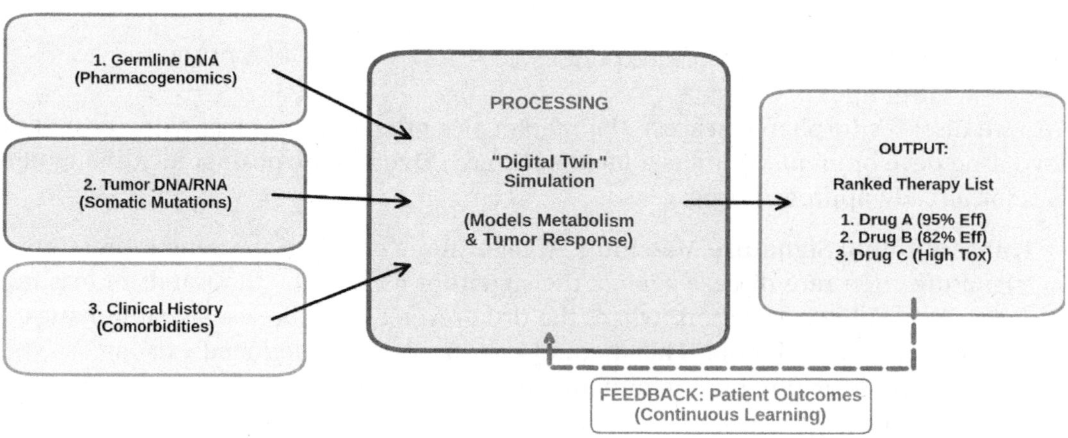

FIGURE 16.4 The AI-enabled precision medicine loop.

16.3.2.2 AI in adaptive clinical trial design

Clinical trials are the bottleneck of medical innovation. AI is optimizing this process through **Simulation and Enrichment**:

- **Patient Enrichment:** AI algorithms analyze EHR data to identify patients who are most likely to respond to the experimental drug (based on biomarkers), ensuring the trial is populated with the "right" participants. This increases the statistical power of the trial, allowing for smaller sample sizes.
- **Synthetic Control Arms:** In rare diseases or aggressive cancers where giving a placebo is unethical, AI generates "Synthetic Control Arms" using historical patient data. The experimental group is compared against this digital cohort rather than a real-life placebo group, accelerating recruitment and reducing ethical concerns.

16.3.3 Remote monitoring and digital biomarkers

The final frontier is moving healthcare from "episodic" (visits every few months) to "continuous" (Fig. 16.5).

16.3.3.1 Utilizing wearable sensor data for early disease prediction

Consumer wearables (smartwatches, rings) and medical-grade patches collect continuous physiological data (Heart Rate Variability, skin temperature, accelerometry). AI transforms this raw noise into **"Digital Biomarkers."**

- *Cardiology*: Deep learning models analyzing photoplethysmography data from smartwatches can detect Atrial Fibrillation (AFib) with high sensitivity, often

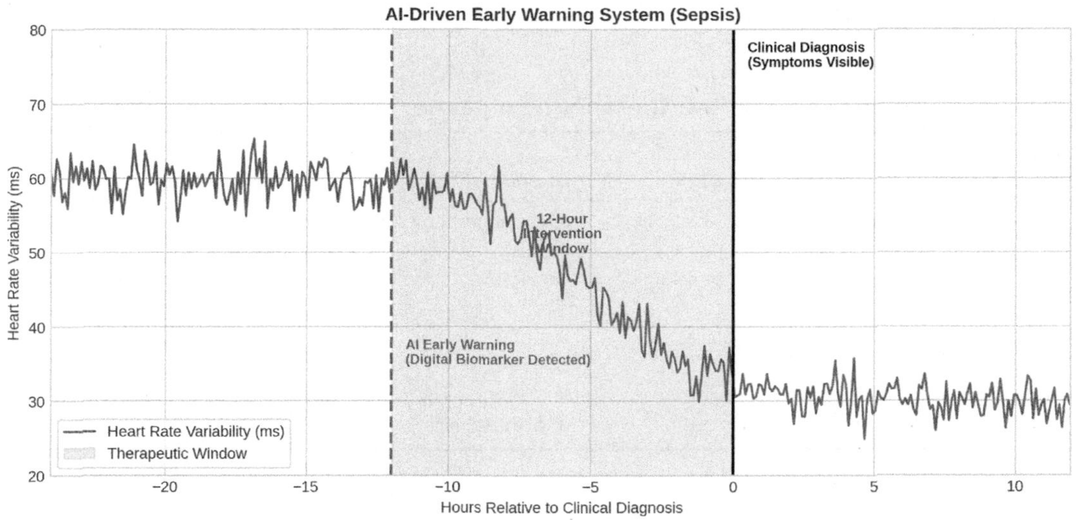

FIGURE 16.5 Continuous prognostic monitoring. The graph illustrates a retrospective analysis of a sepsis patient. The AI model (red dashed line) detects a significant deviation in heart rate variability 12 hours before the onset of clinical symptoms (black vertical line), providing a critical window for early intervention.

identifying asymptomatic episodes that would be missed by a standard 10-second ECG in a clinic.

- *Neurology*: In Parkinson's disease, smartphone accelerometers track "gait quality" and "tremor amplitude" in real-time. AI algorithms detect subtle degradation in motor function weeks before the patient notices, prompting medication adjustments.

16.3.3.2 AI for continuous prognostic monitoring

This continuous stream of data feeds into **Early Warning Systems**.

- **Sepsis Prediction:** In the ICU, AI models analyze streaming vitals (minute-by-minute blood pressure, oxygenation) to predict the onset of sepsis up to 12 hours before clinical shock occurs. This "time-window" allows for early antibiotic intervention, which is critical for survival.
- **Chronic Disease Management:** For heart failure patients, AI algorithms monitor "thoracic fluid content" via wearable sensors. A sudden rise in fluid triggers an alert for the patient to increase their diuretic dose, effectively preventing a hospital readmission for decompensated heart failure (Table 16.2).

16.4 Operational and ethical challenges in implementation

While the scientific potential of AI in medicine is indisputable, the "last mile" of implementation, integrating these algorithms into live clinical workflows, remains fraught with operational and ethical landmines. This section moves beyond the *theoretical* capabilities of AI to address the *practical* barriers preventing widespread adoption: data sovereignty, regulatory flux, algorithmic bias, and the sociology of human-machine interaction.

Table 16.2 Impact of AI on the therapeutic pipeline.

Stage of development	Traditional approach	AI-augmented approach	Impact metric
Target discovery	Literature review, academic serendipity.	Knowledge graphs, network biology.	Identifies "undruggable" targets.
Lead optimization	High-throughput screening (physical).	Generative design (in silico).	Reduces timeline by ~50%.
Clinical trials	Fixed protocols, placebo controls.	Adaptive designs, synthetic controls.	Increases success rate, lowers cost.
Post-market	Passive pharmacovigilance (reporting).	Real-world evidence mining (EHRs/Wearables).	Detects rare side effects faster.

16.4.1 Data Governance and security (HIPAA, GDPR compliance)

The lifeblood of medical AI is data, yet this data is protected by stringent legal frameworks such as the Health Insurance Portability and Accountability Act (HIPAA) in the US and the General Data Protection Regulation (GDPR) in Europe. The tension lies between the need for massive, centralized datasets to train "Foundation Models" and the legal mandate to keep patient data siloed and private.

The Privacy Paradox and Federated Learning: Traditional AI training requires pooling data into a central "data lake," creating a single point of failure for cyberattacks and privacy breaches. "New Applications" are increasingly moving toward **Federated Learning (FL)**:

- *Mechanism*: In FL, the model travels to the data, rather than the data traveling to the model. An untrained algorithm is sent to Hospital A, trains on local patient data, and learns "weights" (mathematical patterns). Only these weights, not the patient data, are sent back to the central server. This process is repeated across Hospitals B, C, and D, allowing the global model to learn from diverse populations without a single patient record ever leaving the hospital firewall.
- *Swarm Learning*: A decentralized evolution of FL using blockchain technology. It eliminates the central server entirely, allowing peer-to-peer sharing of parameters. This architecture is critical for cross-border collaborations where data sovereignty laws prohibit exporting genomic data.

Secure Enclaves and Homomorphic Encryption: For institutions that must share data, new encryption standards like **Homomorphic Encryption** allow algorithms to perform computations on *encrypted* data without ever decrypting it. This mathematically guarantees that even the AI developers cannot see the raw patient health information (PHI) they are analyzing (Fig. 16.6).

16.4.2 Regulatory pathway for evolving AI models (continuous learning systems)

Regulatory bodies like the FDA (US) and EMA (EU) were designed to evaluate static products, drugs that do not change chemical composition after approval. AI challenges this paradigm because optimal AI systems are "Continuous Learning Systems" (CLS) that update themselves as they ingest new data.

From "Locked" to "Adaptive" Algorithms:

- *The "Locked" Era*: Currently, most approved AI devices are "locked." If a developer wants to update the algorithm to improve accuracy, they must submit a new 510 (k) application, a process that takes months. This renders the AI obsolete by the time it is reapproved.
- *The "Predetermined Change Control Plan" (PCCP)*: The FDA's emerging framework for "Software as a Medical Device" (SaMD) introduces the PCCP. This allows

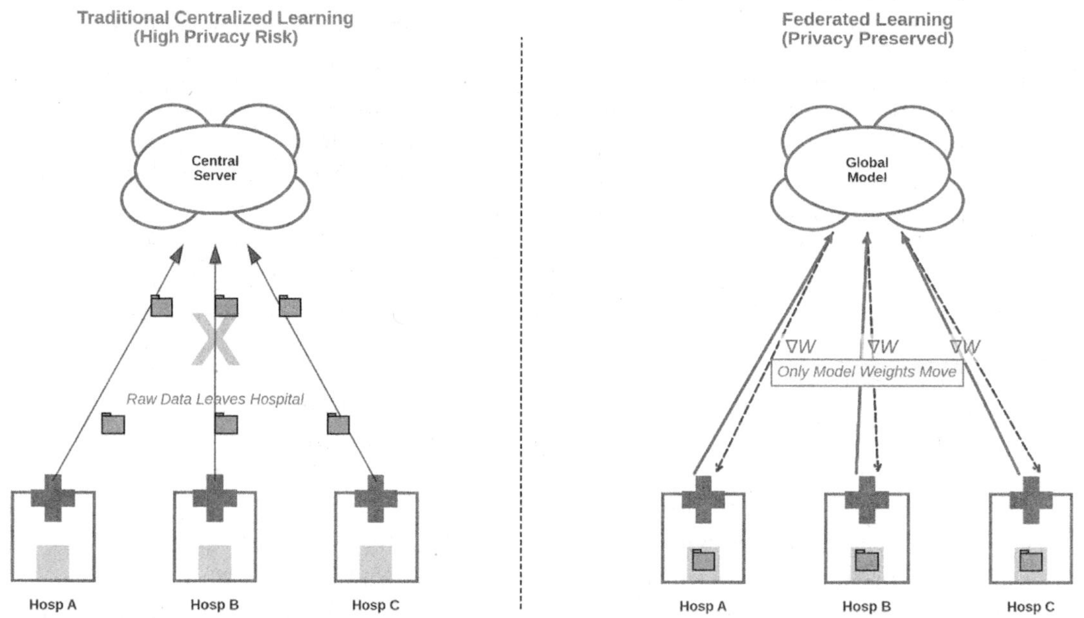

FIGURE 16.6 The federated learning architecture. In contrast to centralized learning (Left), where sensitive data is aggregated in a single location, federated learning (Right) transmits only model updates (gradients/weights), ensuring that raw patient health information (PHI) never leaves the local institution's firewall.

developers to prespecify *how* their model will retrain and *what* performance guardrails are in place. If the AI evolves within these preapproved boundaries, it does not require a new regulatory submission. This shift is essential for deploying AI that adapts to shifting demographics or new viral strains.

Liability and the "Black Box": When an autonomous AI makes an error, for example, missing a stroke on a CT scan, who is liable? The clinician who trusted it? The hospital that purchased it? Or the developer who coded it? Legal frameworks are currently evolving to distinguish between "Assistive AI" (where the physician is the final decision-maker) and "Autonomous AI" (where the algorithm acts independently), with stricter liability standards emerging for the latter.

16.4.3 Addressing algorithmic bias and promoting health equity

If AI models are trained on biased data, they will automate and scale discrimination. This is known as "Algorithmic Bias."

Sources of Bias:

- *Data Representativeness*: Most major biobanks (e.g., The Cancer Genome Atlas) are overwhelmingly composed of data from individuals of European descent. An AI

Table 16.3 The bias cascade. Algorithmic bias is not a single event but a cumulative process that can be introduced at problem definition, data collection, or clinical deployment.

Lifecycle stage	Source of bias	Example scenario	Mitigation strategy
1. Problem formulation	Label bias	Choosing "healthcare cost" as a proxy for "illness severity" (wealthier patients spend more, so AI thinks they are sicker).	Use physiological labels (e.g., Lab values) instead of administrative proxies.
2. Data collection	Selection bias	Training a dermatology AI only on light-skinned images from European biobanks.	**Oversampling:** Deliberately collecting more data from underrepresented groups.
3. Model training	Missingness bias	Patients with low access to care have "missing" data (gaps in EHR), which the model misinterprets as "health."	**Imputation:** Using statistical methods to account for systematic missingness.
4. Deployment	Automation bias	Clinicians trust the AI more than their own judgment, even when the AI is wrong.	**"Human-in-the-loop"** workflows and confidence scoring.

trained on this data may fail to identify disease variants prevalent in Asian or African populations.

- *Label Bias*: Algorithms trained on "healthcare utilization" (cost) as a proxy for "health need" have been shown to discriminate against Black patients. Because systemic inequities mean Black patients historically spend less on healthcare for the same level of illness, the AI falsely concludes they are "healthier" and denies them extra care resources.

Mitigation Strategies:

- *Algorithmic Auditing*: Just as financial audits verify fiscal health, clinical AI must undergo "Algorithmic Fairness Audits" before deployment. These audits distinct subgroups (age, sex, race, and payer status) to ensure **Demographic Parity,** that the error rate is equal across all groups.
- *Synthetic Data Augmentation*: Generative AI can create realistic "synthetic patients" from underrepresented groups to balance training datasets, effectively "filling in the gaps" of historical data exclusion (Table 16.3).

16.4.4 Integration into clinical workflow and physician acceptance

The most common cause of AI failure is not technical inaccuracy but "workflow friction."

The Problem of Alert Fatigue: Early AI systems were "interruptive," popping up alerts for every potential abnormality. Clinicians, overwhelmed by thousands of notifications, developed "Alert Fatigue," ignoring even critical warnings.

- *Solution*: New applications utilize "Silent Mode" testing and "Triage logic." Instead of alerting on every case, the AI prioritizes the worklist. For example, in a radiology queue, the AI places the scan with a suspected brain hemorrhage at the top of the list for immediate review, rather than interrupting the radiologist with a pop-up.

Human-in-the-Loop versus Human-on-the-Loop:

- *Human-in-the-Loop (HITL)*: The AI provides a suggestion, but the human must actively confirm it before any action is taken. This is standard for high-risk decisions like chemotherapy dosing.
- *Human-on-the-Loop (HOTL)*: The AI performs the task autonomously (e.g., screening diabetic retinal scans), and the human only intervenes if the AI flags a "low confidence" case or an anomaly. This model is necessary for scaling AI to population-health levels (Fig. 16.7).

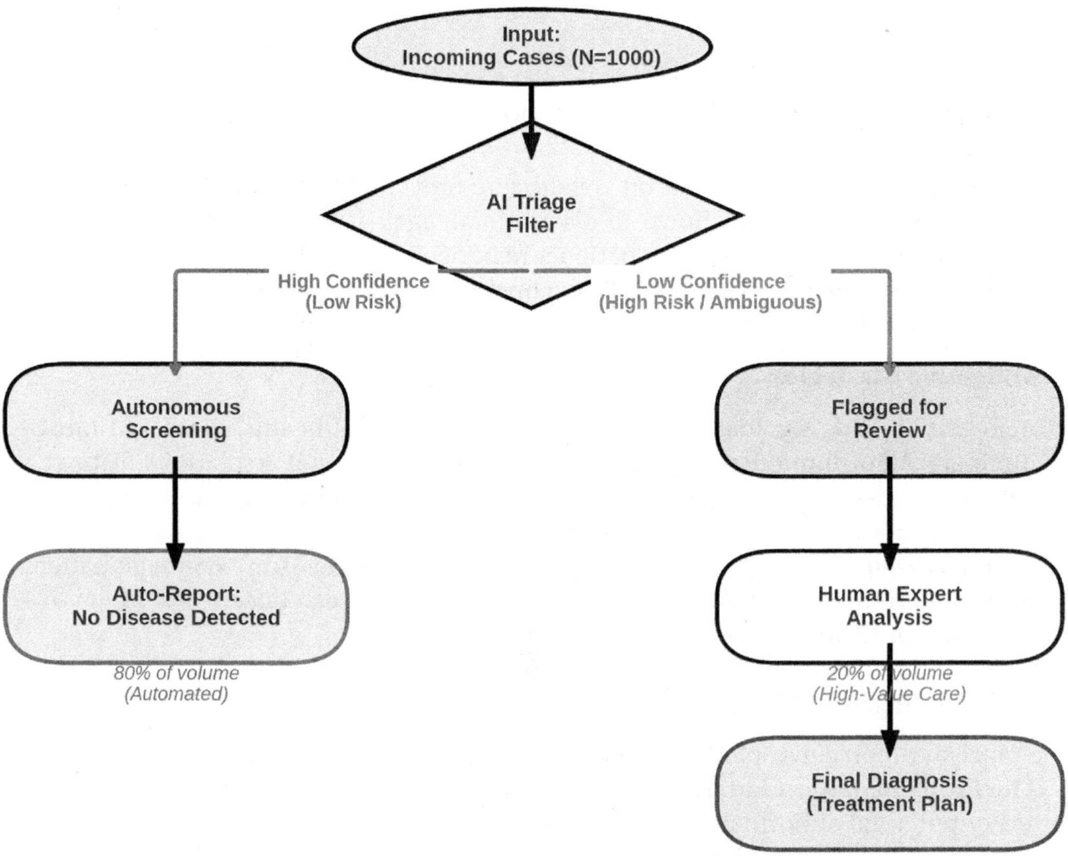

FIGURE 16.7 The "human-in-the-loop" workflow. To mitigate alert fatigue, the AI system acts as a triage engine. It autonomously clears low-risk cases (Path A) while routing complex or ambiguous cases (Path B) to human experts, optimizing clinical resource allocation.

16.5 Conclusion: Future outlook and the AI-augmented clinician

As we stand on the precipice of a new era in medicine, the narrative of "Artificial Intelligence versus Human Intelligence" is increasingly revealed to be a false dichotomy. The future of healthcare lies not in the replacement of the clinician, but in the creation of the **"AI-Augmented Clinician"**, a symbiotic partnership often referred to as the "Centaur Model."

In this model, the AI handles the data-heavy, computational tasks: sifting through terabytes of genomic data, monitoring millisecond-by-millisecond fluctuations in ICU vitals, and memorizing the millions of potential drug interactions. This offloading of cognitive drudgery liberates the human clinician to focus on the tasks that remain uniquely human: the nuanced interpretation of complex social histories, the ethical navigation of end-of-life care, and the empathetic connection that is the cornerstone of healing.

The "New Applications" discussed in this chapter, from multimodal foundation models and generative drug design to digital twins and federated learning, are the tools that will build this future. However, the path forward is not technological alone. It requires a concerted effort to build **Ethical AI** that promotes equity rather than disparity, **Transparent AI** that earns the trust of patients and providers, and **Adaptive AI** that grows alongside our ever-expanding understanding of human biology.

The implementation of these technologies will define the trajectory of 21st-century medicine. We are moving from a reactive, population-based model of care to a proactive, hyper-personalized, and continuous system of health. In this transition, AI is not merely a new tool in the doctor's bag; it is the very infrastructure upon which the next generation of healthcare will be built.

Index

'*Note:* Page numbers followed by "f" indicate figures and "t" indicate tables.'

N

Q

R

X

9780443449154